DIAGNOSIS AND TREATMENT OF CANCER USING THERMAL THERAPIES

MINIMAL AND NON-INVASIVE TECHNIQUES

Editors

Citlalli J. Trujillo-Romero
Researcher in Medical Sciences
National Institute of Rehabilitation – LGII
Mexico City, Mexico

Dora-Luz Flores
Professor, Universidad Autónoma de Baja California
Facultad de Ingeniería, Arquitectura y Diseño
Ensenada, Mexico

CRC Press
Taylor & Francis Group
Boca Raton London New York

CRC Press is an imprint of the
Taylor & Francis Group, an **informa** business
A SCIENCE PUBLISHERS BOOK

First edition published 2024
by CRC Press
2385 NW Executive Center Drive, Suite 320, Boca Raton FL 33431

and by CRC Press
4 Park Square, Milton Park, Abingdon, Oxon, OX14 4RN

CRC Press is an imprint of Taylor & Francis Group, LLC

Library of Congress Cataloging-in-Publication Data (applied for)

ISBN: 978-1-032-37936-4 (hbk)
ISBN: 978-1-032-37937-1 (pbk)
ISBN: 978-1-003-34266-3 (ebk)

DOI: 10.1201/9781003342663

Typeset in Times New Roman
by Radiant Productions

Dedication

To my mentors, family and Tut who have guided and inspired me throughout my scientific journey.

Citlalli J. Trujillo-Romero

To Sofía and Diego.

Dora-Luz Flores

Preface

According to the World Health Organization, cancer is a leading cause of death worldwide. In 2020, cancer caused 10 million deaths; moreover, it is considered an important barrier to increasing life expectancy in every country. Therefore, cancer is a worldwide health hazard of great significance to the scientific community. New, less invasive techniques for diagnosis and cancer treatment have been proposed during the recent past. Recently, techniques such as artificial intelligence, bioimpedance, thermal images, and nanomaterials have been studied for an early cancer diagnosis.

On the other hand, although chemotherapy and radiotherapy are common treatments, they present several side effects. Therefore, new treatments based on the generation of microwaves, radiofrequencies, or ultrasound have been proposed in the last decades. Thermotherapies have proved efficient; however, to be considered a regular treatment, some challenges must be met. One of the main challenges is the design of applicators capable of focusing either the electromagnetic or the mechanical waves on the region of interest without affecting the surrounding healthy tissues; in some cases, nanoparticles have also been designed to achieve better focus. New applicators can be designed with computational models based on numerical methods such as the finite element. However, to efficiently predict the applicator performance, it is crucial to know and include dielectric, thermal and acoustic properties (tissue characterization) in models. Not only healthy tissue but also tumors must be characterized. A powerful tool called patient-specific treatment planning can be developed to implement a safe treatment.

Moreover, tissue properties, as well as the applicator, must be defined. Parameters such as temperature increase and heat pattern must be evaluated to ensure patient safety and treatment success. This book will guide the readers through the most critical aspects of the latest cancer detection and diagnosis techniques and the fundamental aspects of thermal therapies based on radiofrequencies, microwaves, and ultrasound.

Contents

Chapter 1

Main Problems in Cancer Diagnosis and Treatments

Citlalli J. Trujillo-Romero,[1,*] *Genaro Rico Martínez*[2] and *Raquel Martínez Valdez*[3]

Cancer is considered a public health care problem and one of the leading causes of deaths worldwide. One of the main reasons for death is the late diagnosis of the disease when it is in the final stages. Therefore, few options are available for treatment. An early diagnosis and treatment can increase the survival rate of patients significantly. Standard cancer treatment procedures such as chemotherapy, radiotherapy, and surgery present several side effects that lower the quality of life of the patients. Therefore, researchers have been focused on early diagnosis as well as proposing new treatments to reduce side effects. This chapter addresses the main challenges in cancer diagnosis and treatment. Moreover, the main challenges expected in most recently proposed cancer treatments are also addressed.

1.1 Introduction

Cancer is defined as a decontrolled division of tissue cells; moreover, it can invade other tissues. The behavior of cancer cells is different from that

[1] Division of Medical Engineering Research, National Institute of Rehabilitation LGII. Calz. Mexico Xochimilco No. 289, Col. Arenal de Guadalupe, Mexico City, 14389, Mexico.
[2] Bone Tumors Service, National Institute of Rehabilitation LGII. Calz. Mexico Xochimilco No. 289, Col. Arenal de Guadalupe, Mexico City, 14389, Mexico.
[3] Biomedical Engineering Program, Polytechnic University of Chiapas. Carretera Tuxtla Gutiérrez – Portillo Zaragoza km 21+500, Col. Las Brisas, Suchiapa, Chiapas, 29150, Mexico.
Emails: grico@inr.gob.mx; rmartinez@ib.upchiapas.edu.mx
* Corresponding author: cjtrujillo@inr.gob.mx

of healthy cells. Healthy cells mature in different cell types with specific functions. However, cancer cell division is so quick they do not have the opportunity to mature and become specialized in any function. Healthy cells are programmed to die (apoptosis). Nevertheless, the cancerous cells ignore the body's signal to stop dividing once enough cells are present. There are many types of cancer; most of them are named according to the name of the organ where its growth starts.

Cancer is a leading cause of death around the world. A total of 18.1 million cancer cases were estimated for 2020; 8.8 million women and 9.3 million men were affected by this disease. Worldwide, breast and lung cancer are the most common types of cancer; both represent 24.7% of the total new cases diagnosed in 2020 (12.5% and 12.2%, respectively). The third most common type of cancer is colorectal; it represents 10.7% of the total new cases (1.9 million). In women, breast, colorectal, and lung cancer corresponds to a total of 44.5% of all cancer cases. Moreover, cervical cancer represents 6.9% of the new cases of cancer diagnosed in 2020. On the other hand, lung, prostate, and colorectal cancers represent 41.9% of all new cases of cancer diagnosed in men (*Worldwide Cancer Data*, 2022).

Cancer survival depends on several factors after diagnosis; among these factors are the type and stage of cancer, as well as the treatment. Therefore, early cancer detection is highly related to the survival of the patient. In recent decades, several multi-disciplinary researchers have been investigating different ways to improve the screening process and treatments. Unfortunately, these investigations have been restricted to specific cancers, such as breast and liver. Moreover, the development of each country and its healthcare systems make it difficult for all to access opportune diagnosis and treatment. Therefore, lower-income countries present lower survival rates.

1.2 Main challenges in cancer detection

1.2.1 Standard methods for cancer detection

To be able to suggest a treatment plan to a patient with cancer, it is important to have a diagnosis. The main approaches to diagnosing cancer are:

Physical examination: The medical doctor makes an examination looking for lumps, abnormalities, changes in the skin color, or an organ elongation that may be due to the presence of cancer.

Laboratory test: Clinical lab testing like urine and blood tests measure the levels of different components in the body fluids and tissues. The main analyzed components include blood glucose, calcium, potassium, lipids, hormones, enzymes, and proteins.

Imaging tests: They are used to analyze different anatomical structures in a non-invasive way. The images used for cancer diagnoses include X-rays, computerized tomography (CT) scans, magnetic resonance imaging (MRI), positron emission tomography (PET) and ultrasound. Imagine techniques and morphological analysis of cells and tissues (cytology and histopathology) are the most used for cancer diagnosis. Both use CT scans, MRI and X-rays among others , to identify healthy and cancerous cells; however, either imaging techniques or morphological analysis have low sensitivity; moreover, they are not able to detect cancer in its early stages (Choi et al., 2010).

Biopsy: A sample of cells is collected for analysis in the laboratory; in most cases, a biopsy is the only way to diagnose cancer. Cancer detection is possible due to the poor organization of cancer cells, i.e., under the microscope, they look disorganized with different sizes.

Biopsy, imaging tests, and quantification of biomarkers in physiological fluids are the most common diagnostic methods to monitor cancers. The main disadvantage of these techniques is their expense and the time required for their implementation; moreover, they cannot provide an evaluation of the response to the treatment nor an evaluation of the physiological conditions of the patients. Additionally, the lack of specificity and sensitivity of these techniques is also a disadvantage (Sharifianjazi et al., 2022).

1.2.2 Recent methods for cancer detection and diagnosis

Early diagnosis

As described before, cancer is defined as a group of diseases where an uncontrolled growth and spread of abnormal cells is presented. According to the International Agency for Cancer Research (IACR), 17 million new cancer cases and 9.5 million deaths were predicted for 2018 worldwide (American Cancer Society, 2018). However, if an early diagnosis is made, the treatment can be more effective, and consequently, the survival rate of patients increases. Therefore, one of the main challenges to researchers related to cancer is early diagnosis; unfortunately, in several cases, cancer is detected at an advanced stage; therefore, there are fewer treatment options for such patients. The importance of early diagnosis is mainly due to the necessity for increasing patient survival rates. Unfortunately, effective tests for early diagnosis of several kinds of cancer are not possible. Moreover, one of the disadvantages is the possibility of overdiagnosis and overtreatment of patients without cancer. Early detection is important to increase the possibility of a cure. The main strategies for early diagnosis include patient awareness for detecting early symptoms followed by a consultation; therefore, the diagnosis, and consequently, the treatment can be given as soon as possible. The other strategy involves the screening of asymptomatic people to detect either cancer at the early stage or precancerous lesions to make an early diagnosis.

Some countries have implemented a system called *fast track* to improve the early diagnosis of cancer, whose objective is to give access to treatment as soon as possible after the consultation of a suspected cancer patient (Vedsted et al., 2015). However, this system cannot be easily applied in all cancer types, i.e., patients with alarm symptoms such as lumps, and ulcerations, have a higher probability for fast-tracking (breast, oral cancer and others) than those with symptoms of low predictive value such as brain cancer and myeloma (Zhou et al., 2017).

According to Crosby et al., to make the early detection of cancer a reality the following five challenges must be faced (Crosby et al., 2022).

1) A clear understanding of the biological aspects and behavior of the early disease to identify consequential, aggressive, and inconsequential lesions.
2) Determining the risk of developing cancer according to genomic susceptibility, family history, demographic, among others, to generate models to predict who should be tested for cancer and the follow-up process.
3) Identification and validation of early cancer biomarkers.
4) The improvement of recent approaches and the development of new disruptive and innovative detection technologies.
5) To find ways to evaluate different early detection approaches.

In this book chapter, we are mainly focused on challenge four. To give a proper and timely diagnosis, medical doctors and researchers of different topics have been working together for several years. The available technology to detect and monitor different kinds of cancer is strongly related to the advancement and sophistication of the recent analysis methods in different knowledge areas.

Medical images

One of the main problems in cancer is its early detection; usually, cancer is detected at later stages, when it has compromised the function of vital organs and has spread to other parts of the body. Different methods for early detection, such as medical images, represent a huge and active area of research. The most used medical images are X-rays, magnetic resonance imaging (MRI), computed tomography (CT), and ultrasound. Data from medical images are a crucial tool in early cancer diagnosis; moreover, they can help to choose the most appropriate and least invasive treatment for each specific patient.

However, it depends on the interpretation of the medical doctor, which could not be accurate. The use of medical imaging to detect cancer is based on so-called semantic tumor characteristics, such as density, intratumoral cellular, acellular composition, and tumor margin regularities. Moreover, recently, a field called radiomics has worked with image decoding to quantify features, such as descriptors of shape, size, and texture (Bi et al., 2019). In more recent years, a different approach of artificial intelligence (AI) has been used to improve cancer detection, classification, and tumor segmentation for diagnosis by using medical image processing (Hunter et al., 2022). AI represents the ability of machines to perform tasks and solve problems; i.e., to show intelligence. Deep learning a sub-division of AI is a powerful tool for detection, diagnosis, and tumor segmentation because it automatically extracts high-level features from raw images by arranging either linear or nonlinear processing units in a deep architecture (Pacal et al., 2020).

This is a valuable tool for clinicians because they can delimitate the tumor size over time and track multiple tumors at the same time, i.e., AI quantifies the information into the medical images to complement the clinical decision-making. The non invasiveness of using AI as a cancer detection method is one of the main advantages. The patient does not have to go through medical procedures such as endoscopies, biopsies, and MRIs, where contrast agents are added to make the detection of tumors possible. The main goal of medical image analysis is the detection of tumors to make a diagnosis. Moreover, with the development of fully automatic applications, it is possible to perform a complete analysis of medical images without the intervention of a medical doctor.

Although AI is a promising tool to detect, diagnose, and treat cancer, several issues must be considered before being adopted as a regular tool in the clinic. One of the main issues to be addressed for the use of AI in cancer detection is model bias. Although there are different ways to introduce an undesired bias in the AI algorithms, the most common one is the lack of a representative and heterogeneous population sample in the training data set, particularly respecting demographic and other characteristics like ethnicity and sex. It is important to mention that the use of AI is having a great impact on the prediction and diagnosis of medical issues as well as recommending treatments. However, the complexity of the internal structure makes impossible to provide an understandable explanation of how the output was obtained. Moreover, the use of algorithms to treat medical images to help in the clinical-decision making process can lack reliability and interpretability.

On the other hand, due to the large amount of data generated by these applications, it is quite important to ensure that the datasets are well organized, and that access and retrieval are easy. The biggest obstacle in the development of automatic clinical tools is data curation, i.e., the creation, organization, and maintenance of data sets to be accessed and used. Data curation can be expensive due to the necessity of professional training; furthermore, data must be curated in terms of labeling, segmentation, quality assurance, and more (Bi et al., 2019). Another important issue to be faced is the regulatory aspect, i.e., the application based on AI must be tested to ensure quality control and risk assessment. Moreover, data security and privacy are some other concerns.

Recently, hyperspectral images (HSI) have been used for cancer detection because they include a very large number of wavelengths. HSI is a promising tool for detecting and diagnosing cancer due to the amount of information provided by a broader spectral band and higher spectral resolution, i.e., it can capture images under different spectral bands such as visible, near-infrared, and ultraviolet. Moreover, it is a non-invasive technique. HIS has had a big impact on cancer detection; it can be used to detect tumors; moreover, it provides information about the surrounding tissues, which helps in treatment selection. One of the main disadvantages of hyperspectral images is the required technology because it could be very expensive; therefore, there is a necessity of portable and inexpensive equipment required for use in the clinic.

In general, the main challenges in early cancer detection and diagnosis include the following:

1) Cancer imaging techniques must be improved to enhance early cancer detection.
2) To develop noninvasive techniques for accurate detection, diagnosis, and treatment.
3) To combine the current capable techniques to improve early cancer detection and diagnosis.

Biosensors

A biomarker is a "molecule with biologically important intra-or-intercellular functions, whose expression or activities either cause or are specifically altered in response to corresponding pathological conditions" (Wang et al., 2017); therefore, cancer detection and diagnosis by using biomarkers is a very promising and effective technique. In cancer, the main uses of biomarkers are focused on screening for cancer detection, diagnosis, and identification of specific cancers, patient prognosis, and monitoring the cancer's course. Therefore, the development of new techniques to evaluate the physiology and metabolic response of the patients is required (Wang et al., 2017). Biosensors have been one of the most recent advances related to cancer prognosis and diagnosis in the early stages. The main advantage of biosensors is their cost-effectiveness, noninvasiveness, sensitivity, and their fast responses due to their direct relation with the physiological fluids of the patient (Sharifianjazi et al., 2022). The development and use of different nanomaterials, nanoelectronics, optoelectronics, drug delivery and microelectronics have been helping in the development of new electrochemical sensors for cancer detection using biomarkers. These kinds of biosensors are based on the most common nanomaterials such as graphene, carbon nanotubes, Au, Ag, Pt and more.

Recently, new highly sensitive electrochemical biosensors based on the detection of cancer biomarkers at ultra-low levels in clinical samples such as tissues, blood and urine have been developed (Wang et al., 2017). Different biomarkers such as enzymatic tumor markers, protein cancer markers and hormones related to cancer have been detected by these biosensors, which have been so helpful not only in cancer detection in the early stages but also in the process of cancer classification, and treatment selection. For example, the carcinoembryonic antigen (CEA) is considered to be a broad-spectrum multi-tumor biomarker; it is used for the detection of colon, ovarian, breast, and lung tumors (Jiang et al., 2011). Lung cancer has one of the highest indices of lethality due to the lack of early detection techniques. The reduced rate of survival is because lung cancer is usually detected at stage IV when the chances of survival are low (Uday Kumar et al., 2013). Therefore, early detection and diagnosis of lung cancer is important to give opportune treatment. In this sense, the development of biosensors to detect biomarkers related to lung cancer is a great area of opportunity for researchers. Several biosensors for cancer detection of the lung, breast, prostate, and colon have been developed (Yang et al., 2019; Mittal et al., 2017; Pothipor et al., 2019; Ratajczak et al., 2018).

The main problems faced by biosensors are related to the control of the size, shape, and stability of the nanoparticles in a physiological environment. Efforts of the researchers have been focused on the biocompatibility of the nanoparticles used in biosensors. Several works report the surface modification of the iron oxide nanozymes, and the nanoparticles functionalized by either carboxyl or amine groups to improve the immobilization of the target biomolecular probes over their surface (Farhana et al., 2021). Biosensors can face different challenges depending on the kind of cancer to be detected and diagnosed. For example, for breast cancer, due to the low levels of biomarkers related to it, one of the main problems for early detection is the accuracy of the biosensors. Several biosensors have been tested in vivo and in vitro by using animals; however, more real approaches for human beings are still required. Moreover, the most general challenges related to biosensors are stability, precision, accuracy, cost, and toxicity. Therefore, reaching medical standards is still a big challenge (Sharifi et al., 2020).

Bioimpedance spectroscopy

Recently bioimpedance spectroscopy has been under research to develop a tool to diagnose breast, cervical, and prostate cancer, among others, and consequently to avoid the progression of cancer. The bioimpedance principle is based on the response of the biological organism to an external current; it measures the opposition to the flow of the current throughout the tissues. Bioimpedance devices inject a current (normally lower than 1 mA) into the tissue under test; then the voltage generated by the current circulation is measured (Tucker et al., 2013). Bioimpedance is considered a non-invasive method that can be used to characterize and assess the composition of the tissue. Several researchers have found notable differences between healthy tissues and tumors. Bioimpedance of a tumor differs from that of the healthy tissues, mainly because of the increment in cellular water and salts that modify the permeability of the membrane, density, and cell orientation (Sarode et al., 2016). Tissue properties are dependent on the frequency of the applied electric field; moreover, the tissue bioimpedance is also temperature and time-dependent. The main advantages of bioimpedance are its low cost and instant results; moreover, it does not require too much training to be implemented at the clinic (Abdul et al., 2006). Portable devices are available, and they are feasible for any patient, no matter their age and physical state (Mantzorou et al., 2020). The advantages of bioimpedance as a screening method are focused on the ability to repeat an incorrect test immediately without producing anxiety in the patient (Sarode et al., 2016).

Bioimpedance is considered to be a robust method against noise and motion artifacts due to the reference current at a specific frequency or range of frequencies, that generates an easily measured voltage (Naranjo-Hernández et al., 2019). One of the main challenges in bioimpedance is the design of a precise and stable current driver (Kassanos et al., 2014). Moreover, the design of a robust module against noise, to measure voltage is another challenge.

Nanotheranostics

Nanotheranostics is the combination of diagnostic and targeted therapy in a nano system. Its main objective is to find the most effective treatment for a specific patient. This is a recent research field where drug molecules and imaging agents are combined to give simultaneous diagnosis and treatment for cancer (Kelkar et al., 2011). Image-contrast molecules and fluorescent probes are used as diagnosis agents, to have a better visualization of the tumor, while as therapeutic agents, some chemotherapeutic molecules or oligonucleotides are used. The main advantage of nanotheranostics is the possibility of saving time, and money, and avoiding side effects. Moreover, this technique allows monitoring the response to the treatment in real-time and consequently planning the following therapies (Kim et al., 2010). Artificial intelligence is playing an important role in the analysis and processing of a huge amount of data from nanotheranostic systems. Additionally, recent advances in nanomaterials related to size and biocompatibility make the use of nanotheranostics as a tool to diagnose and treat cancer even more promising. The main challenge of nanotheranostics is the customization of treatment according to patient-specific necessities. The most important aspect to be considered is the selection of the most adequate combination of therapy and the imaging technique because each has its issues. Moreover, the inclusion of both in one system is also a challenge. Besides, it is also still important to evaluate its safe use in humans. To improve the use of this technique long-term studies about toxicity and regulatory protocols are still required (Murar et al., 2022).

1.3 Main challenges in cancer treatments

Cancer control is related to the reduction of mortality, morbidity, and incidence of cancer as well as to improve the quality of life of patients with it. Cancer control components are prevention, early detection, diagnosis, treatment, and palliative care. On the other hand, cancer care is related to the different actions to support, assist, and treat patients with cancer, while cancer treatment is an intervention to cure/palliate cancer. Cancer treatments are divided into local/locoregional treatments such as surgery, radiotherapy, and systemic approaches like chemotherapy, hormone and gene therapy, and immunotherapy. Moreover, recently, less invasive, and fewer side-effect therapies such as thermal therapies are under development.

1.3.1 Main cancer treatments vs thermotherapies

The most common cancer treatments around the world are (1) surgery, (2) radio, and (3) chemotherapy. The treatment selection will depend on the patient's condition and necessities. Treatment characteristics depend on the tumor type, size, and location. Each treatment has its benefits and side effects; moreover, each tumor presents a different response to these treatments.

Surgery: It is the most common and effective treatment to eliminate most types of cancers before reaching a metastatic phase. For several years, surgery was the first option to treat cancer; however, due to the increased use of neoadjuvant therapies, surgery has been relegated. Normally, either radio or chemotherapy is also applied as neoadjuvant therapy to reduce the tumor size before the surgery. Moreover, radio and chemotherapy can also be applied as adjuvant therapies to ensure the total elimination of cancer cells and avoid tumor recurrence. Usually, in the surgery, the medical doctor removes the lymph nodes near the tumor to check if cancer has spread to them. If cancer has spread, then the patient must be treated with radio or chemotherapy to avoid tumor recurrence. Not all patients are candidates for surgery; it depends on the cancer stage; moreover, in some cases, the tumor is in an inaccessible site. Therefore, radio and chemotherapy are the best options.

Nowadays, cancer must be treated following a multidisciplinary approach. Surgeons play a key role in different aspects because of their direct access to different kinds of tumors. Surgeons not only perform surgeries to remove the tumor but also participate in the diagnosis and application of neoadjuvant therapies. Moreover, they also play the role of physician-scientist, and play an active part in the research on new therapies such as molecular and immunotherapy (Chang et al., 2006). Surgery as a cancer treatment can be divided into two main types.

Curative surgery: It is mainly used to treat primary cancers, defined as those located at the original site, either organ or tissue, where the cancer originated. The main goal of this surgery is to eliminate the tumor by achieving optimum local control. Local control refers to the stopping of tumor growth at the place of origin. Normally, the efficiency of a treatment outcome is defined as a function of the local control of the tumor. Extirpative surgeries such as mastectomy for breast cancer and amputation for bone cancer maximize local control by avoiding the recurrence of the tumor. During the surgery, one of the main goals is to achieve adequate negative margins around the primary tumor. If they are not achieved, the risk of local recurrence of the tumor increases. For example, for hepatocellular carcinomas, a resection margin of 2 cm reduces the prospective recurrence rate of the tumor (Pilewskie et al., 2018). Moreover, for bone sarcoma, a margin > 2 mm has been considered an acceptable limit to say that the surgical resection is safe (Bertrand et al., 2016). Surgical resection has been more effective, especially when it is combined with radiotherapy (Chang et al., 2006).

Cytoreductive surgery: Its main goal is to reduce the number of cancer cells; therefore, the possibility of destroying the residual tumor cells by radio or chemotherapy increases. This kind of surgery is used to treat ovarian cancer, bone tumors, and tumors in the abdominal region, among others (Mehta et al., 2016).

Moreover, the main complications can be attributed to the surgery itself, the patient's overall health, and the used drugs. Some complications/side effects of surgery are damage to other organs/tissues, pain, bleeding, infections and pain. These side effects can result in slow recovery of the patient and affects the body functions.

Radiotherapy: It refers to the medical use of ionizing radiation to control the growth of tumor cells. Once the ionizing radiation strikes the tumor, the energy needed to un-bind one of the electrons from its atom is achieved. Therefore, the ionization of tumor cells produces irreparable damage to DNA, liposomes and cell membranes; as a consequence the cells lose their reproductive capacity and eventually lead to death. It has been widely used as a local or locoregional treatment of several kinds of cancer, such as breast, lung, liver, bone and prostate (Alfouzan, 2021; Haussmann et al., 2020). As long as the cancerous cells do not escape the locoregional boundaries and metastasize, there is a chance to cure cancer by using a locoregional approach such as radiotherapy (Rosenblantt et al., 2017). In these cases, a systemic approach including chemotherapy, hormonal therapy, or targeted therapy must be applied. One of the main challenges in radiotherapy has been the investigation of how to deliver higher radiation doses over a well-defined volume in the tumor, while minimizing the dosage to healthy tissues. To make it possible, different strategies such as the use of radiosensitizers and radioprotectors, and brachytherapy among others have been utilized. Recent advances in radiotherapy are focused on the use of high precision collimating devices and the radiation fluence modulation of an individual beam (Rosenblantt et al., 2017). One of the main problems of this technique is that the radiation could affect not only tumor cells but also healthy ones. This damage of healthy cells can provoke several side effects in the patients. Moreover, it is important to remark that not all tumor types are highly sensitive to radiotherapy such as bone tumors.

Chemotherapy: It refers to the use of chemicals/drugs to inhibit cancer cells. The drugs are designed to kill or inhibit the target organism with a minimal effect on the host cell (Alam, 2018). The drug is systemically delivered and reaches the abnormal cells to kill them; actually, the conventional chemotherapeutic agents are distributed randomly in the body. Therefore, both cancer and healthy cells are affected by the agent, i.e., all the chemicals used to treat cancer are cytotoxic not only for the cancer cells but also for the healthy ones (Ferrara et al., 2005). Consequently, several side effects such as mucositis, alopecia, myelosuppression, and immunosuppression are provoked (Alam, 2018). Normally, chemotherapy is used under four schemes: (1) to reduce the tumor size before surgery or radiotherapy (neoadjuvant), (2) to destroy remaining cancer cells after surgery or radiotherapy (adjuvant), (3) to help improve the performance of some other treatments and, (4) to treat the recurrence of cancer cells including those spreading to other body parts. Therefore, chemotherapy goals are either to cure, control, or palliate cancer. The type of subminister drug will depend on several factors such as the type and stage of cancer, age, and the overall health of the patient among others. Medical doctors must carefully determine the chemotherapy dose; especially because these drugs are too strong, and the incorrect dose can either not be enough to treat cancer or can be too much to produce life-threatening side effects.

Thermotherapies: In the last years, new techniques to treat cancer have been developed. Thermotherapy is a technique used to induce heat and achieve therapeutic benefits. As a cancer treatment, it can be divided into two main categories:

Hyperthermia: The tumor is exposed to temperatures around 40°C–45°C for periods of approximately 60 min. The tumor temperature increase is due to the absorption of energy generated by an applicator. Literature reports that a temperature increase can damage and kill the tumor cells while producing minimal damage to the surrounding healthy tissue (Habash et al., 2006). Hyperthermia can be divided into local, regional, and whole body, depending on the region to be treated (Beck et al., 2015; Gautherie, 1992; Szasz et al., 2013). Oncological hyperthermia has been successfully applied in several kinds of cancer including breast, lung, head & neck, and bone tumors, among others (Huilgol et al., 2010; Mitsumori et al., 2007; Paulides et al., 2010; Staruch et al., 2012; Van Der Zee et al., 2002).

Thermal ablation: The tumor is exposed to temperatures between 60°C–100°C for short periods (minutes/seconds) causing coagulative necrosis. The main goal is to destroy the tumor by burning the cancer cells. Thermal ablation is considered to be a minimally invasive treatment because the applicators are a kind of needle inserted over the tumor.

Table 1.1 shows the main side effects produced by the standard cancer treatments, as well as by the thermotherapies. As can be observed, the main challenges in most of the cancer treatments include reduction of side-effects, cost and specificity.

1.4 Main challenges in the clinical application of new treatments

Despite the recent advances in medicine, to improve the results of cancer treatments many issues must be addressed. Therefore, the most recent advances have been focused on finding less invasive, more precise, and effective cancer treatments; moreover, the reduction of side effects produced by common

Table 1.1. Side effects and cost of the most common cancer treatments.

	Effect		Cost	
	Short term	**Long term**	**Equipment**	**Treatment**
Surgery	Removal of the tumor (best case), Limb loss, breast loss, phantom limb pain, and more.	Psychological trauma.	–	–
Radiotherapy (Han et al., 2009; Icrp, 1990; Therapy et al., 2014)	Skin issues, fatigue, hair loss, nausea, vomiting, loss of appetite, and changes in blood counts.	Depending on the irradiated area, it could produce heart, lungs, intestinal, and infertility issues, among others.	Radiotherapy equipment: linear accelerators 175,000–300.000 dls (the oldest systems) 750,000–1,500,00 dls (the latest systems) (Linear Accelerator (LINAC) Price Guide & Costs, 2021).	The median cost of radiation therapy (Paravati et al., 2015): Breast: \$8,600 (\$7,300 to \$10,300), Lung: \$9,000 (\$7,500 to \$11,100), Prostate: \$18,000 (\$11,300 to \$25,500).
Chemotherapy	Nausea, vomiting, reduced appetite, hair loss, mouth ulcers, diarrhea, constipation, and more.	Heart, lung, hormonal and digestive problems, secondary cancer, fatigue, and more.	–	The average cost of care per episode (Health, 2012): Office-managed chemotherapy: \$19,640 Hospital Outpatient Department (HOPD)-managed chemotherapy: \$26,300.
Thermotherapy (RF y MW)	Bleeding, perforation of a vein or artery, and burns that could cause irreversible damage.	-	–	Cost per session (Trujillo-Romero et al., 2018) (4–16 sessions). \$225.76 dls \$516.29 dls
Thermotherapy (US)	Toxicity, peripheral nerve damage, ligament weakness.	-	–	–

treatments has also been an important issue. Nowadays, several new kinds of cancer treatments are under evaluation in research laboratories and clinical trials; some others have already been introduced into clinical practice. Among the most recent proposed cancer treatments are nanomedicine, targeted therapy, gene therapy, immunotherapy, thermal ablation, hyperthermia and magnetic hyperthermia. However, all these cancer treatments pose many challenges to be adoptable as regular treatments. According to the database of clinical trials (https://www.clinicaltrials.gov/), immunotherapy, gene therapy, and targeted therapy are the most highlighted and popular therapies under investigation due to their promising feasibility as cancer treatments. Some other therapies such as hyperthermia and thermal ablation are still under investigation. Figure 1.1 presents the number of clinical studies per treatment registered in the database of clinical trials. Moreover, Table 1.2 shows the main issues to be solved in each treatment.

Although so much cancer treatment development research is ongoing, the effectiveness of the treatment outcome for all patients is still a challenge, i.e., the effectiveness of the treatments for specific kinds of cancer is limited.

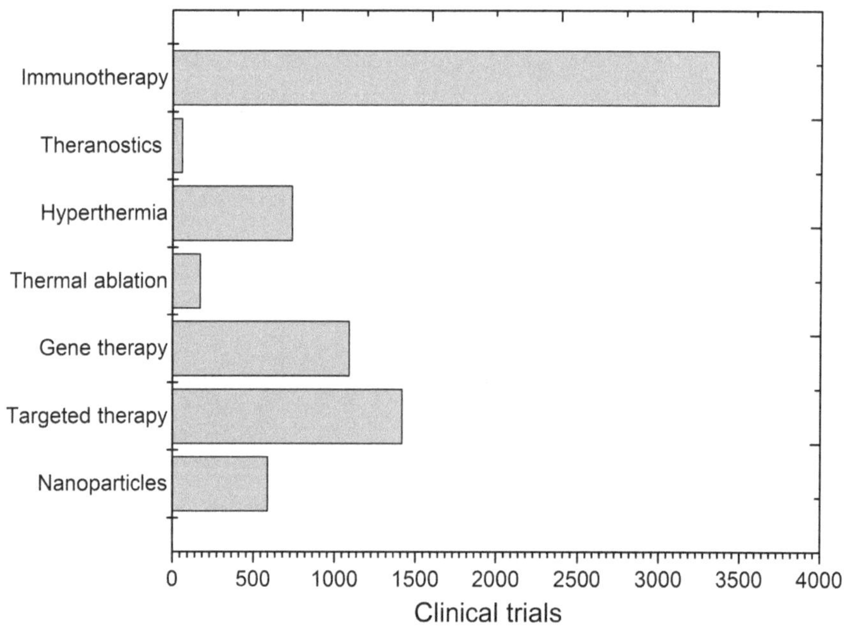

Figure 1.1. Number of registered clinical trials per treatment.

Table 1.2. Main advantages and disadvantages according to the treatment.

Technique	**Advantage**	**Disadvantage**
Nanoparticles	• Biocompatibility, bioavailability, stability, and specificity.	• Safety and tolerance of the new modalities must still be inspected. • Just a few nanoparticles are used in hospitals, and clinical trials; most of them are under development in research laboratories. • Topics related to toxicity, drug-delivery as well as cellular and physiological factors need regulation.
Targeted therapy	• Fewer side effects. • High level of specificity.	• Insufficient information about long-term side effects. • Cancer cells can become resistant to the therapy. • Drugs for some targets are hard to develop due to the target's structure and function in the cell. • Presence of side-effects such as liver problems, blood pressure and skin problems. • Sometimes, a hole can be formed through the wall of the stomach, rectum and intestine.
Gene therapy	• It could be a cure in cases where no other treatment has worked. • Just one application is needed. • Its positive effects can pass through generations.	• The approval of gene therapy products has been strongly regulated by the FDA to ensure their efficiency and safety; therefore, access to them can be delayed (Ross et al., 2015). • Ethical issues. • Manufacturing and distribution of drugs is also a challenge. • The number of certified hospitals to give this treatment is limited. • Treatments can be very expensive. • Gene therapy is a relatively new approach to treating cancer; therefore, delayed adverse events are still known.

Table 1.2 contd. ...

...Table 1.2 contd.

Technique	Advantage	Disadvantage
Radiofrequency (RF) thermal ablation	• Relatively low rates of complications (0–12%) (Riadh et al., 2007). • Mortality and morbidity rates are reduced when compared to those from surgeries. • Possibility of treating patients who are not candidates for surgery. • Can be applied through three processes: (a) open surgery, (b) laparoscopy, and (c) percutaneously. • Well tolerated by patients. • Can be applied in combination with other therapies. • The patient recovery is faster. • Relatively inexpensive.	• Burns a small volume (if only one applicator is used). • Impossibility of treating regions close to blood vessels or essential structures that could be damaged by high temperatures. • Difficulty in seeing the volume burned. • Need for an imaging system to properly insert the antenna into the tumor. • Generation of skin burns or appearance of sores.
Microwave (MW) thermal ablation	Same advantages of RF ablation as well as: • Higher temperature rises. • Larger treatment volumes. • Reduction of treatment times.	Same disadvantages of RF ablation.
Ultrasound (US) thermal ablation	• Non-invasive nature. • Possibility of repeating the treatment until the clinical objective is reached.	• Need to vary application times and energy levels to achieve greater penetration of mechanical waves.

Reduction of short and long-term side effects produced by most of these treatments is also an important issue to be solved before their adoption as regular treatments. Therefore, more research is required to ensure their safety and effectiveness.

Acronym/Abbreviation

AI	Artificial Intelligence
CEA	Carcinoembryonic Antigen
CT	Computerized Tomography
FDA	Food and Drug Administration
HSI	Hyperspectral Images
IARC	International Agency for Research Cancer
MRI	Magnetic Resonance Imaging
MW	Microwave

PET Positron Emission Tomography
RF Radiofrequency
US Ultrasound

Reference list

Abdul, S., Brown, B.H., Milnes, P. and Tidy, J.A. (2006). The use of electrical impedance spectroscopy in the detection of cervical intraepithelial neoplasia. *International Journal of Gynecological Cancer*, 16(5): 1823–1832. https://doi.org/10.1111/j.1525-1438.2006.00651.x.

Alam, A. (2018). Chemotherapy treatment and strategy schemes: A review. *Open Access Journal of Toxicology*, 2(5). https://doi.org/10.19080/oajt.2018.02.555600.

Alfouzan, A.F. (2021). Radiation therapy in head and neck cancer. *Saudi Medical Journal*, 42(3): 247–254. https://doi.org/10.15537/SMJ.2021.42.3.20210660.

American Cancer Society. (2018). *Global Cancer - Facts & Figures 4th Edition*. American Cancer Society. https://www.cancer.org/research/cancer-facts-statistics/global.html.

Beck, M., Ghadjar, P., Weihrauch, M., Burock, S. and Budach, V. (2015). Regional hyperthermia of the abdomen, a pilot study towards the treatment of peritoneal carcinomatosis. *Radiation Oncology*, 10(1): 157. https://doi.org/10.1186/s13014-015-0451-3.

Bertrand, T.E., Cruz, A., Binitie, O., Cheong, D. and Letson, G.D. (2016). Do surgical margins affect local recurrence and survival in extremity, nonmetastatic, high-grade osteosarcoma? *Clinical Orthopaedics and Related Research*, 474(3): 677. https://doi.org/10.1007/S11999-015-4359-X.

Bi, W.L., Hosny, A., Schabath, M.B., Giger, M.L., Birkbak, N.J., Mehrtash, A., Allison, T., Arnaout, O., Abbosh, C., Dunn, I.F., Mak, R.H., Tamimi, R.M., Tempany, C.M., Swanton, C., Hoffmann, U., Schwartz, L.H., Gillies, R.J., Huang, R.Y. and Aerts, H.J.W.L. (2019). Artificial intelligence in cancer imaging: Clinical challenges and applications. *CA: A Cancer Journal for Clinicians*, 69(2): 127–157. https://doi.org/10.3322/caac.21552.

Chang, A.E., Hayes, D.F., Pass, H.I., Stone, R.M., Ganz, P.A., Kinsella, T.J., Schiller, J.H. and Strecher, V.J. (2006). Principles of surgical therapy in oncolgy. In *Oncology: An Evidence-Based Approach* (Issue January, pp. 1–2005). https://doi.org/10.1007/0-387-31056-8.

Choi, Y.E., Kwak, J.W. and Park, J.W. (2010). Nanotechnology for early cancer detection. *Sensors*, 10(1): 428–455. https://doi.org/10.3390/s100100428.

Crosby, D., Bhatia, S., Brindle, K.M., Coussens, L.M., Dive, C., Emberton, M., Esener, S., Fitzgerald, R.C., Gambhir, S.S., Kuhn, P., Rebbeck, T.R. and Balasubramanian, S. (2022). Early detection of cancer. *Science*, 375(6586). https://doi.org/10.1126/SCIENCE.AAY9040/ASSET/F415B967-E1FC-41BB-A566-FB27F9720D77/ASSETS/IMAGES/LARGE/SCIENCE.AAY9040-FA.JPG.

Farhana, F.Z., Umer, M., Saeed, A., Pannu, A.S., Shahbazi, M., Jabur, A., Nam, H.J., Ostrikov, K., Sonar, P., Firoz, S.H. and Shiddiky, M.J.A. (2021). Isolation and detection of exosomes using Fe_2O_3 nanoparticles. *ACS Applied Nano Materials*, 4(2): 1175–1186. https://doi.org/10.1021/acsanm.0c02807.

Ferrara, N. and Kerbel, R.S. (2005). Angiogenesis as a therapeutic target. *Nature*, 438(7070): 967–974. https://doi.org/10.1038/NATURE04483.

Gautherie, D.M. (ed.). (1992). *Whole Body Hyperthermia: Biological and Clinical Aspects*. Springer Berlin Heidelberg. https://doi.org/10.1007/978-3-642-84596-3.

Habash, R.W.Y., Bansal, R., Krewski, D. and Alhafid, H.T. (2006). Thermal therapy, Part 2: Hyperthermia techniques. *Critical Reviews in Biomedical Engineering*, 34(6): 491–542. https://doi.org/4644b8e8318701ed,20ba3f104d3b2090 [pii].

Han, W. and Yu, K.N. (2009). Response of cells to ionizing radiation. p. 320. *In*: Tjong, S.C. (ed.). *Advances in Biomedical Sciences and Engineering*. Bentham Science. https://doi.org/10.2174/978160805040610901010204.

Haussmann, J., Corradini, S., Nestle-Kraemling, C., Bölke, E., Njanang, F.J.D., Tamaskovics, B., Orth, K., Ruckhaeberle, E., Fehm, T., Mohrmann, S., Simiantonakis, I., Budach, W. and Matuschek, C. (2020). Recent advances in radiotherapy of breast cancer. *Radiation Oncology*, 15(1): 1–10. https://doi.org/10.1186/s13014-020-01501-x.

Health, A. (2012). *Total Cost of Cancer Care by Site of Service: Physician Office vs Outpatient Hospital March*.

Huilgol, N.G., Gupta, S. and Dixit, R. (2010). Chemoradiation with hyperthermia in the treatment of head and neck cancer. *International Journal of Hyperthermia*, 26(1): 21–25. https://doi.org/10.3109/02656730903418283.

Hunter, B., Hindocha, S. and Lee, R.W. (2022). The role of artificial intelligence in early cancer diagnosis. *Cancers*, 14(6). https://doi.org/10.3390/cancers14061524.

Icrp. (1990). Annals of the ICRP 60. In *International Commission on Radiological Protection*. https://doi.org/10.1016/j.icrp.2006.06.001.

Jiang, W., Yuan, R., Chai, Y., Mao, L. and Su, H. (2011). A novel electrochemical immunoassay based on diazotization-coupled functionalized bioconjugates as trace labels for ultrasensitive detection of carcinoembryonic antigen. *Biosensors and Bioelectronics*, 26(5): 2786–2790. https://doi.org/10.1016/j.bios.2010.10.042.

Kassanos, P., Constantinou, L., Triantis, I.F. and Demosthenous, A. (2014). An integrated analog readout for multi-frequency bioimpedance measurements. *IEEE Sensors Journal*, 14(8): 2792–2800. https://doi.org/10.1109/JSEN.2014.2315963.

Kelkar, S.S. and Reineke, T.M. (2011). Theranostics: Combining imaging and therapy. *Bioconjugate Chemistry*, 22(10): 1879–1903. https://doi.org/10.1021/bc200151q.

Kim, K., Kim, J.H., Park, H., Kim, Y.S., Park, K., Nam, H., Lee, S., Park, J.H., Park, R.W., Kim, I.S., Choi, K., Kim, S.Y., Park, K. and Kwon, I.C. (2010). Tumor-homing multifunctional nanoparticles for cancer theragnosis: Simultaneous diagnosis, drug delivery, and therapeutic monitoring. *Journal of Controlled Release*, 146(2): 219–227. https://doi.org/10.1016/j.jconrel.2010.04.004.

Linear Accelerator (LINAC) Price Guide & Costs. (2021). Systems Radio Oncology. https://www.oncologysystems.com/resources/linear-accelerator-guides/used-linac-price.

Mantzorou, M., Tolia, M., Poultsidi, A., Pavlidou, E., Papadopoulou, S.K., Papandreou, D. and Giaginis, C. (2020). Can bioelectrical impedance analysis and BMI be a prognostic tool in head and neck cancer patients? A review of the evidence. *Cancers*, 12(3): 1–16. https://doi.org/10.3390/cancers12030557.

Mehta, S.S., Bhatt, A. and Glehen, O. (2016). Cytoreductive surgery and peritonectomy procedures. *Indian Journal of Surgical Oncology*, 7(2): 139–151. https://doi.org/10.1007/s13193-016-0505-5.

Mitsumori, M., Zeng, Z.-F., Oliynychenko, P., Park, J.H., Choi, I.B., Tatsuzaki, H., Tanaka, Y. and Hiraoka, M. (2007). Regional hyperthermia combined with radiotherapy for locally advanced non-small cell lung cancers: A multi-institutional prospective randomized trial of the international atomic energy agency. *International Journal of Clinical Oncology*, 12(3): 192–198. https://doi.org/10.1007/s10147-006-0647-5.

Mittal, S., Kaur, H., Gautam, N. and Mantha, A.K. (2017). Biosensors for breast cancer diagnosis: A review of bioreceptors, biotransducers and signal amplification strategies. *Biosensors and Bioelectronics*, 88: 217–231. https://doi.org/10.1016/j.bios.2016.08.028.

Murar, M., Albertazzi, L. and Pujals, S. (2022). Advanced optical imaging-guided nanotheranostics toward personalized cancer drug delivery. *Nanomaterials*, 12(3): 1–17. https://doi.org/10.3390/nano12030399.

Naranjo-Hernández, D., Reina-Tosina, J. and Min, M. (2019). Fundamentals, recent advances, and future challenges in bioimpedance devices for healthcare applications. *Journal of Sensors*, 2019. https://doi.org/10.1155/2019/9210258.

Pacal, I., Karaboga, D., Basturk, A., Akay, B. and Nalbantoglu, U. (2020). A comprehensive review of deep learning in colon cancer. *Computers in Biology and Medicine*, 126(September). https://doi.org/10.1016/j.compbiomed.2020.104003.

Paravati, A.J., Boero, I.J., Triplett, D.P., Hwang, L., Matsuno, R.K., Xu, B., Mell, L.K. and Murphy, J.D. (2015). Variation in the cost of radiation therapy among Medicare patients with cancer. *Journal of Oncology Practice*, 11(5): 403–409. https://doi.org/10.1200/JOP.2015.005694.

Paulides, M.M., Bakker, J.F., Linthorst, M., van der Zee, J., Rijnen, Z., Neufeld, E., Pattynama, P.M.T., Jansen, P.P., Levendag, P.C. and van Rhoon, G.C. (2010). The clinical feasibility of deep hyperthermia treatment in the head and neck: new challenges for positioning and temperature measurement. *Physics in Medicine and Biology*, *55*(9): 2465–2480. https://doi.org/10.1088/0031-9155/55/9/003.

Pilewskie, M. and Morrow, M. (2018). Margins in breast cancer: How much is enough? *Cancer*, 124(7): 1335. https://doi.org/10.1002/CNCR.31221.

Pothipor, C., Wiriyakun, N., Putnin, T., Ngamaroonchote, A., Jakmunee, J., Ounnunkad, K., Laocharoensuk, R. and Aroonyadet, N. (2019). Highly sensitive biosensor based on graphene–poly (3-aminobenzoic acid) modified electrodes and porous-hollowed-silver-gold nanoparticle labelling for prostate cancer detection. *Sensors and Actuators, B: Chemical*, 296(March): 126657. https://doi.org/10.1016/j.snb.2019.126657.

Ratajczak, K., Krazinski, B.E., Kowalczyk, A.E., Dworakowska, B., Jakiela, S. and Stobiecka, M. (2018). Optical biosensing system for the detection of survivin mRNA in colorectal cancer cells using a graphene oxide carrier-bound oligonucleotide molecular beacon. *Nanomaterials*, 8(7): 13–19. https://doi.org/10.3390/nano8070510.

Rosenblantt, E. and Zubizarreta, E. (eds.). (2017). *Radiotherapy in Cancer Care: Facing the Global Challenge*. https://doi.org/10.1016/S0168-8278(94)80112-6.

Ross, J.S., Dzara, K. and Downing, N.S. (2015). Efficacy and safety concerns are important reasons why the FDA requires multiple reviews before approval of new drugs. *Health Affairs*, 34(4): 681–688. https://doi.org/10.1377/hlthaff.2014.1160.

Sarode, G.S., Sarode, S.C., Kulkarni, M., Karmarkar, S. and Patil, S. (2016). Role of bioimpedance in cancer detection: A brief review. *International Journal of Dental Science and Research*, 3(1): 15–21. https://doi.org/10.1016/j.ijdsr.2015.11.003.

Sharifi, M., Hasan, A., Attar, F., Taghizadeh, A. and Falahati, M. (2020). Development of point-of-care nanobiosensors for breast cancers diagnosis. *Talanta*, 217(April): 121091. https://doi.org/10.1016/j.talanta.2020.121091.

Sharifianjazi, F., Jafari Rad, A., Bakhtiari, A., Niazvand, F., Esmaeilkhanian, A., Bazli, L., Abniki, M., Irani, M. and Moghanian, A. (2022). Biosensors and nanotechnology for cancer diagnosis (lung and bronchus, breast, prostate, and colon): A systematic review. *Biomedical Materials (Bristol)*, 17(1). 12002. https://doi.org/10.1088/1748-605X/ac41fd.

Staruch, R., Chopra, R. and Hynynen, K. (2012). Hyperthermia in bone generated with MR imaging–controlled focused ultrasound: control strategies and drug delivery. *Radiology*, 263(1): 117–127. https://doi.org/10.1148/radiol.11111189.

Szasz, A., Iluri, N. and Szasz, O. (2013). Local Hyperthermia in Oncology – To Choose or not to Choose? In *Hyperthermia* (pp. 1–82).

Therapy, R. and Cdks, B. (2014). The science behind radiation therapy. *American Cancer Society*, 15.

Trujillo-Romero, C.J., Rico-Martínez, G. and Gutiérrez-Martínez, J. (2018). Thermal ablation: An alternative to bone cancer. *Investigación En Discapacidad*, 7: 36–47.

Tucker, A.S., Fox, R.M., Member, S. and Sadleir, R.J. (2013). Improved howland current source with lead-lag compensation. *IEEE Transactions on Biomedical Circuits and Systems*, 7(1): 63–70.

Uday Kumar, S., Bharat, B., Poornima, D., Matai, I., Abhay, S. and Gopinath, P. (2013). Emerging applications of nanoparticles for lung cancer diagnosis and therapy. *International Nano Letters*, 2013: 1–17. https://doi.org/10.1155/2013/148578.

Van Der Zee, J., Rietveld, P.J.M., Broekmeyer-Reurink, M.P., Wielheesen, D.H.M. and van Rhoon, G.C. (2002). Hyperthermia in recurrent breast cancer: from experimental oncology to standard practice. *Experimental Oncology*, 2002(24): 45–50.

Vedsted, P. and Olesen, F. (2015). A differentiated approach to referrals from general practice to support early cancer diagnosis – The Danish three-legged strategy. *British Journal of Cancer*, 112: S65–S69. https://doi.org/10.1038/bjc.2015.44.

Wang, L., Xiong, Q., Xiao, F. and Duan, H. (2017). 2D nanomaterials based electrochemical biosensors for cancer diagnosis. *Biosensors and Bioelectronics*, 89: 136–151. https://doi.org/10.1016/j.bios.2016.06.011.

Worldwide Cancer Data. (2022). World Cancer Research Fund International. https://www.wcrf.org/cancer-trends/worldwide-cancer-data/.

Yang, G., Xiao, Z., Tang, C., Deng, Y., Huang, H. and He, Z. (2019). Recent advances in biosensor for detection of lung cancer biomarkers. *Biosensors and Bioelectronics*, 141(April): 111416. https://doi.org/10.1016/j.bios.2019.111416.

Zhou, Y., Mendonca, S.C., Abel, G.A., Hamilton, W., Walter, F.M., Johnson, S., Shelton, J., Elliss-Brookes, L., Mcphail, S. and Lyratzopoulos, G. (2017). Variation in "fast-track" referrals for suspected cancer by patient characteristic and cancer diagnosis: Evidence from 670 000 patients with cancers of 35 different sites. *Nature Publishing Group*, 118. https://doi.org/10.1038/bjc.2017.381.

Chapter 2

Bioimpedance and Cancer Detection

Rafael Gonzalez-Landaeta[1,*] and *César Antonio González-Díaz*[2]

Biological tissues have electrical properties linked to their constitution, size, and shape. Tissues that contain cancer cells present differences in electrical properties due to the structural and histological characteristics of such cells, which differ from healthy cells. This chapter describes the role of Bioimpedance in the estimation of the electrical properties of malignant tissues. The first part describes the fundamentals of Bioimpedance as a useful technique to estimate the electrical properties of tissues/cells; that is, conductivity, resistivity, and permittivity. Secondly, the electrical properties of tissues that contain malignant cells are presented, including some studies where these properties have been estimated in different tissues, namely, breast, prostate, and skin. Thirdly, a short description of the electrodes and the electronic instrumentation used in Bioimpedance for the detection of cancerous tissues/cells is presented. Finally, applications of Electrical Impedance Spectroscopy (EIS) in the detection of Circulating Tumoral Cells (CTC) and breast cancer, as well as an introduction to the Magnetic Induction Spectroscopy (MIS) as a valuable non-contact technique to detect neoplasms in breast tissue have been presented.

[1] Departamento de Ingeniería Eléctrica y Computación, Instituto de Ingeniería y Tecnología, Universidad Autónoma de Ciudad Juárez. Av. del Charro 450 nte. Ciudad Juárez, Chihuahua, México. C.P. 32310.

[2] Escuela Superior de Medicina. Departamento de Posgrado e Investigación Instituto Politécnico Nacional. Plan de san Luis esquina con Diaz Miron, s/n, colonia casco de santo Tomás. Ciudad de México. C.P. 11320.

Email: cgonzalezd@ipn.mx

* Corresponding author: rafael.gonzalez@uacj.mx

2.1 Bioimpedance basics

Since the middle of the last century, the electrical properties of tissues have been studied using Bioimpedance. In the first works done in this field, the intention was to estimate the water content in different tissues. This opened the way to new investigations where electrical properties have been obtained to detect several pathologies, both at the tissue and cellular levels. In this sense, Bioimpedance has been used in the development of various devices ranging from small biosensors to electrical impedance tomographs. Among the diversity of applications, Bioimpedance has been widely used in detecting malignant tissues/cells through *ex vivo* and *in vivo* measurements, where the different electrical properties are evaluated at the microscopic and macroscopic levels. In this sense, here is a brief description of the electrical properties that can be obtained using Bioimpedance, and how these properties allow differentiation between healthy and malignant tissues/cells. On the other hand, the measurement methods and the electronic instrumentation involved in the detection of malignant tissues/cells are also described.

2.1.1 Electrical properties of the cells

Electrical impedance is the term used to estimate the resistance of a given system to electric current when an electric field is applied to it. It is estimated by the ratio between the voltage (V) and the current (I); that is $Z = V/I$. In biological tissues, it is called Electrical Bioimpedance, or simply Bioimpedance, where a voltage is produced due to the distribution of the current injected into the tissue. The Bioimpedance of tissue is determined by the electrical conductivity (σ), the dielectric permittivity (ε), and geometrical factors (length, L, and area, A) of the tissue (see Equation 2.1) and can be used to estimate the electrical behavior of the tissues from a macroscopic point of view (Rigaud et al., 1996).

$$Z^* = \frac{1}{\sigma + j\omega\varepsilon}\frac{L}{A} \tag{2.1}$$

Z^* indicates a complex quantity (combination of real, R, and imaginary, X, components), and $\varepsilon = \varepsilon_r\varepsilon_0$, where ε_r is the relative permittivity and ε_0 is the permittivity of vacuum (8.85×10^{-12} F/m). The conductivity determines the mobility of charge carriers (ions in biological tissues), and the dielectric permittivity is related to the polarization of media as it determines the mobility of bound charges when an electric field is applied (Grimnes and Martinsen, 2015). The permittivity is used to assess the dielectric properties of biological materials. The conductivity and the permittivity of tissues depend on their composition and structure and are independent of geometrical constraints (Ivorra, 2003).

Considering only one cell suspended in an ionic solution, the ionic solution outside the cell is called the extracellular medium. Inside the cell, there also exists an ionic medium where the metabolic processes occur, and it is called the intracellular medium. Both media are separated by the cell membrane, which acts as a barrier to the ionic flow. The conductivity of both media is higher than that of the cell membrane. So, electrically, the extracellular and intracellular media can be modeled like resistors, and the cell membrane can be modeled like a capacitor.

Biological tissues are not perfect conductors or perfect capacitors. They could be considered conductors with capacitive properties and capacitors with conduction properties. Due to these non-ideal properties, once an electric field is applied to a biological tissue (assume a step function), a certain time is required for it to reach a new electrical equilibrium: this process is called *relaxation* and depends on the polarization (perturbation of the charge distribution) mechanism (Grimnes and Martinsen, 2015). Since suspended cells can be emulated as RC circuits, it is feasible to say that the electrical properties of tissues are frequency-dependent.

Dispersions represent the *relaxation* mechanisms in the frequency domain. Three zones of change from a specific value to another specific value define the dispersions called alpha (α), beta (β), and gamma (γ). At low frequencies (< 1 kHz), the α-dispersion describes the accumulation of opposite charges between both sides of the cell membranes, creating a large dipole due to the electric field applied (surface polarization) (Pethig, 1984; Scholz and Anderson, 2000). In the radiofrequency region (1 kHz – 100 MHz), the β-dispersion is attributed to the charging and discharging mechanism of the cell membrane (Pethig, 1984; Rigaud et al., 1996). The γ-dispersion, which occurs at higher frequencies (≥ 1 GHz), is attributed to the relaxation of the water molecules (Schwan, 1957).

The terms "complex conductivity" and "complex permittivity" are mainly used to estimate the electrical properties of the cells from a microscopic point of view (Rigaud et al., 1996), and are defined by Equations 2.2 and 2.3, respectively:

$$\sigma^* = \sigma + j\omega\varepsilon_0\varepsilon_r \tag{2.2}$$

$$\varepsilon^* = \varepsilon_r - j\frac{\sigma}{\omega\varepsilon_0} \tag{2.3}$$

From Equation (2.2), as the frequency increases the conductivity increases, which is attributed to the fact that the cell membrane is short-circuited, so its contribution to the distribution of the injected current is negligible. In the case of the complex permittivity (Equation (2.3)), it becomes prominent at lower frequencies, where the capacitive effect of the cell membrane prevails. So, when a multifrequency current is injected into a tissue, the path for higher frequency currents includes both intracellular and extracellular media because the capacitive effects of the cell membrane disappear (Grimnes and Martinsen, 2015; Webster, 1990).

The complex conductivity and permittivity can be obtained indirectly by measuring Z^*. However, the geometry of the tissue could achieve misleading results. For this, a scaling factor called "cell constant" is used to relate the complex impedance and the conductivity (or the permittivity) of the tissue (Ivorra, 2003). The cell constant is $K = A/L$, being A and L the area and length of the measurement cell, respectively.

2.1.2 Electrical properties of malignant tissues

Cancer cells have structural differences compared with healthy cells and have histological characteristics that can be reflected in their electrical properties, namely the conductivity, resistivity ($\rho = 1/\sigma$) of the intracellular and extracellular media, and the permittivity of the cell membrane. Bioimpedance has become a useful technique for detecting malignant tissues, where measurements can be performed *in vivo*, and *in vitro* in controlled environments. In the literature, most of the studies are focused on breast, prostatic and, skin cancers (Sarode et al., 2016). However, other experimental studies have used Bioimpedance to study the electrical properties of cancer cells in muscles (Blad and Baldetorp, 1996), lymph nodes (Malich et al., 2002), and colon (Haemmerich et al., 2003).

In vitro measurements of breast cancer have revealed differences in the conductivity and the permittivity, which have been attributed to the structural and cellular inhomogeneities of tumoral tissue (Surowiec et al., 1988). In the range of 499 Hz and 1 MHz, an increment in the low-frequency-limit resistance of carcinoma in comparison with the mammary gland has been revealed (Jossinet, 1998). Nevertheless, these results disagree with those obtained by Surowiec et al. (1988), who obtained an increase in the conductivity and the permittivity in the radiofrequency range. The discrepancies between these two studies can be attributed to the measurement conditions, which may alter the results. In the case of *in vitro* measurement, the conditions of the environment where the cells are immersed define the reliability of the technique. Living tissues have temperature coefficients that affect the conductivity and the permittivity of the cells (Pethig, 1984), so temperature changes could distort the values of the conductivity and the permittivity. Besides, the electrical properties of the excised tissues differ from those of the living tissues because of the lack of blood irrigation. However, in the case of breast tumors, these differences are considered negligible within the first 30 minutes after tissue removal since the passive electrical properties of the cells remain stable during this time (Heinitz and Minet, 1995).

In vivo measurements are less common but provide more reliable results. Morimoto et al. (1990) used a coaxial needle electrode inserted into the breast to estimate the bioimpedance of breast tumors based on the three-electrodes method in the range of 0–200 kHz. The impedance (Z), the extracellular (R_e) and intracellular (R_i) resistances, and the membrane capacitance (C_m) were obtained. R_e and R_i of malignant tumors were considerably higher than benign tumors, while C_m was considerably lower. Regarding C_m, these results are opposite to those obtained by Singh et al. (1979), where higher electrical permittivity values were obtained in breast tumors in the range of 0.1 Hz – 1 kHz. The discrepancy between these two studies can be explained by the fact that in the study of Singh et al. (1979), the measurements were made with ECG electrodes attached to the skin of the breast, so the effects of polarization at the skin interface could affect the values of C_m. In addition, the measurements were not made directly on the breast tumor.

Bioimpedance has also been used to detect skin cancer based on the estimation of four different indexes, which have been used as indicators of the electrical impedance in the complex space (Emtestam et al., 1998). These indexes not only can be used to detect basal cell carcinoma (BCC), but also allergic reactions, irritation, and hydration of the skin (Beetner et al., 2003). Such indexes are described by Equations 2.4–2.7:

$$\text{MIX} = \frac{abs(Z_{20\text{ kHz}})}{abs(Z_{500\text{ kHz}})} \quad \text{Magnitude Index} \tag{2.4}$$

$$\text{PIX} = \arg(Z_{20\text{ kHz}}) - \arg(Z_{500\text{ kHz}}) \quad \text{Phase Index} \tag{2.5}$$

$$\text{RIX} = \frac{\text{Re}(Z_{20\text{ kHz}})}{abs(Z_{500\text{ kHz}})} \quad \text{Real part Index} \tag{2.6}$$

$$\text{IMIX} = \frac{\text{Im}(Z_{20\text{ kHz}})}{abs(Z_{500\text{ kHz}})} \quad \text{Imaginary part Index} \tag{2.7}$$

Intact skin yields a straight-line plot spectrum between 10 kHz and 1 MHz; however, 20 kHz and 500 kHz were considered sufficient to characterize most of the information in this frequency range (Nicander et al., 1996). The most significant difference found between BCC and normal skin is manifested in the MIX and IMIX values (Beetner et al., 2003; Emtestam et al., 1998), where both indexes were lower for BCC. Distinction between nonmelanoma skin cancer and benign nevi by electrical bioimpedance is also feasible. Aberg et al. (2004) demonstrated the possibility of differentiating skin cancer from benign nevi with 75% of specificity in the range of 1 kHz – 1 MHz. However, the reasons why the impedance values are different have not been investigated, although it is believed to be due to the histopathological difference between both tissues.

In the case of prostate tissue, significant differences in permittivity have been observed between prostate cancer (CaP) and benign prostatic hyperplasia (BPH). Halter et al. (2009) observed a significantly higher permittivity in CaP at 100 kHz, which is attributed to the variability of the glandular and cellular content. The mean conductivity of CaP tissue was slightly higher than that of BPH between 0.1 kHz and 100 kHz, and significantly lower than that of non-hyperplastic glandular tissue and Stroma, specifically at 0.1 kHz. This is attributed to the fact that the extracellular space has been reduced by the proliferation of cancer cells causing neoplastic cells to infiltrate the Stroma (Halter et al., 2008).

Bioimpedance measurements not only detect the differences in electrical properties between malignant and benign tissues but have also shown their usefulness in estimating the stage of cancer. Qiao et al. (2010) have shown that the relaxation frequency increases as the stage of cancer advances. A possible hypothesis that can explain this, is based on the fact that the capacitance in cancer cells is lower than in normal cells, and this value decreases as cancer progresses (Han et al., 2007). Therefore, this makes the tissue less polarizable, causing the relaxation frequency to increase (Gregory et al., 2012).

2.1.3 Electrodes and probe

In biological tissues, ions are the charge carriers, giving rise to ionic currents. In wires and electronic circuits, the charge carriers are electrons, giving rise to electronic currents. In that sense, there must be an interface that allows the interaction between the biological tissues and the electronic instrumentation. The electrodes allow the exchange of charge carriers because they transform ionic currents into electronic currents and vice versa. It is a transduction function, and it can be carried out with or without galvanic contact. Although it is not the intention of this chapter to discuss the electrode theory in depth, it is important to understand the fundamentals of the electrode transduction mechanism.

When a metal comes into contact with an electrolyte, double layers are formed near the electrode surface where electric charges of opposite polarity are distributed across the interface, acting as a capacitor (Rigaud et al., 1996). Depending on the rates of diffusion of ions into and out of the metal, an equilibrium is reached that gives rise to a half-cell potential. The value of this potential depends on the material of the electrode, the ionic concentration in the electrolyte, the temperature, and other second-order factors (Webster, 2009), and it is produced under conditions of no current flowing through the interface. When a current crosses the interface, the half-cell potential is altered. If such alteration is considerable, the electrode is considered polarizable. If such alteration is negligible, then the electrode is considered non-polarizable (Grimnes and Martinsen, 2015). The former acts as a capacitor and does not allow charge exchange at the metal-electrolyte interface. In this kind of electrode, no DC currents are available, unless a high dc voltage is applied. The latter acts as a resistor and it does allow an unhindered transfer of charges between the metal and the electrolyte. For applications that involve the injection of AC signals, platinum electrodes are often used as they have a lower polarization impedance than electrodes made of silver and stainless steel (Rigaud et al., 1996). According to Ivorra (2003), electrodes made of noble metals (platinum, gold) or stainless steel are used in bioimpedance measurements to avoid tissue damage or electrode degradation. In Electrical Impedance Spectroscopy (EIS), when using polarizable electrodes, the effect of the double layer can be notorious at low frequencies (< 10 kHz). To compensate for this effect, different approaches have been proposed (Ishai et al., 2013).

Cancer detection by electrical impedance can be carried out by *in vivo* measurements, which can be invasive, minimally invasive, or non-invasive. *In vitro* measurements can also be performed, which are carried out on a piece of excised tissue, on a single cell, or on cells in suspension. When measuring in tissues (*in vivo* or *ex vivo*), it is common to use arrangements made up of multiple electrodes to cover a larger measurement area of the tissue and, in some cases, to be able to reconstruct images from impedance values of the tissues involved. By performing multiplexing between the different injection electrodes and the voltage detection electrodes it is possible to perform tetrapolar measurements. Bioimpedance measurements through the surface of the skin involve several factors that can mask impedance measurements, e.g., skin conditions and movement artifacts. To avoid this, invasive *in vivo* measurements are preferable, where needle electrodes that penetrate the tissue are used.

For *ex vivo* measurements, stainless steel needles with an exposed pair of electrodes have been proposed. In this way, it is possible to measure the electrical properties of tissues simultaneously with the extraction of the tissue (Mishra et al., 2013). For this, an 18-gauge stainless steel needle has been used. It consists of two concentric elements separated by 54 μm of air. Both elements are insulated by a polyimide coating adhered to the inner element, allowing only two electrodes to be exposed for tissue contact. With this probe, measurements are made using a bipolar configuration.

For the detection of skin cancer by EIS, many studies have used the spectrometer manufactured by SciBase AB (Hudinge, Sweden). This system initially used a probe made up of four concentric circular gold electrodes deposited on a ceramic. The inner electrode is a sink electrode, and the next electrode is a guard, helping to reduce the effects of surface currents. The two outermost electrodes are the source electrodes (Beetner et al., 2003). Between these two electrodes, a virtual electrode is created which can be "moved" by changing the distribution of current between the source electrodes (Aberg et al., 2004). This probe allows measurement at five different depths by varying the position of the virtual electrode. A recent model changed the probe configuration to a square array made up of five disposable electrodes covered with gold microinvasive pins. Similarly, four levels of depth can be achieved by alternating between the different electrode bars (Braun et al., 2017).

Cancer detection by electrical bioimpedance can also be performed at cellular level. One way to do this is by using microelectrodes to measure the impedance of a single cell. Another way is by measuring impedance on cells in suspension, which minimizes errors such as variability in the size and shape of each cell, polarization effects of the electrodes, and heating effects from high current density (Guofeng Qiao et al., 2012). To measure cells in suspension, chambers have been proposed that allow controlling factors such as temperature, volume, and cell distribution within the chamber.

2.1.4 Instrumentation

To extract the electrical properties of tissues and/or cells, the response to a certain stimulus containing a specific range of frequencies must be measured, hence the term EIS. The stimulus can be a step function, electronic noise, a sinusoidal signal, or a chirp signal, among others (Bertemes-Filho, 2018; Xu and Hong, 2020). There are different techniques based on the number of electrodes used to apply the stimulus and measure the response (Figure 2.1a). The bipolar (2 electrodes), the tripolar (3 electrodes), and the tetrapolar techniques (4 electrodes) have been used. Each of these techniques has its advantages and disadvantages, which have been well described in the literature (Grimnes and Martinsen, 2015; Grossi and Riccò, 2017; Kassanos, 2021; Morimoto et al., 1990).

In EIS, the most common technique is tetrapolar since it reduces the effects of the contact impedance of the electrodes and the effects of polarization. In this technique there are four electrodes connected to the tissue/cell (Figure 2.1b): two electrodes (Z_{HC}, Z_{LC}) inject current (or voltage) and the other two electrodes (Z_{HP}, Z_{LP}) measure the response to that current (or voltage). Each block of the injection and the detection systems must have

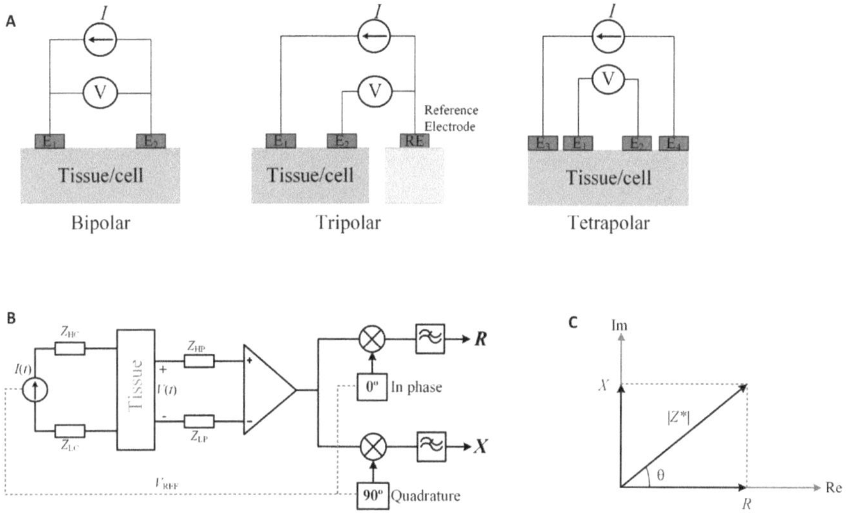

Figure 2.1. (A) Different techniques based on the number of electrodes used in Bioimpedance measurements, (B) Block diagram of the instrumentation involved in Bioimpedance measurements, (C) Phasor diagram where the real and imaginary parts of Z^* are depicted.

certain characteristics to reduce the errors that could cause misinterpretations of the impedance measurement. Briefly:

Oscillator: It is responsible for generating the waveform. In EIS applied to cancer detection, it is necessary to be able to generate a wave (usually sinusoidal) in a frequency range between 100 Hz and 1 MHz. For this, the oscillator must have high linearity (low THD[1]) over a wide frequency range, which affects the precision of the EIS system. There are oscillators implemented with off-the-shelf components and embedded on a chip.

Current source: The signal coming from the oscillator is an alternating voltage. To transform that voltage into a current that is injected into the target tissue/cell, a circuit known as a V-I converter is required, specifically a Voltage Controlled Current Source (VCCS). This circuit must have a high input impedance (Z_i), to reduce the loading effects when connecting it to the oscillator. It must also have a high output impedance (Z_S) to reduce drifts in the injected current when the biological material is connected to the VCCS output. However, the values of Z_i and Z_S are frequency dependent, so it is a challenge to keep the values of these impedances high in multifrequency applications.

[1] Total Harmonic Distortion: is a measurement of the harmonic content of a signal that indicates the level of waveform distortion.

Front-end: This stage is responsible for measuring the voltage response due to the current injected into the tissue. *V*(*t*) is a signal modulated in amplitude by the impedance (Z_X) of the tissue. The amplitude level of this signal is between 100 μV and 100 mV (Bertemes-Filho, 2018), so it is susceptible to errors from common mode signals, electrode contact impedance noise, and movement artifacts.

Demodulator: The measurement system must be able to extract the real part and the imaginary part of Z^*. The electrical impedance of the tissue not only modulates the amplitude (AM) of the injected current but also causes a phase shift (PM). Synchronous demodulation is a technique that has proven to be robust in bioimpedance measurements. It offers a high dynamic range and rejects non-synchronous signals. For this, the AM and PM signal is multiplied by a reference wave that is synchronous with the injected current, but can be in-phase (0°) or in quadrature (90°) (Figure 2.1b) with respect to the reference wave. The measurements in-phase (I) allow to extract the real part of Z^*, and the measurements in quadrature (Q) allow to extract the imaginary part of Z^*. From these components, it is possible to estimate the magnitude and phase of Z^* (Figure 2.1c). However, any mismatch between the IQ channels must be avoided to reduce phase errors (Yang et al., 2006).

2.2 Bioimpedance applications

In 1959, the Nobel laureate Richard Feynman, pronounce an emblematic speech… "There's Plenty of Room at the Bottom"… in that time, Feynman wanted to encourage his colleagues to think about the possibility of developing very small devices at the micro and nano scales, for instance, he mentioned the possibility of storing the British encyclopedia on the head of a pin, even more, he mentioned that there are no physical reasons that make it impossible to store information at the molecular level. It seems that Feynman's message was not fully understood by the scientific community till the last two decades. Now, several researchers worldwide work on the design and implementation of micro-devices at micro and nanoscale dimensions, and micro and nano biosensors are the current challenges in lab-on-a-chip development.

2.2.1 Circulating tumoral cells detection (CTC)

Breast cancer (BC) is an important health problem worldwide. In general, BC is the cause of high levels of mortality due to the absence of early identification, lack of response to therapy, and metastasis. Through several investigations in which the process of metastasis has been studied, the possibility to detect the presence of Circulating Tumoral Cells (CTC) exist, which could be a relevant signal of dissemination of the disease, prognosis, and response to treatment

(Samia et al., 2012) (Morán et al., 2014) (Giuliano et al., 2014). Metastasis causes high morbidity/mortality associated with BC, and 30% of the patients with BC develop micro-metastases in connective tissue. Approximately 50% of women with localized BC, treated with surgery show high recurrence and poor outcomes by oncological treatment, such as radiotherapy and chemotherapy, often ineffective because the status of CTCs in the body is not well known (Bray et al., 2018) and (Sua et al., 2011).

Currently, several techniques for the detection of CTC based on biological and physical properties have been proposed, since their detection might represent a non-invasive technique to monitor the evolution of cancer patients. However, the amount of CTC in peripheral blood is very low compared to the total blood cells (Markou et al., 2011). So, improvements in these technologies are continually being sought, which require expensive equipment and specialized processes, which require emerging technological proposals that may represent CTC detection alternatives that can be used in low-income healthcare levels. An ideal solution should be a method with very high sensitivity, good reproducibility, intuitive, easy to use, as well as practical implementation in the clinical field. Research focused on new techniques to deter these drawbacks to achieve low-cost and easily accessible technology is still necessary (Sua et al., 2011) (Adams et al., 2015).

In a study reported by Xu et al. (2011), the feasibility of a technique to perform the separation of CTC from whole blood by immunomagnetic techniques is shown. Han et al. (2007) showed the feasibility of differentiating tumor cells from normal ones through the characterization of their electrical properties in a study (Han et al. 2007). CTC detection through EIS measurements assisted with magnetic nanoparticles (MnP) bound to an antibody (Ab) to recognize proteins in the cell surface has been proposed (Silva et al., 2014). The main idea is to add nanoprobes "Magnetic Nanoparticle-Antibody" (MnP-Ab) on the CTC membrane and then, through a microfluidic system, separate immunomagnetically by the effect of a permanent magnet on the surface of a microelectrode array, thus perform EIS measurements for detection.

This subsection presents the advances in the development of a new CTC biosensor based on EIS measurements. Different types of cancer cells commonly used in the study of BC (MCF-7, MDA-MB-231 and SK-BR-3) were employed as potential targets for the use of an MnP-Ab nanoprobe. The three cell lines were quantified and incubated with the corresponding doses of bioconjugate Mnp-Ab according to their predominant surface protein (Huerta-Nuñez et al., 2019). Immunomagnetic isolation of the cells and subsequent measurement by multifrequency electrical bioimpedance was developed by the use of a ScioSpec-ISX3 impedance measurement system. The biosensor designed for isolation/detection of BC cells by the use of a microfluidic system and magnetic nanoparticle-assisted with EIS measurements is technically

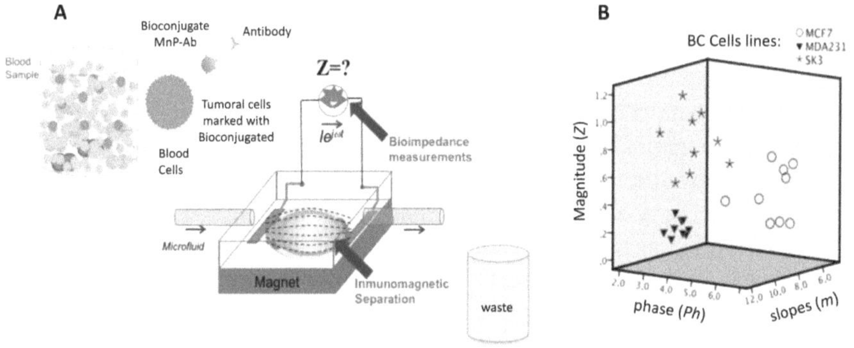

Figure 2.2. (A) Schematic representation of CTC detection by EIS measurements assisted with bioconjugated MnP-Ab. A microfluid system allows immunomagnetic anchoring by the use of a permanent magnet. (B) Three-dimensional representation of principal components for detection and classification of different types of BC cell lines.

viable to detect low concentrations of cancer cells and magnetic nanoparticles (Huerta-Nuñez et al., 2019).

The EIS magnitude parameter at low frequencies seems to represent the best option to get relevant sensitivity. A main factors analysis allows for identifying sensitive components as representative elements of the EIS spectra; thus, components for Magnitude (Z), phase (Ph) and slopes (m) were selected as a three-dimensional representation for cell lines characterization. The results indicate that the detection and classification of CTC through the proposed technique is feasible and allows supporting the operation principle of a biosensor for detection of CTC based on EIS measurements assisted with nanotechnology (González-Díaz and Golberg, 2020). Figure 2.2 shows the schematic representation of CTCs detection by EIS measurements assisted with MnP-Ab, as well as a three-dimensional representation of the main factor for classification of BC cell lines.

2.2.2 Breast cancer detection by EIS

BC is an uncontrolled malignant proliferation. In the search for diagnostic methods for this disease, it has been found that bioimpedance measurements of tumor tissues reveal important data. The commonly accepted concept of electrical bioimpedance is the opposition that a body presents to the passage of an alternating current through it, where the body is a biological tissue.

The structure of the tissues has to be considered since it affects the bioimpedance and the electrical properties of biological tissue change according to the frequency of the current; at high frequencies, the conductivity of the tissue increases until it becomes constant at frequencies from

10 to 100 MHz. The opposite is observed with the permittivity, which decreases at higher frequencies (Schwan, 1957).

Currently, the use of nanomaterials for the diagnosis of numerous diseases is somewhat feasible. These materials also work as drug carriers for the treatment of some pathologies and have also been used coupled with antibodies to have greater specificity when releasing the drug. Magnetic nanoparticles coupled to an antibody that recognize an overexpressed antigen as a receptor in breast cancer cells could help in the diagnosis/treatment of the disease, modifying the cellular structure and increasing its conductivity through nanomaterials.

The infusion of "magnetic nanoparticle - anti Her2" (MnP-Ab) inside of a biological organism and its physiological effect in tumor tissue through electrical bioimpedance measurements at different frequencies has been evaluated to show changes in tissue electrical conductivity. BC was induced in rats of the Sprague Dawley strain with 1-Methyl 1-Nistrosurea (MNU) intraperitoneally (30 mg/kg), according to the model reported by Thompson et al. (1983). Five rats with tumor development were anesthetized with acepromazine (300 mg/100 ml), butorphanol (10 mg/50 ml), and ketamine (1000 mg/10 ml), and a central venous catheter was placed by jugular venous dissection for systemic MnP-Ab infusion. The "magnetic nanoparticle–anti Her-2" bioconjugate was made according to a previous report (Silva et al., 2015). An impedance analyzer (Agilent: mod. 4294 A) was configured to perform impedance measurements in a broad range of 100 Hz to 100 MHz at 181 logarithmically spaced steps in three conditions: Basal (before MnP-Ab infusion), t = 0 hrs (immediately after MnP-Ab infusion), and t = 24 hrs (24 hours post- MnP-Ab infusion). The placement of a magnet in the tumor region was used in order to attract the magnetic nanoparticles to the region of interest. In all cases, bioimpedance measurements were normalized concerning their contralateral value in healthy tissue. The systemic injection of MNP-Ab that recognizes an antigen in cancer cells promotes a lower bioimpedance of cancer tissue at low frequencies and 24 hours post-infusion; such a condition allowed higher tissue electrical conductivity values. Specific descriptions of the findings could be found in a study reported by Silva et al. (2015).

2.2.3 Breast cancer detection by MIS

In general, limitations in the treatment of BC are associated with a late diagnosis. The scarcity or non availability of modern imaging systems at the primary level of medical attention promotes the risk of BC in women candidates since early detection opportunities in the early stages of the disease by screening studies were lost and it became impossible to continuously monitor the therapeutic response. Currently, no portable, non-invasive and low-cost technology has the potential to assist in early diagnosis, as well as give timely alarms in remote places that do not have modern imaging systems.

Hyper-vascularization commonly found in malignance represents important changes in the tissue's electrical properties. Such a condition suggests the use of the bioimpedance technique in order to detect cancer. Hutten et al. (1998) reported the use of EIS to produce relevant bioimpedance parameters associated with abnormal conditions. Newell et al. (1996) and Holder et al. (1999) proposed the Tomography through Electrical Impedance (EIT) for monitoring neoplasms in different organs/tissues. EIT depends on an adequate electrode-skin galvanic coupling, and this is frequently affected by sweating or skin hydration levels, thus it represents the main disadvantage. On the other hand, Griffiths et al. (1999) and Griffiths (2001) have shown the technical feasibility of using mono-frequency magnetic fields through non-contact coils for monitoring the status of organs and/or tissues. Al-Zeiback (1993) and Korzhenevskii (1997 and 1999), proposed the use of Magnetic Induction Tomography (MIT) to produce contactless bioimpedance imaging. The potential use of EIT and MIT to assist the diagnosis of BC is a feasible technique because the conductivity in a malignant tissue is different with respect to a healthy condition. Burdette (1982), reported that normal and BC tumors have different conductivities, for instance, 4 $mS \cdot cm^{-1}$ and 8 $mS \cdot cm^{-1}$, respectively.

Bioelectrical measurements through magnetic induction at multiple frequencies have been proposed as a relevant technique for non-contact detection of neoplasms in breast tissues. An inductor-sensor system to produce low intensity magnetic fields at non ionizing frequencies, adapted in an ergonomic form to the human breast shape was designed; the prototype works based on the Magnetic Induction Spectroscopy (MIS) technique.

The induction of currents in conductive materials through oscillating magnetic fields (of a certain frequency) is explained by the Faraday's law of induction (alternating magnetic fields induce electric potentials in conductive mediums). This potential in turn induces a flow of electric currents (Eddy currents) inside the medium proportional to its conductivity and the electric potential induced, so the greater the conductivity of the material, the greater the energy that the material must absorb. It has been documented that the presence of neoplasms in breast tissue increases its volumetric conductivity (Kim et al., 2007), so the influence of magnetic fields of a certain frequency in a volume of normal breast tissue should induce a lower amount of Eddy currents than in those breast tissues with mammographic findings.

A volume of breast tissue placed between an inductor coil and a sensor coil was considered as a case study (see Figure 2.3). An alternating current

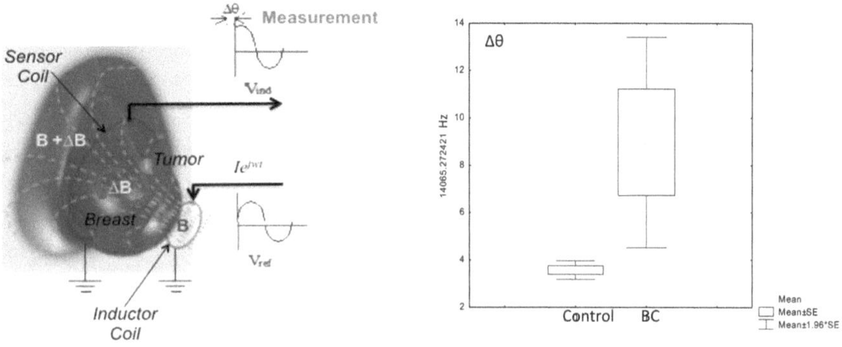

Figure 2.3. (A) Biophysical concept for non-invasive detection of malignant neoplasms in breast tissue by MIS measurements. (B) Inductive phase shift (Δθ) at a specific frequency for Control and BC groups (mean values and dispersion).

is injected into the inductor coil, and it induces a magnetic field **B** in the second coil (sensor). The phase of **B** is tied to the phase of the voltage (V_{ref}) in the field coil itself as a reference. The bulk of the tissue is considered non-magnetic with conductive properties. The induced current in the breast volume under study causes a modification Δ**B** in the initial **B** (Griffiths et al., 1999). The final result is a composite field **B** + Δ**B** and it is detected in the secondary coil out of phase with respect to **B** by an angle θ (González-Díaz et al., 2017). Thus, the phase shift ($\Delta\theta$) between V_{ref} and the voltage induced in the secondary coil (V_{ind}), can be calculated by Equation 2.8:

$$\Delta\theta = \theta(V_{ind}) - (V_{ref}) \qquad (2.8)$$

The presence of neoplasia modifies the electrical properties of the breast volume, thus the conductivity increases (Martinsen and Grimnes, 2011) (González-Díaz et al., 2017). It means that conductivity increase promotes that the influence of a magnetic field is reflected as a selective induction of currents in the neoplastic region, which in turn causes a greater disturbance of **B** and consequent increase in Δ**B**, thus volumetric changes in $\Delta\theta$ are detected.

Women who were assisted with mastography screening were enrolled as volunteer patients separated into control and BC groups. All volunteers were explored by the MIS technique to determine multifrequency $\Delta\theta$ in breast tissue in the bandwidth of 0.001 to 100 MHz. In addition, patients were auscultated by experienced radiologists through conventional mastography. Statistical comparison of the two independent groups (Control vs BC) shows significant differences. Specific details regarding experimental design are described in (González-Díaz et al., 2017).

The observations indicate that inductive $\Delta\theta$ increments in the β-dispersion region might correlate with differences in morphological patterns such as cell structure (Martinsen and Grimnes, 2011). Inductive $\Delta\theta$ increments could be used to detect tissue abnormalities such as BC (González-Díaz et al., 2017).

In conclusion, it seems necessary to search for new low-cost technologies and easy access molecular biomarker detection methods. Biosensors of different biological analytes designed based on bioimpedance measurements have been increased as an important emerging inexpensive technology; such developments require specialized knowledge of instrumentation for designing bioimpedance meters and constraints.

BC detection by EIS measurements assisted with nanotechnology as well as MIS exploration is an important approach for designing innovative diagnostic tools. Also, bioimpedance micro-devices integrated with different measuring techniques seem to be complementarily linked, thus, a combination of different techniques assisted with bioimpedance measurements seems to be the future key in biosensor designs.

Acronym/Abbreviation

AC	Alternating Current
AM	Amplitude Modulation
BC	Breast Cancer
BCC	Basal Cell Carcinoma
BPH	Benign Prostatic Hyperplasia
CaP	Prostate Cancer
CTC	Circulating Tumoral Cells
DC	Direct Current
ECG	Electrocardiography
EIS	Electrical Impedance Spectroscopy
IMIX	Imaginary part Index
MIS	Magnetic Induction Spectroscopy
MIX	Magnitude Index
PIX	Phase Index
PM	Phase Modulation
RIX	Real part Index

Reference list

Aberg, P., Nicander, I., Hansson, J., Geladi, P., Holmgren, U., Ollmar, S. et al. (2004). Skin cancer identification using multifrequency electrical impedance-a potential screening tool. *IEEE Transactions on Biomedical Engineering*, 51(12): 2097–2102. https://doi.org/10.1109/TBME.2004.836523.

Adams, D.L., Stefansson, S., Haudenschild, C., Martin, S.S., Charpentier, M. et al. (2015). Cytometric characterization of circulating tumor cells captured by microfiltration and their correlation to the cell search VR CTC test. *Cytometry Part A*, 87A: 137–144. https://doi.org/10.1002/cyto.a.22613.

Al-Zeiback. (1993). A feasibility study of *in vivo* electromagnetic imaging. *Physics in Medicine & Biology*, 38: 151–160. https://doi.org/.

Beetner, D.G., Kapoor, S., Manjunath, S., Zhou, X., Stoecker, W.V. et al. (2003). Differentiation among basal cell carcinoma, benign lesions, and normal skin using electric impedance. *IEEE Transactions on Biomedical Engineering*, 50(8): 1020–1025. https://doi.org/10.1109/TBME.2003.814534.

Bertemes-Filho, P. (2018). Electrical impedance spectroscopy. *Bioimpedance in Biomedical Applications and Research*, pp. 5–27. Springer.

Blad, B. and Baldetorp, B. (1996). Impedance spectra of tumour tissue in comparison with normal tissue; a possible clinical application for electrical impedance tomography. *Physiological measurement*, 17(4A): A105. https://doi.org/10.1088/0967-3334/17/4A/015.

Braun, R.P., Mangana, J., Goldinger, S., French, L., Dummer, R. et al. (2017). Electrical impedance spectroscopy in skin cancer diagnosis. *Dermatologic Clinics*, 35(4): 489–493. https://doi.org/10.1016/j.det.2017.06.009.

Bray, F., Ferlay, J., Soerjomataram, I., Siegel, R.L., Torre, L.A. et al. (2018). Global cancer statistics 2018: GLOBOCAN estimates of incidence and mortality worldwide for 36 cancers in 185 countries. *CA: A Cancer Journal for Clinicians*, 68(6): 394–424. https://doi.org/10.3322/caac.21492.

Burdette, E.C. (1982), Electromagnetic and acoustic properties of tissues. pp. 105–150. *In*: Nussbaum, G.H. (ed.). *Physical Aspects of Hyperthermia, AAPM Medical Physics Monographs*, 8. https://doi.org/.

Emtestam, L., Nicander, I., Stenström, M. and Ollmar, S. (1998). Electrical impedance of nodular basal cell carcinoma: A pilot study. *Dermatology*, 197(4): 313–316. https://doi.org/10.1159/000018023.

Giuliano, M., Giordano, A., Jackson, S., De Giorgi, H. and Mego, M. (2014). Circulating tumor cells as early predictors of metastatic spread in breast cancer patients with limited metastatic dissemination. *Breast Cancer Research*, 16: 440. https://doi.org/10.1186/s13058-014-0440-8.

González-Díaz, C.A., Uscanga-Carmona, M.C., Lozano-Trenado, L.M., Ortíz, J.L., González, J.A. et al. (2017). Clinical evaluation of inductive spectrometer to detect breast cancer. *In*: Torres, I., Bustamante, J. and Sierra, D. (eds.). *VII Latin American Congress on Biomedical Engineering CLAIB 2016, Bucaramanga, Santander, Colombia, October 26th–28th, 2016. IFMBE Proceedings*, vol 60. Springer, Singapore. https://doi.org/10.1007/978-981-10-4086-3_170.

González-Díaz, C.A. and Golberg, A. (2020). Sensitivity analysis of electrical bioimpedance patterns of breast cancer cells labeled with magnetic nanoparticles: Forming the foundation for a biosensor of circulating tumor cells. *Physiological Measurement*, 41(6): 064001. https://doi.org/10.1088/1361-6579/ab9377.

Gregory, W., Marx, J., Gregory, C., Mikkelson, W., Tjoe, J. et al. (2012). The Cole relaxation frequency as a parameter to identify cancer in breast tissue. *Medical Physics*, 39(7Part1): 4167–4174. https://doi.org/10.1118/1.4725172.

Griffiths, H., Stewart, W.R. and Gough, W. (1999). Magnetic induction tomography - A measuring system for biological materials. *Annals of the New York Academy of Sciences*, 873: 335–345. https://doi.org/10.1111/j.1749-6632.1999.tb09481.x.

Griffiths, H. (2001). Magnetic induction tomography. *Measurement Science and Technology*, 12: 1126–31. https://doi.org/10.1088/0957-0233/12/8/319.

Grimnes, S. and Martinsen, O.G. (2015). *Bioimpedance and Bioelectricity Basics* (3rd. ed.). Academic Press.

Grossi, M. and Riccò, B. (2017). Electrical impedance spectroscopy (EIS) for biological analysis and food characterization: A review. *Journal of Sensors and Sensor Systems*, 6(2): 303–325. https://doi.org/10.5194/jsss-6-303-2017.

Haemmerich, D., Staelin, S.T., Tsai, J.-Z., Tungjitkusolmun, S., Mahvi, D.M. et al. (2003). *In vivo* electrical conductivity of hepatic tumours. *Physiological Measurement*, 24(2): 251. https://doi.org/10.1088/0967-3334/24/2/302.

Halter, R.J., Schned, A., Heaney, J., Hartov, A., Schutz, S. et al. (2008). Electrical impedance spectroscopy of benign and malignant prostatic tissues. *The Journal of Urology*, 179(4): 1580–1586. https://doi.org/10.1016/j.juro.2007.11.043.

Halter, R.J., Schned, A., Heaney, J., Hartov, A., Paulsen, K.D. et al. (2009). Electrical properties of prostatic tissues: I. Single frequency admittivity properties. *The Journal of Urology*, 182(4): 1600–1607. https://doi.org/10.1016/j.juro.2009.06.007.

Han, A., Yang, L. and Frazier, A.B. (2007). Quantification of the heterogeneity in breast cancer cell lines using whole-cell impedance spectroscopy. *Clinical Cancer Research*, 13(1): 139–143. https://doi.org/10.1158/1078-0432.CCR-06-1346.

Heinitz, J. and Minet, O. (1995). Dielectric Properties of Female Breast Tumors. *In Ninth International Conference on Electrical Bio-Impedance*, pp. 356–359. Heidelberg.

Holder, D.S., Gonzalez-Correa, C.A., Tidswell, T., Gibson, A., Gusick, G. et al. (1999). Assessment and calibration of a low-frequency system for Electrical Impedance Tomography (EIT), optimized for use in imaging brain function in ambulant human subjects. *Annals of the New York Academy of Sciences*, 873: 512–519. https://doi.org/10.1111/j.1749-6632.1999.tb09500.x.

Huerta-Nuñez, L.F.E., Gutierrez-Iglesias, G., Martinez-Cuazitl, A., Mata-Miranda, M.M., Alvarez-Jiménez, V.D. et al. (2019). A biosensor capable of identifying low quantities of breast cancer cells by electrical impedance spectroscopy. *Scientific Reports*, 9: 6419. https://doi.org/10.1038/s41598-019-42776-9.

Hutten, H., Scharfetter, H., Ninaus, W., Puswald, B., Petrova, G.I. et al. (1998). Inductively coupled wideband transceiver for bioimpedance spectroscopy (IBIS). *Proceedings of the 20th Annual International Conference of the IEEE Engineering in Medicine and Biology Society*, 20(4): 1791–1794. https://doi.org/10.1109/IEMBS.1998.7469.

Ishai, P.B., Talary, M.S., Caduff, A., Levy, E., Feldman, Y. et al. (2013). Electrode polarization in dielectric measurements: A review. *Measurement Science and Technology*, 24(10): 102001. https://doi.org/10.1088/0957-0233/24/10/102001.

Ivorra, A. (2003). Bioimpedance monitoring for physicians: An overview. *Centre Nacional de Microelectrònica Biomedical Applications Group*, 11(17).

Jossinet, J. (1998). The impedivity of freshly excised human breast tissue. *Physiological Measurement*, 19(1): 61. https://doi.org/10.1088/0967-3334/19/1/006.

Kassanos, P. (2021). Bioimpedance sensors: A tutorial. *IEEE Sensors Journal*, 21(20): 22190–22219. https://doi.org/ 10.1109/JSEN.2021.3110283.

Kim, B.S., Isaacson, D., Xia, H., Kao, T.-J., Newell, J.C. et al. (2007). A method for analyzing electrical impedance spectroscopy data from breast cancer patients. *Physiological Measurements*, 28: S237. https://doi.org/10.1088/0967-3334/28/7/S17.

Korjenevsky, A.V. and Cherepenin, A. (1999). Progress in realization of magnetic induction tomography. *Annals of the New York Academy of Sciences*, 873: 346–352. https://doi.org/10.1111/j.1749-6632.1999.tb09482.x.

Korzhenevskii, A.V. and Cherepenin, A. (1997). Magnetic induction tomography. *Journal of Communications Technology and Electronics*, 42(4): 469–474.

López-Carrillo, L., Torres-Sánchez, L., Blanco-Muñoz, J., Hernández-Ramírez, R.U. and Knaul, F.M. et al. (2014). Utilización correcta de las técnicas de detección de cáncer de mama en mujeres mexicanas. *Salud Pública de México*, 56(5): 538–546.

Malich, A., Boehm, T., Facius, M., Freesmeyer, M., Azhari, T. et al. (2002). Electrical impedance scanning of lymph nodes: Initial clinical and technical findings. *Clinical Radiology*, 57(7): 579–586. https://doi.org/10.1053/crad.2001.0927.

Markou, A., Strati, A., Malamos, N., Georgoulias, V., Lianidou, E.S. et al. (2011). Molecular characterization of circulating tumor cells in breast cancer by a liquid bead array hybridization assay. *Clinical Chemistry*, 57(3): 421–430. DOI: https://doi.org/10.1373/clinchem.2010.154328.

Mishra, V., Schned, A., Hartov, A., Heaney, J., Seigne, J. et al. (2013). Electrical property sensing biopsy needle for prostate cancer detection. *The Prostate*, 73(15): 1603–1613. https://doi.org/10.1002/pros.22695.

Morán, M.E., Rodríguez, G.J., Lara, L.M., Piña, L.C., Thompson, B.M.R. et al. (2014). Células tumorales circulantes en cáncer de mama: un posible riesgo biológico. *Revista de Especialidades Médico-Quirúrgicas*, 19(1): 45–51.

Morimoto, T., Kinouchi, Y., Iritani, T., Kimura, S., Konishi, Y. et al. (1990). Measurement of the electrical bio-impedance of breast tumors. *European Surgical Research*, 22(2): 86–92. https://doi.org/10.1159/000129087.

Newell, J.C., Edic, P.M., Ren, X., Larson-Wiseman, J., Danyleiko, M. et al. (1996). Assessment of acute pulmonary edema in dogs by electrical impedance imaging. *IEEE Transactions on Biomedical Engineering*, 43: 133–138. https://doi.org/10.1109/10.481982.

Nicander, I., Ollmar, S., Eek, A., Rozell, B.L., Emtestam, L. et al. (1996). Correlation of impedance response patterns to histological findings in irritant skin reactions induced by various surfactants. *British Journal of Dermatology*, 134(2): 221–228. https://doi.org/10.1111/j.1365-2133.1996.tb07605.x.

Pethig, R. (1984). Dielectric properties of biological materials: Biophysical and medical applications. *IEEE Transactions on Electrical Insulation*, (5): 453–474. https://doi.org/10.1109/TEI.1984.298769.

Qiao, G., Duan, W., Chatwin, C., Sinclair, A., Wang, W. et al. (2010). Electrical properties of breast cancer cells from impedance measurement of cell suspensions. *Journal of Physics: Conference Series*. https://doi.org/10.1088/1742-6596/224/1/01208.

Qiao, G., Wang, W., Duan, W., Zheng, F., Sinclair, A.J. et al. (2012). Bioimpedance analysis for the characterization of breast cancer cells in suspension. *IEEE Transactions on Biomedical Engineering*, 59(8): 2321–2329. https://doi.org/10.1109/TBME.2012.2202904.

Rigaud, B., Morucci, J.-P. and Chauveau, N. (1996). Bioelectrical impedance techniques in medicine part I: Bioimpedance measurement second section: impedance spectrometry. *Critical Reviews™ in Biomedical Engineering*, 24(4-6). https://doi.org/10.1615/CritRevBiomedEng.v24.i4-6.20.

Samia A. Ebeed, Nadia A. Abd El-Moneim, Ahmed Saad, Ebtsam R.E. Zaher et al. (2012). Diagnostic and prognostic value of circulating tumor cells in female breast cancer patients. *Alexandria Journal of Medicine*, 48(3): 197–206. https://doi.org/10.1016/j.ajme.2012.02.005.

Sarode, G.S., Sarode, S.C., Kulkarni, M., Karmarkar, S., Patil, S. et al. (2016). Role of bioimpedance in cancer detection: A brief review. *International Journal of Dental Science and Research*, 3(1): 15–21. https://doi.org/10.1016/j.ijdsr.2015.11.003.

Scholz, B. and Anderson, R. (2000). On electrical impedance scanning-principles and simulations. *Electromedica*, 68(1): 35–44.

Schwan, H.P. (1957). Electrical properties of tissue and cell suspensions. *Advances in Biological and Medical Physics*, 5: 147–209. https://doi.org/10.1016/B978-1-4832-3111-2.50008-0.

Silva, J.G., Cárdenas, R.A., Quiróz, A.R., Sánchez, V., Lozano, L.M. et al. (2014). Impedance spectroscopy assisted by magnetic nanoparticles as a potential biosensor principle for breast cancer cells in suspension. *Physiological Measurement*, 35(6): 931–941. https://doi.org/10.1088/0967-3334/35/6/931.

Silva, J.G., López, J., Sánchez, V., Lozano, L.M., González, C.A. et al. (2015). Breast cancer tissue marked selectively by magnetic nanoparticles in an experimental animal model. *Journal of Nanoscience and Nanotechnology*, 15: 1–6. https://doi.org/10.1166/jnn.2015.10610.

Singh, B., Smith, C. and Hughes, R. (1979). *In vivo* dielectric spectrometer. *Medical and Biological Engineering and Computing*, 17(1): 45–60. https://doi.org/10.1007/BF02440953.

Sua, L.F., Silva, N.M., Vidaurreta, M., María, L., Maestro, M. et al. (2011). Detección inmunomagnética de células tumorales circulantes en cáncer de mama metastásico: nuevas tecnologías. *Revista Colombiana de Cancerología*; 15(2): 104–109. https://doi.org/10.1016/S0123-9015(11)70073-7.

Surowiec, A.J., Stuchly, S.S., Barr, J.R. and Swarup, A. (1988). Dielectric properties of breast carcinoma and the surrounding tissues. *IEEE Transactions on Biomedical Engineering*, 35(4): 257–263. https://doi.org/10.1109/10.1374.

Thompson, H.J. and Meeker, L.D. (1983). Induction of mammary gland carcinomas by the subcutaneous injection of 1-methyl-1-nitrosourea. *Cancer Research*, 43: 1628–1629. https://doi.org/10.1016/0304-3835(95)03761-K.

Webster, J.G. (1990). *Electrical Impedance Tomography*. CRC Press.

Webster, J.G. (2009). *Medical Instrumentation: Application and Design*. John Wiley & Sons.

Xu, J. and Hong, Z. (2020). Low power bio-impedance sensor interfaces: Review and electronics design methodology. *IEEE Reviews in Biomedical Engineering*, 15: 23–25. https://doi.org/10.1109/RBME.2020.3041053.

Xu, H., Aguilar, Z.P., Yang, L., Kuang, M., Duan, H. et al. (2011). Antibody conjugated magnetic iron oxide nanoparticles for cancer cell separation in fresh whole blood. *Biomaterials*, 32(36): 9758–9765. DOI: https://doi.org/10.1016/j.biomaterials.2011.08.076.

Yang, Y., Wang, J., Yu, G., Niu, F., He, P. et al. (2006). Design and preliminary evaluation of a portable device for the measurement of bioimpedance spectroscopy. *Physiological Measurement*, 27(12): 1293. https://doi.org/10.1088/0967-3334/27/12/004.

Chapter 3

Thermal Images

Towards Cancer Detection

Rafael Bayareh-Mancilla

Temperature is a physical quantity that measures the average kinetic energy which represents hot and cold. Nonetheless it is one of the most frequently measured physical quantities in science and healthcare. Accurate and precise measurement is relevant for the health and safety monitoring of patients, and in the manufacturing and processing of materials with clinical applications. The correct determination of temperature depends on the selection and knowledge of the chosen sensor since currently, some different forms and methods vary in cost, temperature scale, and accuracy according to the needs of the study. One of the most current methods is the implementation of infrared (IR) sensors through thermography. This method has allowed not only the detection of temperature asymmetries but more currently presents an open door for quantitative studies of temperature differences by regions in paired organs. This chapter presents the implementation of thermal imaging as a tool for cancer detection, from technical aspects to clinical protocols of medical IR thermography.

3.1 Introduction

In the third century BC, Hippocrates introduced the detection of disease through fever in ancient medicine. He proposed to cover an ill person with wet

Electrical Engineering Department, Bioelectronics Section, CINVESTAV-IPN, Mexico City 07360, Mexico.
Email: rafael.bayareh@cinvestav.mx

mud, in such a way that the rate of drying in certain regions was faster than others. This discovery led to the correlation of temperature asymmetries with the presence of tumours. However, for nearly four centuries, fever detection remained relatively qualitative until Galileo introduced a tubular-shaped glass tube resembling a thermometer in the sixteenth century. Currently, clinical thermometry has been universally accepted; it was introduced by Carl Wunderlich in 1868 with the invention of the first thermometer. This concept states that internal body temperature measurements would hover around 37°C; meaning that it can maintain a constant temperature that differs from the environment.

By 19th century, Herschel introduced the term "thermal spectrum" when he discovered the IR energy as a form of heat radiation by exposing samples of crystals to light. Nowadays, while glass thermometers have evolved, technology and knowledge of new materials have brought new ways to measure body temperature in clinical practice. These advances allowed the creation of sensors based on liquid crystals for non-contact temperature measurement. These crystals were encapsulated to create the first sheet detector. Current sensors are based on heat transfer by radiation, which is of great value in medicine, known as Medical Infrared Thermography (MIT). MIT is a non-invasive, non-contact technique to obtain thermal images. Currently, the U.S. Food and Drug Administration (FDA) recommends that MIT should not be a standalone diagnostic tool but a complement to mammography. However, the implementation as a first contact tool makes thermography an attractive technique for research. The goal of modern thermal imaging is to promote quantitative and digital studies for a full integration into clinical centres and telemedicine computer networks (Ring and Ammer, 2012).

This chapter will discuss the principles of MIT as a tool for breast cancer detection. The content is structured by the principles of thermal imaging, protocols, criteria in cancer detection studies, and the perspectives of this promising technique.

3.2 Thermal images principles and applications

Thermography principles are based on the fact that everybody emits infrared radiation when the surface temperature rises above absolute zero. Infrared radiation belongs to an electromagnetic band spectrum, whose wavelength λ and intensity Q_λ are used to characterize the sensors. The IR spectrum is between 1 μm (near-infrared) and 1 mm (far-infrared). In terms of temperature, a wavelength near 1 mm can reflect a temperature close to 0 K, while a wavelength near 1 μm reflects a temperature close to 10 000 K. Thus, the IR spectrum can be divided into three regions: near-IR (NIR), mid-IR (MIR), and far-IR (FIR). NIR spectral band is 0.8 μm to 2.5 μm. NIR absorption is less intense concerning light, due to red color spectrum proximity. MIR has a spectral range between 3 μm to 5 μm. Lastly, FIR has a range between 8 μm to 14 μm. The physical principles are currently described by several radiation laws postulated by Plank, Stefan-Boltzman, and Kirchhoff. The emitted IR radiation is described by the intensity of the wavelength, which is introduced by Planck's law in Equation (3.1).

$$Q_\lambda = A/\lambda^5\left(e^{(B/\lambda T)} - 1\right) \tag{3.1}$$

where Q_λ is the emitted radiation [W], λ is the wavelength [m], T is the surface temperature [K], and the Plank constants A = 3.742 × 10^8 W μm^4 m^{-2} and B = 1.439 × 10^4 μm K.

Equation (3.1) generates a collection of curves that correlate the amount of radiation emitted for each wavelength λ. An approximation of a body radiation was presented by Stefan-Boltzmann by integrating Equation (3.1) over the wavelength range. This physical property is known as emissivity, described by Equation (3.2).

$$\varepsilon = e'\sigma T^4 \tag{3.2}$$

where ε is the radiant emittance $[\frac{W}{m^2}]$, e' is the effective emissivity (a value between 0 and 1), σ is the Stefan-Boltzmann constant ($\sigma = 5.67 \times 10^{-8} \frac{W}{m^2 k^4}$), and T is the absolute temperature [K].

Later, Kirchhoff postulated the principle of emissivity, which states that thermal equilibrium can be achieved if the power radiated by an object is proportional to the power absorbed. This implies that emissivity can be expressed as an absorption rate, which introduces the concept of the black body radiator. All materials emit specific radiation at a constant rate with a wavelength distribution, that is determined by the object temperature and spectral emissivity ε. A material that can perfectly absorbs IR radiation is considered a blackbody, which is a theoretical model $\varepsilon = 1$. However, most materials are not blackbodies. Only a fraction of the energy incident on the object can be reflected or transmitted. The ratio of the emitted radiation W' and the radiation emitted from the energy source W of the black body at the same temperature is known as emissivity (see Equation (3.3)).

$$\varepsilon(\lambda, T) = \frac{W'}{W} = \frac{\varepsilon(\lambda_1, T)_{source}}{\varepsilon(\lambda_2, T)_{blackbody}} \tag{3.3}$$

where $\varepsilon(\lambda, T)$ is the ratio of the exitance between the source $\varepsilon(\lambda_1, T)_{source}$ and the blackbody $\varepsilon(\lambda_2, T)_{blackbody}$, which is dimensionless between [0–1].

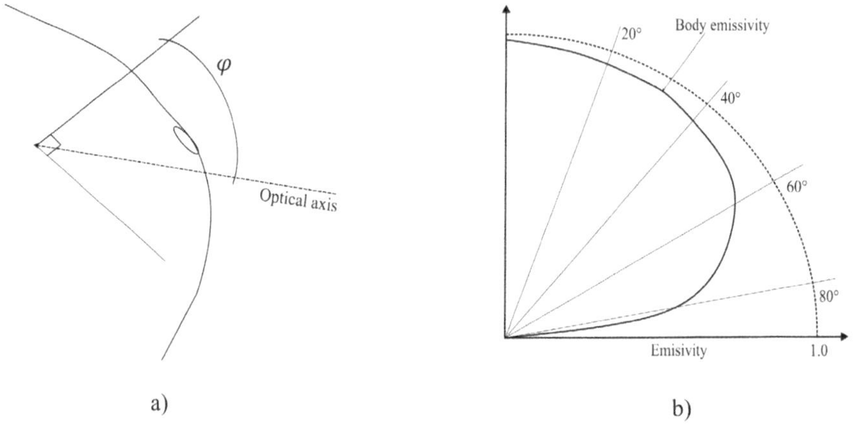

Figure 3.1. Effect of angular emissivity, (a) represents the optical axis to the normal angle, (b) polar graph representing the emissivity attenuation close to the normal axis (90°).

Human skin can be considered as a blackbody radiator with $\varepsilon \approx 0.98$ (Hardy and Muschenheim, 1936). Nevertheless, emissivity is dependent on the viewing angle between the optical axis and the normal surface (Figure 3.1). Several studies point out that the emissivity drops significantly when the viewing angle is extended above 60° (Arnon et al., 2017).

Several models evaluate the angular effects to compute an accurate temperature reading. An estimation from Equation (3.4), approximates an error dependent on the angle of capture (Cheng et al., 2012).

$$\Delta T = \frac{T_n - T(\theta)}{T_n - T_a} \tag{3.4}$$

where ΔT is a normalized temperature level with a range between [0,1], T_n is the temperature viewed at a normal angle, at which the viewing angle concerning the camera axis is 0°. T_a is the room temperature, and $T(\theta)$ is the temperature recorded at a specific angle.

Other approaches have been proposed based on artificial neural networks (ANN) (Muniz et al., 2014). An experimental ANN model and linear regression are proposed for error correction . This approach could be suitable when there is a database of measurements such that the ANN can compensate.

3.2.1 Infrared sensors, readings, and data arrays

Currently, IR sensors can be classified based on photon detectors or bolometers. Photon detectors are wavelength-dependent requiring cryogenic-

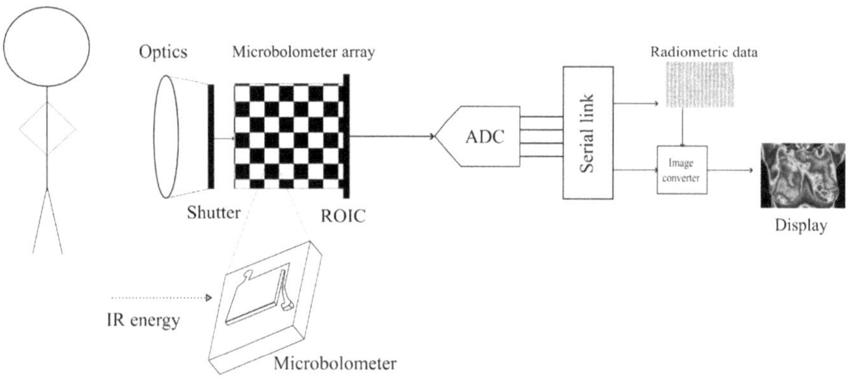

Figure 3.2. Schematic of a sensor based on a microbolometer array. The detector array sits on a Read Out Integrated Circuit surface (ROIC) board. When the temperature changes over the filament, it changes the electrical charge which is digitized by the ROIC.

based cooling. These types of sensors are not intended for temperature measurement. On the other hand, bolometer-based detectors are lower cost than photon detectors and are not wavelength dependent. Figure 3.2 illustrates the schematic of a bolometer array sensor. The pinnacle of IR sensors led to uncooled Focal Plane Array (FPA) cameras, which are lower cost and more efficient than their predecessors. Currently, most prototyping sensors, as well as thermal imaging cameras, are based on FPA technology. These signals are obtained as a matrix of values commonly scaled to a resolution of 14 or 16 bits. The extracted information from the sensors is known as radiometric data.

In medical studies, it is relevant to store radiometric data, since they can be preserved, shared, used for the development of detection algorithms, or simply for further studies. Depending on the thermographic equipment, the information may contain metadata besides the IR radiation record such as room temperature, set emissivity and relative humidity. Commonly, the radiometric data arrays are stored in a TXT file extension to be post-processed.

3.2.2 Thermal camera characterization

Typically, the characterization is performed by the manufacturer in such a way that the device can interpret the IR signals as temperature. The characterization should provide a mathematical model that predicts the temperature in each pixel of the image, regardless of the quality of the sensor. Several thermal devices are designed to detect temperature variations rather than map a temperature distribution to which it was calibrated. In medical applications, assessing blood circulation based on thermal imaging involves searching for cold regions that should normally be warm, to quantify thermal differences.

Even with periodic calibrations, the devices can still lose lifetime causing a shift in calibration due to age. Calibration methods can be classified as internal and external. Internal calibration refers to the ability to adjust equipment parameters such as emissivity, relative humidity, and ambient temperature. On the other hand, external calibration refers to the characterization of the sensors against a black body, so it is recommended to contact the manufacturer for calibration at least once a year. If the situation is appropriate, accuracy and precision can be verified against gold-standard instruments.

Calibration targets may use an epoxy resin-loaded iron absorber coating such that the metal structure maintains a homogeneous temperature distribution. The coating should be thick enough to absorb the radiation but at the same time, it should be thin enough to be attached enough to the metal (Cui and Xing, 2022). However, even air fluctuations can lead to changes in the measurements. An approach is usually direct compensation in the thermographic equipment; nevertheless, this practice would affect the measurement since the emissivity would be inaccurate or the compensation would not correspond to the study. The most common designs are cone-shaped geometries with emissivity $\varepsilon \approx 0.9$ (Nunak et al., 2015). The hypothesis of proposing a conical geometry is that the probability of the light beam being reflected is less than or equal to 0.1 due to the multiple reflections within the inside of the cavity, even if the material is characterized by 0.9 emissivity and 0.1 reflection. The calibration will be given by comparing the measured temperature inside the cavity against the temperature recorded by the sensors. Several studies have suggested different methods to compensate the temperature difference, among them the most common one being linear regression (Fidali and Mikulski, 2008). Some parameters that may be involved in the calibration and characterization are equivalent noise temperature difference, dynamic range, spatial resolution, pixel size, and minimum resolvable temperature difference (Pellegrino et al., 2006).

The noise equivalent temperature difference (NETD) is relevant for classifying LWIR sensors. It is a signal-to-noise ratio in which the temperature difference is calculated in the signal produced by temporal noise. Sensors can be susceptible to different types of noise, including spatial and temporal noise. These errors can be represented in a space in which the sensor is stimulated by a uniform temperature distribution. Continuous samples are taken from the system (typically $n = 100$ (Pellegrino et al., 2006)) such that the FOV is directed at the temperature reference and the response to the stimulus is measured directly from the sensor. When the samples are piled, in theory, a uniform temperature "cube" is obtained. The temperature difference should be minimal, however, in practice, the heat radiated from the object and the noise cannot be distinguished from the heat source. For each sample, the voltage standard deviation of the sensor is computed. Then, the response matrix is obtained which is computed by the difference between the sets of samples at different temperatures. The NEDT is obtained by the ratio of the temporal noise to the response matrix (see Equation (3.5)). A lower NETD indicates an 'excellent' quality (NETD $\approx$ 40 mK) while an 'acceptable' quality has a NETD $\approx$ 80 mK (Diakides et al., 2006).

$$NEDT = \frac{1}{n}\sum_{i=1}^{n} T\_noise_{mxm} \Big/ R_{mxm} \ [\text{mK}] \tag{3.5}$$

where $NEDT[mK]$ is the noise equivalent temperature difference, computed by the average of the rate between the temporal noise array and the responsivity array given n samples. T_noise_{mxm} $[V]$ is the temporal noise array (size $m \, x \, m$) computed by the standard deviation of each pixel at the median selected temperature. R_{mxm} $[V/K]$ is the responsivity matrix computed by the difference between the first dataset and the last dataset.

The dynamic range is the rate given by the maximum input signal and the minimum signal. Particularly for thermography applications, the dynamic range can be computed by the temperature differences reaching a maximum and minimum saturation with respect to the NEDT (Equation (3.6)).

$$Dynamic\ range = 20\log_{10}\left(\frac{\Delta T_{max} - \Delta T_{min}}{NEDT}\right)[\text{dB}] \tag{3.6}$$

Another parameter to be considered in the calibration is the spatial resolution, which refers to the size of the image in pixels. Although there are techniques to resize images, the spatial resolution cannot be modified so it is an intangible property of the thermal camera. It is important to consider the spatial resolution when choosing the imager model, particularly when it is required as an input for pattern detection. The number of pixels can define the spatial resolution of the image; however, the pixel size can define the resolution of the temperature difference detection. Thermal cameras map a colour scale concerning the hottest point in the captured scene. A characteristic that defines the performance is the minimum resolvable temperature difference (MRTD), which characterizes the thermographic equipment between spatial resolution and thermal sensitivity. This rate computes the minimum detectable temperature. The technical test consists of measuring different targets with different spatial frequencies. Then, the temperature of a black body must be measured, and the temperature must be adjusted until the slightest change is detected (Equation (3.7)).

$$\text{MRDT} = \frac{\Delta T(s)}{F(s)} \tag{3.7}$$

where $\Delta T(s)$ is an array of temperature differences and $F(s)$ is an array of spatial frequencies. A lower MRDT rate indicates the equipment has better resolution.

3.2.3 Thermal imaging in medicine

The fundamental concept of MIT is the correlation of surface temperature with physiological processes under the skin, acting as a third eye for the physician to detect asymmetries and temperature anomalies. Among the most cited and attractive advantages is that it is a non-invasive, noncontact technology with no ionization risks like other medical imaging techniques. Suggestions include training of personnel in data acquisition and interpretation, standardization of data, promoting quantitative studies, finding a better correlation between temperature patterns and physiological processes, and characterization of patterns (Lahiri et al., 2012).

Thermal imaging is currently employed with FPA-type cameras so that a temperature mapping is acquired instead of a spot temperature. These records can be of the temperature distribution on a specific moment or the temperature variation over time. However, the recorded information is the vascular response that can decrease or increase with the presence of disease or inflammatory processes. Different protocols have been established to try to standardize medical IR thermography studies. However, regardless of whatever protocol is chosen, a controlled and regulated environment is required. Thermal interferences (or thermal artifacts) that are caused by external temperature changes such as sun exposure, cosmetics, antiperspirants, air currents, lamps, and the personnel themselves must be avoided prior to properly preparing for the study.

There are two ways of acquiring thermal images, passive and active. Active thermography refers to direct temperature stimulation of the patient through heat sources or lotions that stimulate metabolism, while passive thermography refers to the homogenization of the patient's temperature with the environment. Active thermography is also known as Active Dynamic Thermal IR-imaging (ADT) since the temperature is time-varying. The application of paraffins, ultrasound, hot-packs, and shock waves are some ways to externally induce heat (Pasquali, 2020). The fundamental idea of ADT is to observe the biofeedback mechanism or the thermal recovery response of the patient. If the patient has no vascular problems, the temperature should regulate near 37°C, however, difficulty in thermoregulation may indicate the presence of disease (Renkielska et al., 2006). However, there are technical difficulties in ADT that give passive thermography an advantage in medical applications. The fundamental problem is that temperature distribution is difficult to manage, and optical excitation is frequently non-uniform. Furthermore, the temperature must not exceed 42°C; any higher temperature will damage cells. Normally, boundary conditions cannot be controlled when tissues decrease in temperature by external stimulation. For these reasons, passive thermography is more adjustable for medical purposes (Edis et al., 2014). The protocol for passive thermography indicates that the patient should be acclimatized in 15 minutes to homogenize the body temperature. If the patient presents difficulties in thermoregulation or vascular complications, regions with temperature alterations should be observed. Otherwise, temperature variations will have slow and uniform changes. At this point, thermal interferences can be minimized or eliminated due to the homogenization of the target temperature and scene. This protocol is more suitable for laboratory studies so that neutral data can be obtained without interferences that may be better for interpretation (Ring and Ammer, 2012). Some applications are listed in Table 3.1, as a summary of the most common medical applications. Some applications are listed in Table 3.1, as a summary of the most common medical applications.

In conclusion, IR thermography has evolved into a diagnostic and medical follow-up tool. The interpretation led to the observation of regions with temperature changes due to ischemia or inflammation. However, it should be noted that this imaging technique cannot replace other studies that require

Table 3.1. Most common medical applications with methods and conditions.

Application	Method
Breast cancer	Passive or active
Diabetic foot detection	Passive or active
Mechanisms of Thermoregulation in Mammals	Passive
Diabetes (vascular study)	Active
Assessment of Burn Depth and Healing Potential	Passive
Neurology	Passive or active
Burns (dermatology)	Passive
Pain	Active
Respiratory diseases	Passive
Rehabilitation medicine	Passive or active
Arthritis/rheumatism	Active

tomography inspection. Although MIT is an assistive tool, it is limited to the detection of superficial temperature changes. Due to the mentioned features and advantages, MIT has potential in the medical field. The following section introduces the correlation between the thermal patterns and physiology.

3.2.4 Correlation between physiology and thermal images

A thermal image can provide a qualitative result but does not provide quantitative information. Therefore, the interpretation of thermal images requires a basic understanding of correlation between physiology and thermal imaging. Since temperature measurement is superficial, it is important to consider the thermal properties of the skin as it plays a role in temperature regulation such that it can change drastically to maintain a range between 25°C (environmental temperature) to 45°C (cell denaturation) due to adaptive metabolic processes in response to injury, hormones, fever, and disease. However, temperature changes are not only produced internally but can also be induced externally. For example, intense physical activity and exposure to heat increase blood flow to dissipate heat through the evaporation of sweat. Under a cold environment, the skin develops insulating properties by reducing blood perfusion to maintain a stable body temperature and reducing convective heat transfer (hypothermia) (Magalhaes et al., 2018).

Temperature changes are qualitatively and quantitatively detectable by thermal cameras. Even in a controlled environment, regional temperature changes on breasts may be detected, leading to the possibility of tumours. Currently, thermal cameras can detect temperatures greater than 0.05°C. Regardless of whether the skin is exposed to external stimulation, the sensors must be able to record such changes for further quantifiable study. This is the principle of detecting breast cancer when it is not yet visible to the physician, or the patient has no symptoms. The following section presents several processing techniques that provide reliability in diagnostics.

3.2.5 Thermal data and image processing

Radiometric data processing has several advantages that can be exploited prior to thermal imaging. One of the recurring problems in thermography is thermal interference, which generally occurs in the scene. The simplest form of segmenting is the homogenization of the scene, that in which even thermal interference is eliminated. Although this process can also be performed with digital image processing techniques, the contrast of the colour mapping can be modified only within the region of interest (RoI). Other advantages of radiometric data processing are the adjustment of gain conditions such that thermal interferences are eliminated due to small variations; quantitative studies can be performed in the temperature domain such as computing averages, standard deviations and maxima/minima. Some thermal cameras with advanced radiometric processing allow compensation of parameters such as ambient temperature and emissivity. In this way, the operator can normalize the results to enhance radiometric processing.

Some methods describe detection of features based on the average, standard deviation, and asymmetry. Temperature asymmetry is the most desirable characteristic, particularly in the detection of cystic diseases, infections, vascular complications, and cancer. The detection of abnormal temperature through the thermal signature is based on the study of histograms. Other techniques are transformations such as wavelets, Fourier analysis, and linear filtering to complement the classification. An initial suspicion would be to find a tendency in temperatures near 40°C even after temperature homogenization. However, detection in paired organs is based on the comparison between breasts. Figure 3.3 shows a comparison between histograms with possible cancer detection due to the temperature differences.

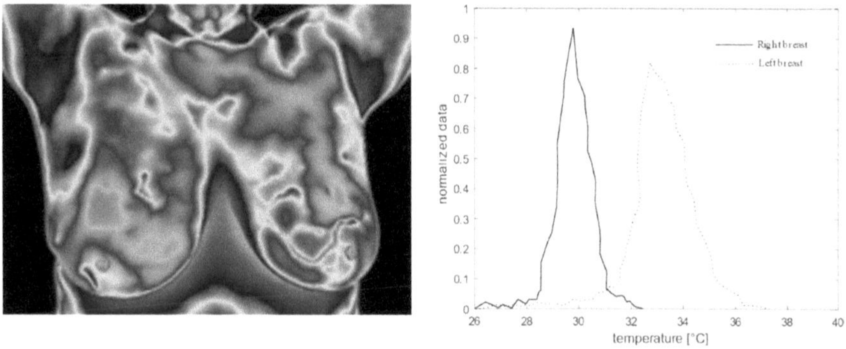

Figure 3.3. Comparison of left vs right breast temperature, (a) thermal image of a patient with a history of fibrocystic change in the left breast was retrieved from (Merino González et al., n.d.) (assuming corporal average temperature as 27°C), (b) histograms of temperature between each breast, with the solid line being the right breast and the dotted line being the left breast.

So far, only statistical analysis of the temperature difference by region has been presented. Before the analysis, segmentation of both breasts separately is required. Segmentation by radiometric signal processing can be effective for upper or lower limbs, however, it can be challenging to segment the breasts due to the thorax. Therefore, several studies rely on advanced signal and image processing to validate the results:

- Second-order parameters (co-occurrence of a matrix), which compare the intensity distribution in different regions concerning neighbouring pixels.

If the distribution is homogeneous, the variance should tend to be low, otherwise, a disease could be suspected (Hao et al., 2021).

- Classification is based on an ANN, such that the system is trained to differentiate a thermal image that could predict disease and a database of images with no pathological background (Jafari-Marandi et al., 2018).
- Wavelet transform is used as a system of linear high-pass and low-pass filters. The fundamental idea is to find descriptive features to produce sub-images at different scales that can denote sudden temperature variations (Zhou et al., 2020).
- Comparison of segmented regions by contour detection based on the Canny technique. The result is to obtain separated regions in such a way that statistical calculations can be performed on that sub-image automatically (Anjum et al., 2021).
- Support Vector Machines (SVM), the basis of which is the classification of images from previous trainings and that this can predict the presence of tumours (Lim et al., 2021).

In conclusion, thermography along with signal and image processing can be a method of medical assistance capable of detecting pathologies when they are not visible to the physician. However, for breast cancer detection, it can present a challenge since the changes in physiology are multifactorial. It is therefore relevant to validate the results with statistical methods that support the hypothesis. The following chapter introduces the conditions that should be sought in MIT for breast cancer detection.

3.3 Breast cancer thermal imaging

The application of MIT has been adopted in the medical field more frequently in the 21st century. However, thermal imaging studies on breasts date back to the middle of the 20th century. These studies can be traced back to the contributions of Lawson in 1956 when he observed that the skin temperature in a cancerous region in the breast was higher than its symmetrical pair. Since then, distinct studies have focused on finding the correlation between the presence of breast tumours and abnormal surface temperature. The first time the FDA approved this technique as a complementary tool in the diagnosis of breast cancer was in 1982. Since then, there have been significant advances due to the implementation of pattern recognition and artificial intelligence (AI). Currently, the detection specificity is around 90% (Ng et al., 2007). Currently, the availability of thermographic cameras and prototyping sensors has grown exponentially. Thus, a criterion must be established for medical studies since these devices are usually for general or industrial applications.

3.3.1 Device criteria

Most radiometric sensors on the market are built for industrial applications where substantial temperature differences are expected and the temperature of the object is well over 100°C. The temperature difference between both breasts may be roughly a tenth of a degree. Several studies suggest the simultaneous implementation of MWIR and LWIR sensors to cover a band spectrum that can provide confidence in tumor detection. Also,there are devices with microbolometer-FPA sensors that have the feature of operating in both spectra simultaneously (Diakides and Bronzino, 2008). Features such as spatial resolution and thermal sensitivity are important in the selection of a thermal camera since the success of signal or image processing depends critically on the quality of the information. Even considering the refresh rate is relevant since there are temperature variations over time. Currently, thermal cameras can be portable, fixed, or simply contain the sensor array together with digital communication electronics. Handheld thermal imagers are commonly used in oncology applications so that the device can sample at different locations without the patient moving. In this way, images can be appropriately obtained at different distances, angles, and heights. However, precaution must be advised with the acquisition protocol as different parameters compromise reproducibility, making tumour detection difficult. On the other hand, fixed thermographic cameras (mounted on stands or tripods) may be adaptable in studies that require standardization. For some camera models, the visualization of the thermal images is observed on a computer, which can even be separated from the control room (temperature and relative humidity). Contrary to portable cameras, reproducibility is guaranteed since distances and angles are always constant. However, particularly in thermography, the comparison between two sequential images over time can be different. This is due to the movement of the patient when ordered to rotate to obtain frontal and lateral images (position errors).

There are studies involving advanced signal and image processing that require a considerable amount of sample volume. Such studies could be regarding classifiers such that they require image collection or even 3D thermal modelling (Bayareh Mancilla et al., 2021). Furthermore, there may be requirements for the placement of different sensors at different positions and angles for simultaneous examination. For these kinds of studies, radiometric sensors are preferred. These sensors can be connected to a computer, an embedded system, or a phone (Netten et al., 2017). These sensors do not have

Table 3.2. Essential characteristics to be considered for studies for breast tumour screening.

Feature	**Specification**	**Units**
Imager type	Fixed (preferable) or imager arrays	
Information type	Thermal images and radiometric data	
Sensitivity	0.05	°C
FPA array size	80 × 60 – 640 × 480	Pixels
Spectral band	6–14 (LWIR)	μm
FOV	20–30	°
Digitalization range	14 (minimum)	bits
Accuracy	±0.3	°C

a display screen, so focusing and alignment is entirely done by the computer. Some models are not factory characterized, so it is strongly recommended to test or calibrate the sensor in advance according to the study. In conclusion, Table 3.2 summarizes the minimum characteristics that thermographic equipment should have for breast cancer screening studies.

3.3.2 Procedures for routine tests and preparing patients

This stage is a sample acquisition protocol, so the environment and the patient must be properly prepared to ensure correct measurements. The study room should maintain a stable temperature and avoid wind currents, otherwise, a small room may change the environment temperature due to the presence of laboratory personnel, lamps, or electronic equipment. Adequate space should be considered since it can be difficult to control ambient temperature and air fluctuations in a large room. The size of the room does not directly impact the quality of the thermal images, but it does alter the quantitative information that is required for signal processing. The dimensions of the room should consider the freedom of movement between the patient, personnel, and equipment for the study, in such a way that sufficient space can be provided for the patient and the devices to be correctly positioned.

The recommendations for breast imaging studies are minimum dimensions of 4 m by 4 m with a height of 2.5 m. The distance from the thermographic camera to the patient can be between 1 m to 2.5 m to ensure a FOV of 25° (Schwartz, 2021). In addition, the floor should be carpeted to isolate contact with the cold floor, and the ceiling and walls should also be insulated and preferably painted with a matte surface finish. This reduces the possibility of thermal interference the camera could be sensitive to. Patient comfort is crucial, the room should have a relaxing environment that could also contain drinking water, towels, and even music; the aim is to avoid patient stress. However, mirrors and windows (sunlight may reflect on human skin) should be avoided as they may reflect IR energy. Internal shutters may be recommended to block reflections from reflections due to glasses or lamps. The recommended light is fluorescent or LED bulbs since gas or filament light is a strong IR source. If fixed equipment or radiometric sensors are chosen, it is preferred to place the control equipment in a separate adjacent room so that there are no other personnel surrounding the sensors. A standard thermometer and hygrometer should constantly monitor the temperature and relative humidity of the room, as this information can be critical for signal processing (Diakides et al., 2006). A way to control these conditions is the implementation of air conditioning. The recommendation is to maintain a temperature between 20°C to 25°C without a variation of more than 1°C during testing (Schwartz, 2021). The temperature during the tests should not cause the patient to shiver or sweat and avoid condensation in the room that may produce humidity. The air conditioner, as well as the air duct, should be placed behind the thermographic system. Water condensation attenuates IR radiation and increases surface heat during routine tests.

Routine passive thermography examinations require the patient to be acclimatized to room temperature. Some studies recommend 21°C room temperature so that the vascular system would be more highly visible. At this temperature, it is possible to produce vasoconstriction so that inflamed tissue or vasodilation patterns can be observed due to the presence of tumours. Because temperature is an important factor, geographic regions can have an impact. If temperature control cannot be guaranteed, it is suggested to acclimatize the patient and perform the study; a correction should then be made in a post-processing stage (The American Academy of Thermology, 2021). Relatively, an advantage can be taken from this situation since the temperature asymmetry between breasts can reach approximately 5°C (Francis and Sasikala, 2013).

The last consideration is the preparation and position of the patient. It is recommended to avoid exposure to hot environments, physical activity, or physical therapy before the study. Regarding intake, hot beverages, alcohol, or spicy food should be avoided. Avoid applying creams, lotions, or make-up on the torso. Bathing is permitted at least one hour before the examination, but hot tubs should be avoided. Tanning baths or prolonged sun exposure should be avoided at least 24 hours before the examination. Regardless of whether an active or passive technique is implemented, the patient should undress in the thermography room and acclimatize by sitting up for at least 15 minutes to produce vasoconstriction. To avoid thermal interference, the patient should keep arms and hands away from the torso and avoid touching the breasts. In this way, the acclimatization process could be accelerated. Once the acclimatization stage has been reached, the patient should stand up or remain seated with the arms behind the head or on the hips (whichever is more comfortable for the patient) (Campbell and Mead, 2022). The first image captured is frontal to the torso, which is the common reference. Some classifiers require an analysis of only the frontal view, however, oblique captures with a 45° orientation or lateral views can be captured for more robust analysis or 3D modelling. Capturing images with orientation must consider curved emissivity to ensure correct quantitative analysis (see Section 3.2). However, frontal images are the most informative, as they can confirm correct vasoconstriction. In this view, any temperature asymmetry should be visible (Amalu et al., 2006).

3.3.3 Image analysis and interpretation

After the examination, the images and radiometric information are analysed to verify the quality and then recorded in the patient's file in the database. Currently, tumours less than 10 mm can be detected, which implies that the patient has an 85% chance of survival (Ng and Sudharsan, 2009). Cancer cells can correlate to warm regions on thermal imaging because they are highly metabolic and angiogenic and thus tend to have a higher temperature than the surrounding cells. Even if tumours are small, their quick spread leads to an increase in temperature that can be seen in a MIT study, which is an advantage over mammography which only detects tumours between 1.2 cm to 1.66 cm. Other studies claim that thermography can detect tumours up to 10 years before mammography can detect advanced tumours (Borchartt et al., 2013).

The fundamental analysis of breast thermography is based on contralateral image cross-matching and slightest temperature asymmetries (Francis and Sasikala, 2013). However, limitations of sensors or image acquisition errors can compromise the visual identification of asymmetries. Therefore, in the last decade, studies on automatic detection based on neural classification and computer vision have been increasing (Farooq and Corcoran, 2020). The first step for automatic analysis and interpretation is the segmentation of the RoI, the most popular techniques being region detection, edge detection, or thresholding. Although these techniques are based on computer vision, the thresholding method can also be applied to the processing of radiometric signals prior to generating an image. Once the RoIs have been segmented, statistical methods such as histogram comparison could be considered adequate for the conclusion. However, when the data are similar, identification by unsupervised methods can be performed. The fundamentals of supervised classification algorithms are based on a dataset training outcome prediction. On the other hand, the unsupervised classification does not contain a training dataset. The most basic algorithm is K-means, in which the shape of the histogram is studied and the breast that is asymmetric with respect to the opposite pair is determined. However, when histogram shape studies are not enough, other image features such as edges, colours, shapes, textures, and spatial and transformation features can be exploited. Some methods for classification of malignant or benign tumours are:

- Cascade forward and Feed-forward back-propagation network (Saini and Vijay, 2015).
- K-nearest neighbour (K-NN) (AhmedMedjahed et al., 2013).
- Naive Bayes classifier (Karabatak, 2015).
- Deep learning (Abdel-Zaher and Eldeib, 2016).

Although diagnosis by asymmetry currently has an accuracy of 98%, reproducibility of results is not always guaranteed among a set of images. That has given breast cancer research based on thermography a wide scope and future to achieve 100% accuracy. Classification algorithms are becoming more robust in a way that minimizes the issue that images are not standardized due to factors such as the quality of the thermographic equipment or the selected acquisition protocol.

3.3.4 Risk and prognostic diagnosis

Currently, breast thermography studies have reported an average accuracy and specificity of 90%. There are even other studies that report a prediction of 84% and a negative prediction of 100% (Campbell and Mead, 2022). Unlike mammography or X-rays which may not detect tumours due to breast density, thermography can detect small tumours based on the temperature that generates the spread of cancer cells, so density does not affect thermography studies. Furthermore, the specificity of thermography is higher in dense breasts than in fat tissue.

The patient's history should include menstrual history, breast implants, medications, symptoms such as pain, lumps or discoloration, previous surgeries, lactation history, previous monographs, ultrasound, or magnetic resonance imaging (MRI). Risk assessment based on thermography alone determines whether the patient is at a risk of developing breast cancer during a lifetime. A thermal image or classification with an unexplained abnormal result indicates a 10% risk of breast cancer. Thermography analysis can detect tumours smaller than 5 mm, which may not be detected on an ultrasound. However, thermography can indicate the possibility of the presence of a tumour but cannot determine its location. Therefore, caareful interpretation is a must because a thermographic study can give a false-positive result while a mammogram can indicate a false negative. The efficiency of MIT is not currently known and cannot be determined with short-term studies. For this reason, considering thermography is recommended for breast imaging as a complementary tool and the interpretation of the results should never be absolute. Only a biopsy can confirm the presence of cancer with absolute certainty.

3.3.5 A comparison of thermography and other medical techniques

Medical imaging has been applied as a complementary tool for the detection and prognosis of breast cancer. The advantage of MIT over other imaging techniques is the detection of tumors in dense breasts that mammography may miss. In addition, thermography does not require mechanical trauma and can monitor patients with previous lesions or therapies such as lumpectomy, mastectomy and reconstruction. On the other hand, mammography is not recommended in young patients since it is an ionizing radiation technique, considered to be one of the factors accelerating the growth of malignant cells (Borchartt et al., 2013).

Figure 3.4 shows a flow chart summarizing the treatment conditions that are appropriate based on the circumstantial condition of the patient (Schwartz, 2021).

3.4 Current advances and perspectives

The number of scientific studies regarding thermal thermography for early detection of breast cancer has been a trend in the last decade. A trend that can be properly exploited in multimodal combination with other sensors, AI implementation, and medical observation. The success of thermography is largely due to improved technologies and standardization of acquisition protocols.

Even with the latest technologies and standardization, the interpretation of thermal images requires training. Not only can images concerning the patient vary over time, the positioning and post-processing algorithms can alter the results and matching. Even colour palette selection can affect the interpretation or 3D reconstruction steps (descriptive feature detection). On the other hand, temperature scales indicate relative and not absolute temperature. Preparation and certification of personnel could reduce errors or increase diagnostic

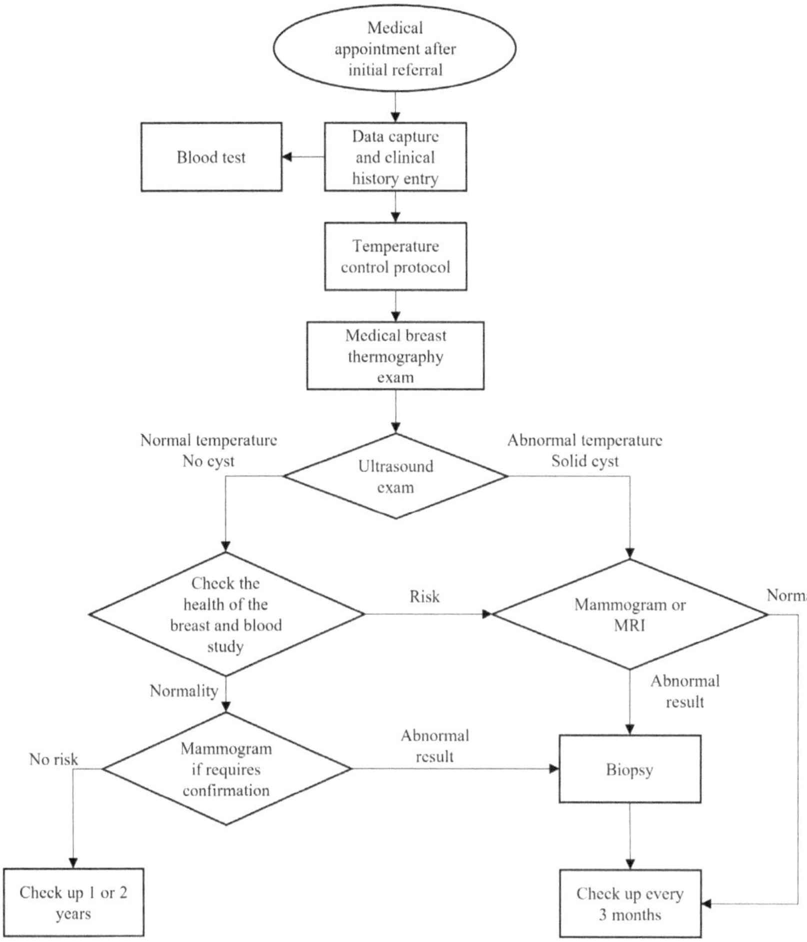

Figure 3.4. Flowchart describing the protocol for breast cancer studies based on thermography and mammography as complementary studies to biopsy.

assertiveness. That is, interpretation should be certified by professionals in the field, and collaboration with physiologists is essential to confirm breast cancer. The information by specialists could create a solid database to improve AI systems, with the prospect of finding a specific relationship between temperature signatures and breast tumours (Mashekova et al., 2022). The literature of the last decade has a clear trend towards the adoption of AI, in such a way that the computer or devices are a kind of medical assistant. In this sense, the training information must be preserved, i.e., the radiometric information which is the core of the processing. The format of the radiometric data should be universalized so that it can be processed by any device or research laboratory (Campbell and Mead, 2022).

Other studies have suggested a multimodal combination of sensors and medical imaging techniques in such a way that the information can be complemented for a more extensive diagnosis. Some hypotheses are the implementation of 3D thermographic models based on CT scans and MRIs (Sanches et al., 2007). 3D breast modelling is a technological advance that can provide morphological information in a single entity rather than a set of images that must be analysed individually. In addition to the detection of temperature signatures, tumour sizes and shapes can be classified in a treatment monitoring scenario. Also, the evaluation in search of malignant tumors can be conducted from different points of view and depth . The tumours can be localized, classified, and form part of a database for training (Pasquali, 2020).

3.5 Conclusions

MIT advantage lies in a non-invasive, non-contact technique study with an early detection perspective. Several clinical and technical studies have provided confidence through statistical results to trust this technique as an auxiliary method for breast detection. In the previous decades, MIT has been exploited to benefit early detection of breast cancer based on technological systems, signal and image processing algorithms, and AI implementation. However, it should be noted that this technique is an auxiliary tool since only biopsy can detect cancer with absolute certainty. MIT can be complemented with tomography techniques to observe anatomical structures among metabolism signs. Continuous improvement in processing, protocols, and technology, not to mention the relative ease of acquiring thermographic equipment due to low prices, will surely lead to increasing acceptance and adoption.

Considering the demonstrated contribution of MIT in this chapter and the prospects of the technique, further clinical and technical research should be promoted for acceptance and incorporation for breast cancer detection.

Acronym/Abbreviation

AI	Artificial Intelligence
FDA	Food and Drug Administration
FIR	Far-IR
FPA	Focal Plane Array
IR	Infrared
K-NN	K-nearest Neighbour
MIR	Mid-IR

MIT	Medical Infrared Thermography
MRI	Magnetic Resonance Imaging
MRTD	Minimum Resolvable Temperature Difference
NETD	Noise Equivalent Temperature Difference
NIR	Near-IR
RoI	Region of Interest
ROIC	Read Out Integrated Circuit surface

Reference list

Abdel-Zaher, A.M. and Eldeib, A.M. (2016). Breast cancer classification using deep belief networks. *Expert Systems with Applications*, 46: 139–144. https://doi.org/10.1016/J.ESWA.2015.10.015.

AhmedMedjahed, S., Ait Saadi, T. and Benyettou, A. (2013). Breast cancer diagnosis by using k-nearest neighbor with different distances and classification rules. *International Journal of Computer Applications*, 62(1): 1–5. https://doi.org/10.5120/10041-4635.

Amalu, W.C., Hobbins, W.B., Head, J.F. and Elliott, R.L. (2006). Infrared imaging of the breast—An overview. *Medical Devices and Systems*, 25-1-25–21. https://doi.org/10.1201/9781420003864.ch25.

Anjum, R., Dipti, R.R., Rashid, H.O. and Ripon, S. (2021). An efficient breast cancer analysis technique by using a combination of hog and canny edge detection techniques. *Proceedings of the 5th International Conference on Trends in Electronics and Informatics, ICOEI 2021*. https://doi.org/10.1109/ICOEI51242.2021.9453095.

Arnon, B., Oria, K. and Arieli, Y. (2017). Correction of the angular emissivity of human skin for clinical thermal imaging. *Imaging Medicine*, 9(4): 103–108.

Bayareh Mancilla, R., Tấn, B., Daul, C., Gutiérrez Martínez, J., Leija Salas, L. et al. (2021). Anatomical 3D modeling using IR sensors and radiometric processing based on structure from motion: Towards a tool for the diabetic foot diagnosis. *Sensors*, 21(11): 3918. https://doi.org/10.3390/s21113918.

Borchartt, T.B., Conci, A., Lima, R.C.F., Resmini, R., Sanchez, A. et al. (2013). *Breast Thermography from an Image Processing Viewpoint—A Survey*. 93: 2785–2803.

Campbell, J.S. and Mead, M.N. (2022). *Human Medical Thermography*. CRC Press. https://www.routledge.com/Human-Medical-Thermography/Campbell-Mead/p/book/9781032251400.

Cheng, T.Y., Deng, D. and Herman, C. (2012). Curvature effect quantification for *in-vivo* IR thermography. *ASME International Mechanical Engineering Congress and Exposition, Proceedings (IMECE)*, 2(2): 127–133. https://doi.org/10.1115/IMECE2012-88105.

Cui, S. and Xing, J. (2022). Research on calibration method of infrared temperature measurement system near room temperature field. *Frontiers in Physics*, 9. https://doi.org/10.3389/FPHY.2021.786443.

Diakides, N.A. and Bronzino, J.D. (2008). *Medical Infrared Imaging*.

Diakides, N.A., Diakides, M., Lupo, J., Paul, J.L., Balcerak, R. et al. (2006). Advances in medical infrared imaging. In *Medical Devices and Systems* (pp. 19-1-19–14). CRC Press. https://doi.org/10.1201/b12938-2.

Edis, E., Flores-Colen, I. and de Brito, J. (2014). Passive thermographic detection of moisture problems in façades with adhered ceramic cladding. *Construction and Building Materials*. https://doi.org/10.1016/j.conbuildmat.2013.10.085.

Farooq, M.A. and Corcoran, P. (2020, June 1). Infrared imaging for human thermography and breast tumor classification using thermal images. *2020 31st Irish Signals and Systems Conference, ISSC 2020*. https://doi.org/10.1109/ISSC49989.2020.9180164.

Fidali, M. and Mikulski, M. (2008). An inexpensive blackbody model. *9th International Conference on Quantitative InfraRed Thermography*.

Francis, S.v. and Sasikala, M. (2013). Automatic detection of abnormal breast thermograms using asymmetry analysis of texture features. *Journal of Medical Engineering and Technology*, 37(1): 17–21. https://doi.org/10.3109/03091902.2012.728674.

Hao, Y., Qiao, S., Zhang, L., Xu, T., Bai, Y. et al. (2021). Breast cancer histopathological images recognition based on low dimensional three-channel features. *Frontiers in Oncology*, 11. https://doi.org/10.3389/fonc.2021.657560.

Hardy, J.D. and Muschenheim, C. (1936). Radiation of heat from the human body. v. the transmission of infra-red radiation through skin. *Journal of Clinical Investigation*, 15(1): 1. https://doi.org/10.1172/JCI100746.

Jafari-Marandi, R., Davarzani, S., Soltanpour Gharibdousti, M. and Smith, B.K. (2018). An optimum ANN-based breast cancer diagnosis: Bridging gaps between ANN learning and decision-making goals. *Applied Soft Computing Journal*, 72. https://doi.org/10.1016/j.asoc.2018.07.060.

Karabatak, M. (2015). A new classifier for breast cancer detection based on Naïve Bayesian. *Measurement*, 72: 32–36. https://doi.org/10.1016/J.MEASUREMENT.2015.04.028.

Lahiri, B.B., Bagavathiappan, S., Jayakumar, T. and Philip, J. (2012). Medical applications of infrared thermography: A review. *Infrared Physics and Technology*, 55(4): 221–235. https://doi.org/10.1016/j.infrared.2012.03.007.

Lim, T.S., Tay, K.G., Huong, A. and Lim, X.Y. (2021). Breast cancer diagnosis system using hybrid support vector machine-artificial neural network. *International Journal of Electrical and Computer Engineering*, 11(4). https://doi.org/10.11591/ijece.v11i4.pp3059-3069.

Magalhaes, C., Vardasca, R. and Mendes, J. (2018). *Classifying Skin Neoplasms with Infrared Thermal Images*. https://doi.org/10.21611/QIRT.2018.013.

Mashekova, A., Zhao, Y., Ng, E.Y.K., Zarikas, V., Fok, S.C. et al. (2022). Early detection of the breast cancer using infrared technology—A comprehensive review. In *Thermal Science and Engineering Progress* (Vol. 27). https://doi.org/10.1016/j.tsep.2021.101142.

Merino González, J.A., Hernández Gómez, L.E., Sánchez Méndez, P.A., Juárez Aburto, J.A. and Hernández Santiago, K.A. (n.d.). Thermy – Soluciones inteligentes con termografía. Retrieved February 20, 2023, from https://thermy.mx/?fbclid=IwAR04fA6WkAU2JZtdWA3FwG2SyBX5a80s9BXr73q-HIiNoSfZUbn92k003FE.

Muniz, P.R., Cani, S.P.N. and da S. Magalhaes, R. (2014). Influence of field of view of thermal imagers and angle of view on temperature measurements by infrared thermovision. *IEEE Sensors Journal*, 14(3): 729–733. https://doi.org/10.1109/JSEN.2013.2287003.

Netten, J.J.V., Clark, D., Lazzarini, P.A., Janda, M., Reed, L.F. et al. (2017). The validity and reliability of remote diabetic foot ulcer assessment using mobile phone images. *Scientific Reports*, 7(1): 1–10. https://doi.org/10.1038/s41598-017-09828-4.

Ng, E.Y.K., Acharya, U.R., Keith, L.G. and Lockwood, S. (2007). Detection and differentiation of breast cancer using neural classifiers with first warning thermal sensors. *Information Sciences*, 177(20). https://doi.org/10.1016/j.ins.2007.03.027.

Ng, E.Y.K. and Sudharsan, N.M. (2009). Numerical computation as a tool to aid thermographic interpretation. *Http://Dx.Doi.Org/10.1080/03091900110043621*, 25(2): 53–60. https://doi.org/10.1080/03091900110043621.

Nunak, T., Rakrueangdet, K., Nunak, N. and Suesut, T. (2015). Thermal image resolution on angular emissivity measurements using infrared thermography. *Lecture Notes in Engineering and Computer Science*, 1(March 2015): 323–327.

Pasquali, P. (2020). History of medical photography. *Photography in Clinical Medicine*, 47–72. https://doi.org/10.1007/978-3-030-24544-3_4.

Pellegrino, J.G., Zeibel, J., Driggers, R.G. and Perconti, P. (2006). Infrared camera characterization. *Medical Devices and Systems*, 38-1-38–10. https://doi.org/10.1201/B12938-5.

Renkielska, A., Nowakowski, A., Kaczmarek, M. and Ruminski, J. (2006). Burn depths evaluation based on active dynamic IR thermal imaging-A preliminary study. *Burns*, 32(7): 867–875. https://doi.org/10.1016/j.burns.2006.01.024.

Ring, E.F.J. and Ammer, K. (2012). Infrared thermal imaging in medicine. In *Physiological Measurement* (Vol. 33, Issue 3). https://doi.org/10.1088/0967-3334/33/3/R33.

Saini, S. and Vijay, R. (2015). Mammogram analysis using feed-forward back propagation and cascade-forward back propagation artificial neural network. *Proceedings - 2015 5th International Conference on Communication Systems and Network Technologies, CSNT 2015*, 1177–1180. https://doi.org/10.1109/CSNT.2015.78.

Sanches, I.J., Brioschi, M. and Traple, F. (2007). 3D MRI/IR imaging fusion: A new medically useful computer tool. *InfraMation 2007 Proceedings, May*. https://www.researchgate.net/publication/259195497_3D_MRIIR_imaging_fusion_a_new_medically_useful_computer_tool.

Schwartz, R.G. (2021). *Guidelines for Breast Thermology*. The American Academy of Thermology. https://aathermology.org/wp-content/uploads/2018/04/AAT-Breast-Guidelines-2021v2.pdf.

The American Academy of Thermology. (2021). *Guidelines For Breast Thermology - Examination Guideline 3.1*.

Zhou, J., Lu, J., Gao, C., Zeng, J., Zhou, C., Lai, X. et al. (2020). Predicting the response to neoadjuvant chemotherapy for breast cancer: Wavelet transforming radiomics in MRI. *BMC Cancer*, 20(1). https://doi.org/10.1186/s12885-020-6523-2.

Chapter 4

Artificial Intelligence and Cancer Detection

Aldo Rodrigo Mejía-Rodríguez,[1,*] *Dora-Luz Flores*[2] and *Nelly Gordillo-Castillo*[3]

Artificial Intelligence (AI) has recently been used to support medical diagnosis, developing tools to increase diagnostic efficiency and decrease response times, particularly in cancer detection with digital images. Algorithms developed based on AI are accurate; however, external or independent validation is needed to ensure that these algorithms are generalizable. This chapter describes the steps to build models based on AI, which include data acquisition (with clinical data acquired by electronic health records or provided by the patient himself and clinical images by instruments such as X-Rays, CT scans, and MRIs, among others). Next, some techniques for pre-processing the acquired data are shown to ensure the data quality. Afterward, some methods for data processing are shown, including feature extraction, classification, and segmentation methods. Some recommendations are also demonstrated for visualizing and presenting the findings in medical images. Finally, the performance

[1] School of Science, Autonomous University of San Luis Potosí, San Luis Potosí, S.L.P. México.
[2] Facultad de Ingeniería, Arquitectura y Diseño, Universidad Autónoma de Baja California, Ensenada, Baja California, México.
[3] Department of Electrical and Computer Engineering, University of Ciudad Juárez, Ciudad Juárez, México.
Emails: dflores@uabc.edu.mx; nelly.gordillo@uacj.mx
* Corresponding author: aldo.mejia@uaslp.mx

metrics must be considered to evaluate the model or algorithms developed, such as recall, accuracy, dice score, sensitivity, and so far.

4.1 Introduction: Artificial Intelligence and its clinical relevance

From a general perspective, Artificial Intelligence (AI) implies that intelligent behavior and critical thinking could be modeled using a computer with minimal human intervention. In recent years, AI has been an exciting topic in many areas related to business, society, technology, and healthcare. However, the concept of AI dates back to 1950, when Alan Turing first described it in his book Computing Machinery and Intelligence by asking can machines think?. He also described a simple test, known as the "Turing test," where a human interrogator tries to distinguish between a computer and human text response to assess the computer's ability to generate responses similar to those a human would give. Later in the same decade, John McCarthy 1956 defined the term AI as the science and engineering of making intelligent machines (Kaul et al., 2020).

Although AI was defined more than sixty years ago, acceptance and applicability were not accessible due to two main drawbacks at the time: (1) limited computation capabilities and (2) limited information available to learn how to solve complex problems. Therefore, AI applications started as a simple "if-then" rule series. With the advance of technology over several decades, it was possible to implement more complex algorithms to emulate the human brain's performance.

It is essential to mention that there are many subsets of AI (see Figure 4.1), such as machine learning (ML), deep learning (DL), natural language processing (NLP), and computer vision (CV), that look to define

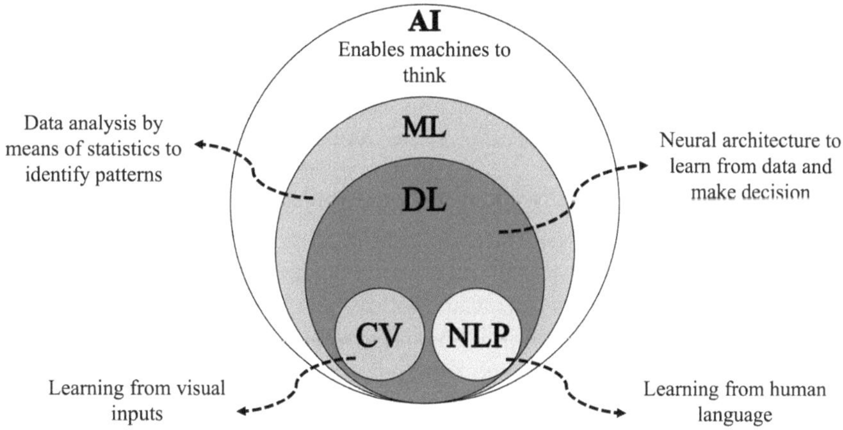

Figure 4.1. Diagram of AI concept and its corresponding subsets.

approaches to enable machines (computers) to think. ML provides models based on statistics to explore and analyze data to look for pattern identification; in this case, devices can improve their performance with experience acquired from provided data sets. In DL, a multi-layer neural architecture is defined, enabling machines to learn from provided data and make decisions independently. On the other hand, NLP allows computers to extract data from human language and make decisions based on that information. Finally, the CV defines processes by which a computer gains meaningful knowledge from visual inputs (a series of images or videos).

With the capability of AI systems to analyze data based on complex algorithms and self-learning, the clinical relevance of AI is now a reality considering that it could be applied to clinical practice through the creation of rule-based expert systems for clinical decision-making, risk assessment models, improvement of diagnosis and treatment applications, and medicine workflow efficiency. In this chapter, we will focus on AI applications for cancer detection, one of AI's major clinical applications.

4.2 Data acquisition

4.2.1 Clinical data

To give a patient an accurate clinical diagnosis, medical personnel must evaluate several types of clinical data. Typically, this information is derived from a patient's clinical information, such as laboratory test results, medical imaging, omics data, and physical symptoms. Additional data may consist of a patient's medical history, previous prescription records, and a history of their current ailment. Several illnesses may also contain significant genetic components; consequently, the family medical history of a patient may also be relevant. In addition, a patient's socioeconomic status, geographic region, nutrition, eating habits, and lifestyle can all play an essential role in identifying their medical problems. These clinical datasets are typically derived from electronic health records (EHR), patient or disease registries (Yanase and Triantaphyllou 2019), and patient-generated health data (PGHD). Patients (or their families or other caregivers) may produce, record, or collect PGHD including, but are not limited to, biometric data, symptoms, and dietary and lifestyle choices (HealthIT 2018). AI systems have been studied in many clinical situations spanning multiple medical specialties, aiding medical practitioners in retrieving relevant patient data from EHR systems with speed and precision (Chi et al., 2021). It is vital to integrate into AI systems diverse clinical data effectively and efficiently to comprehend complex diseases and accomplish correct diagnosis and future treatment. In addition to computational resources, algorithms, models, and data, it is crucial to have

Table 4.1. Overview of clinical data sources for AI systems.

Sources	**Clinical data**
Electronic health records	Lab test results, medical imaging, omics data, physical symptoms, medical history, previous medication records, and history of current disease.
Patient or disease registers	Genetic components, family medical history, socioeconomic standing, geographic location, nutrition, eating patterns, and lifestyle.
Patient-generated health data	Biometric information, symptoms, dietary, and lifestyle decisions.

multiple independent, sufficiently large, and high-quality data sets, expert knowledge, and networks that provide relevant biological entity interactions (Holzinger et al., 2019). Table 4.1 summarizes the clinical data sources that AI systems can utilize. The widespread deployment of EHR systems has enabled the large-scale patient data pooling outside clinical studies. Moreover, it is becoming increasingly apparent that large data sets will be required to find and explain cancer's complex genetic and molecular pathways (Jim et al., 2020).

4.2.2 Cancer medical imaging

Imaging is a tool that can help with improved treatment choices, but it is not a cancer treatment in and of itself. The main uses of medical imaging for cancer are (National Cancer Institute 2016):

- Screening: can be used to identify whether a person has suspicious areas or anomalies that could be malignant.
- Diagnosis and staging: can be used to determine the location of malignancy, if it has spread, and its extent. Imaging may guide physicians in obtaining a biopsy.
- Treatment: in various methods, imaging can assist in navigating the delivery of cancer therapy. By carefully targeting tumor therapies, imaging can make cancer treatments less intrusive.
- Follow-up: can determine whether a tumor is shrinking or has transformed, utilizing fewer body resources than before therapy.
- Monitoring: can determine whether a previously treated cancer has returned or is spreading to other areas.

4.3 Preprocessing

Preprocessing refers to the technique of cleaning and organizing raw data so that it can be appropriate for model building and training. Furthermore,

Table 4.2. A summary of cancer imaging techniques and their applications in artificial intelligence.

Imaging	Uses in cancer[1,2,3]	Cancer types[1,2,3]	AI cancer applications
X-Ray	D, M, Sc T: fluoroscopy-guided procedures.	Lung, breast, and bone cancer.	Tumor and lung cancer detection and classification, breast cancer screening and diagnosis, chest anatomy segmentation, radiation exposure at fluoroscopy-guided procedures reduction, musculoskeletal oncological assessment.
CT	D, F, M, Sc, St, T: cryotherapy, radiofrequency ablation, radioactive seed implantation, external-beam radiation therapy, or surgery planning.	Any type of cancer.	Detection, segmentation, and classification of tumors and nodules, cancer staging, treatment follow-up, anatomy segmentation, body composition assessment, prediction of disease outcome, metastases detection, radiation dose reduction, response and prediction response to treatment, nodule discrepancy.
MRI	D, F, M, Sc, St, T: surgery or radiation therapy planning.	Any type of cancer, especially brain, primary bone, soft tissue, spinal cord, and pelvic organ tumors.	Precision and improved diagnosis, lesion, node and tumor detection, segmentation and classification, cancer staging and prognosis, metastases detection, anatomy segmentation, body composition assessment, preoperative simulation, and treatment predictive outcomes.
Molecular and nuclear	D, F, M, St.	Cancer detection when other imaging techniques produce normal results, tiny tumors.	Tumor and node detection, segmentation, and classification, cancer assessment, prediction of disease outcome, response to treatment, metastases detection, biomarkers measurement, and radiation exposure reduction.
Ultrasound	D: biopsy navigation, F, M, Sc, T.	Breast, liver, pancreas, obstetric, and prostate cancer differentiate fluid-filled cysts from solid tumors.	Detection, segmentation, and lesion assessment, cancer diagnosis, classification, and staging, nodule detection, and characterization, nodule malignancy prediction, differential diagnosis, false-positive findings reduction.

[1] 2020 World Health Organization report on cancer.
[2] National Cancer Institute. Cancer Imaging Program. US Department of Health and Human Services.
[3] Cancer Research UK.
D: diagnosis, F: follow-up, M: monitoring, Sc: screening, St: staging, T: treatment.

it is an important activity that increases data quality to extract the most meaningful information from the data set. Real-world data is often incomplete and contains outliers or errors, so this pre-processing step is necessary. There is no single methodology to carry out this task, and it is regularly adjusted to the problem to be solved.

However, before improving the data quality, it is advisable to analyze the data to understand what change is sought. Six characteristics can be used to assess the data quality; others can also help.

1. **Accuracy.** This term refers to the fact that, in most cases, the data is inaccurate due to how it is obtained, the different instruments for data collection, inconsistencies in naming conventions, incorrect values by users, input formats, and errors in data transmission, among others.
2. **Completeness.** A data set is complete if it has all the necessary characteristics or variables. On the contrary, the data is considered incomplete due to the lack of availability of this information. This incompleteness may be due to unintentional deletion while collecting the data, equipment malfunction while the data is being collected, failure to record history, and so far.
3. **Consistency.** It is expressed quantitatively in the different instances stored as a percentage of matching values. If the information stored in several instances is the same, including format inconsistencies or not, then the data is said to be consistent.
4. **Timeliness.** The time the data is recorded also helps or affects the data quality since the information is valuable as long as it is available when needed. If the data is not up to date or changes are made after analyzing the data sets, then the data quality is inadequate.
5. **Believability.** It is described as the trust that users have in the data. If users find that the data has inconsistencies or errors, they are likely to be wary of using it again. So, credibility is also a factor in determining the quality of the data.
6. **Interpretability.** The ease with which the data stored in a data set can be understood is defined as its interpretability. The statistical analysis that can be obtained from it with the same ease also contributes to the quality of the data set and can help its interpretability.

Table 4.3 shows some pre-processing techniques, classified into four steps, associating some tasks for each step.

Table 4.3. Some pre-processing techniques and their specific associated tasks.

Step	Tasks
Data cleaning	Data and metadata acquisition, handling missing values (ignoring the tuple, manually filling missing values, assigning constants to missing values, imputation, replacing missing values), reformatting, attribute conversions, outliers' identification (binning, regression, outlier analysis).
Data integration	This involves integrating metadata from different sources and resolving potential value conflicts, for example, differences in units of measure, discrepancies in representation, among others.
Data reduction	Dimensionality reduction (Wavelet transforms, principal component analysis, random forest, and more), numerosity reduction (parametric, nonparametric), Data compression.
Data transformation and discretization	Normalization, discretization (correlation analysis, binning, clustering, histogram analysis, and decision tree analysis), concept hierarchy generation.

Felix and Lee, 2019 identified several problems related to data sets and the pre-processing techniques used to solve them. For the problem of unbalanced classes, the techniques of rebalancing, reweighting, and selective learning have been used. For the heterogeneity of the data, the techniques of data transformation, filtering, data normalization, metric selection, and grouping have been addressed. When there is a problem of highly skewed data, the automatic/logarithmic transformation technique has been used. When the dataset has irrelevant and redundant features, it has been addressed with feature selection.

Regarding the problem of continuous data, the literature says that discretization techniques are the ones that have been used the most. To address data privacy issues, the technique used is multi-party data sharing. Another critical problem is the collinearity between the metrics and the techniques used. These include eliminating highly correlated metrics, combining variables with the variance inflation factor technique, and applying principal component analysis. Noise in the data was also identified and addressed with outlier removal and noise filtering techniques.

4.4 Processing

4.4.1 Feature extraction and selection methods

For most AI applications in cancer detection, it is common to have access to hundreds (sometimes more) of data inputs, clinical data, or medical imaging, from which it is possible to extract features or variables that represent the patient's medical condition. A common issue in ML applications is to have similar or more features than the number of observations in the dataset, leading to an overfitted model, meaning that available information of features is redundant and requires more computational resources to be processed. Therefore, reducing the problem's dimensionality is necessary, reducing the number of variables used to represent the problem of interest.

The two main types of algorithms used to reduce the dimensionality of a dataset are Feature Extraction (FE) and Feature Selection (FS). FE aims to reduce the number of features by creating new features from a combination of the original ones, meaning that a summarized version of features will substitute the original set of features. On the other hand, in FS, no new features are created; instead, it aims to prioritize the available features and discard the less important ones. Table 4.4 lists the most used approaches for dimensionality reduction for AI cancer applications (Zhang et al., 2018).

Table 4.4. Feature extraction (FE) and feature selection (FS) methods for dimensionality reduction.

Method	Description
Principal Components Analysis (PCA)	An unsupervised learning algorithm for dimensionality reduction uses the original input set of features to find a new combination that best summarizes the original data distribution but with fewer features.
Independent Component Analysis (ICA)	A linear dimensionality reduction approach that considers the input data (features) as a combination of independent components. This technique aims to split and identify every single component and eliminate the unnecessary ones. The independent components are defined by maximizing the statistical independence of the components.
Linear Discriminant Analysis (LDA)	A supervised learning dimensionality reduction method aims to compute the mean of each class of observations, to maximize the distance between each mean, and to minimize data spread within the class, projecting the data in a lower dimensionality space.
Correlation	Regarding feature selection approaches, correlation could be used to assess the relationship between different observations or measures and selecting only the data with a certain level of similarity.
Mutual Information (MI)	MI is a measure of the mutual dependence between two random variables, and in particular, for feature selection purposes, the variables are a feature (input) and a target (output). Therefore, MI tries to quantify the amount of information obtained from the target by observing the feature or a set of features. Therefore, the MI value could be used to select the more representative features for a specific outcome.

4.4.2 Classification methods

Once the set of features that best represent a phenomenon to be studied is settled, it is possible to search for patterns to perform classification. A classification model uses statistical tools to study the provided input features to learn how to assign a class label or a category from the domain problem.

For example, recognize if a tumor is benign (class 1) or malignant (class 2) based on available information.

For cancer applications, ML approaches have proved to have promising results. In this case, a classification predictive model requires a training dataset with several examples of inputs (features) and outputs (class labels) from which to learn; therefore, the training dataset must sufficiently represent the problem of interest and sufficient examples of each class label.

In general, types of classification can be considered depending on the number of inputs and the outputs of interest: (1) binary classification, with two outputs; this model is commonly used for normal and abnormal situations such as the presence or absence of cancer; (2) multi-class classification, refers to situations where the problem has a range of classes (e.g., different stages of cancer); (3) multi-label classification, where more than one label can be predicted in the same sample; and (4) imbalance classification, where the number of examples in each class is not equally distributed (having more examples of non-cancer cases than those with cancer). Table 4.5 describes some of the more commonly used ML classification approaches for cancer (Obaid et al., 2018).

4.4.3 Segmentation methods

For cancer applications, the delimitation of anatomical structures of interest presented in medical imaging is particularly useful, considering that knowing the body's shape, volume, and spatial location is crucial for a cancer diagnosis, treatment, and follow-up. This delimitation, also known as image segmentation, which is the process of individually identifying and labeling every pixel or voxel in an image by looking for features shared between pixels/voxels.

Considering the complexity and irregular anatomy of cancer-affected tissues and organs, clinicians' gold standard for medical image segmentation remains the manual delimitation. However, this is very time-consuming, complicated, and prone to task variability, and the development of proper automatic or semi-automatic segmentation approaches to properly help clinicians are still of interest to the medical community. Traditional automatic segmentation methodologies have established a base on mathematical methods such as edge detection filters. However, consistency in results was insufficient due to the variability of cancer features in medical imaging. With the rise of ML approaches, several algorithms were proposed using hand-crafted features with promising results. However, the design and extraction of these manual features for proper training have been the central issue in developing these ML systems. In recent years, DL segmentation approaches

Table 4.5. Commonly used ML classification approaches in cancer applications.

Classification algorithm	Description
K-nearest neighbors	A non-parametric supervised learning classifier uses proximity to make classifications or predictions about an individual data point based on a similarity measure (e.g., Euclidean distance) concerning the dispersion of a group of data points that share proximity similarities.
Decision Trees	A non-parametric supervised learning approach predicts an outcome based on a set of observations (features) analyzed in a hierarchical tree structure. This tree model consists of a root node (input of observations), branches, internal nodes (decision nodes), and leaf nodes (output classes). It is one of the simplest, easy to understand and implement classification approaches because of its hierarchical structure based on a set of decision rules.
Support Vector Machines (SVM)	A supervised learning approach can analyze and sort data into two categories, making it suitable for binary classification. However, it is possible to use its formulation for multi-class problems effectively. Given a set of classified training observations, the main idea of SVM relies on the construction of a model capable of assigning new observations into one of the two categories; this classification model groups the training data into two regions on a hyperplane and tries to maximize the gap between the classification regions by defining a maximum margin line, then each new observation is placed on the plane and is categorized based on which side of the line it falls. This approach is one of the most used for cancer classification problems.
Random Forest (RF)	RF is an ensemble learning classification approach that combines multiple decision trees (or a forest of decision trees) to reach a single result. RF is the direct evolution of decision trees, and it overcomes the common bias and overfitting issues of decision trees. The main difference between RF and decision trees is the use of feature randomness to create an ensemble of uncorrelated decision trees, meaning that only a fraction of the features are used to ensure low correlation among decision trees.
Artificial Neural Networks (ANN)	An ANN is a computational model that attempts to simulate a human neural network in order to be able to learn and make decisions. This classification model can analyze complex data and find intricate patterns based on experience (training) and integrated (learning) information from new observations. This particularity makes ANN particularly important for ML and DL applications. ANN is constructed with three or more interconnected layers. First, an input layer is used to pass information to the inner layers, also known as neural layers, where units or neurons try to learn and understand complex patterns by weighting the information and passing it to the next layer. In addition, a set of learning rules is defined to backpropagate information from errors and adapt the weights of the neural layer for a better learning process, looking for a minimization of the errors and unwanted results.

(based on network structures) have proved to have significant capabilities for medical image segmentation and are more robust than ML methodologies. Table 4.6 briefly surveys the most recent and promising DL segmentation approaches (Hesamian et al. 2019).

Although DL approaches have considerably improved segmentation accuracy due to their capabilities to handle complex information, the availability of a large number of annotated data to properly train DL networks is still the main problem, especially in cancer applications. Creating big datasets with annotated images for different cancer cases and generating annotations on new images is very tough and time-consuming. For some diseases, it is not even possible. Therefore, in order to overcome this limitation, some solutions are usually adopted for DL segmentation methodologies, such as (1) data augmentation through the application of affine transformations to the

Table 4.6. DL segmentation approaches are commonly used for medical imaging.

DL segmentation approach	**Description**
Convolutional Neural Network (CNN)	A multi-layered feed-forward neural network designed to process structured arrays of data (images) looking for patterns. The input layer is directly connected to an input image, and the subsequent stacked hidden layers, convolutional layers, present the result of convolving filters with the input image to identify features or patterns, making them especially useful for image classification and segmentation.
Fully Convolutional Network (FCN)	FCN is a modification of a CNN where the last fully connected layer is replaced with a fully convolutional layer, allowing dense pixel-wise predictions from the full-sized input image in just one forward pass.
U-Net	In recent years one of the most used architectures for medical image segmentation due to its ability to be trained with fewer images with more precise segmentation results. The U-net is a symmetric U-form architecture that consists of two major parts, a contracting path (general convolutional layers) and an extension path using up sampling operators to increase the resolution of the output, leading to a segmentation map of the U-net.
Convolutional Residual Network (CRN)	Theoretically, deeper networks should have higher learning capabilities; however, the accuracy of deeper networks gets saturated and degrades more rapidly. Therefore, to take advantage of deeper networks, CRN was established by feeding the stacked layers of the network with a residual map instead of a feature map.
Recurrent Neural Network (RNN)	Recently, RNNs were introduced by adding recurrent connections into the architecture that allow the network to memorize patterns from previous inputs. This is particularly useful in medical imaging, considering that regions of interest (ROI) are usually distributed over multiple adjacent slices. Therefore, inter-slice context is extracted from the input slices as sequential data. In RNN, the extraction of intra-slice information is performed with a CNN model, while the inter-slice information is obtained with a RNN.

limited data available in order to increase the size of the samples and boost the performance of the DL approaches; (2) transfer learning, meaning that successful models implemented using existing data from another problem are tuned and transferred to the problem of interest; (3) Sparse annotation, where data sets available have only sparse annotations, and weighted loss functions are applied; among others.

4.5 Visualization and presentation

Before deployment, it is essential to describe, based on the type of tool, task, or pathology, how interaction with the AI tool will occur, what must be managed in the image, and what must be presented. The AI tool is not just the algorithm but also the algorithm integrated within a workflow with accompanying visualizations and user experiences (Makhlouf et al., 2022). The need for user interfaces in clinical workflows for AI algorithms restricts the widespread adoption of AI models in clinical settings (Huang et al., 2021). Establishing efficient user interfaces, vendor-neutral interoperability standards, and infrastructure requirements is crucial to facilitate the adoption of artificial intelligence in clinical practice. The artificial intelligence system can generate a diagnosis, prognosis, or quantitative data from tissue/organ/lesion segmentation, volumetric representations, image fusion, and data over time.

For a better understanding of the complete process of an AI application for cancer detection, Figure 4.2 shows an example of lung cancer segmentation performed on a computed tomography (CT) image. In this case, it is possible to observe that from the input images available on the dataset, a feature extraction phase is performed in order to identify patterns to differentiate lung tissue and cancerous tissue; then, these features are used to classify the two anatomical structures of interest at a voxel level in all the CT slices, leading to a volumetric segmentation of the lungs and the tumor; finally, the segmentation is used to create a tridimensional representation (rendering) that provides a visual perspective of the spatial distribution and dimensions of the lung cancer inside the lungs, which is information of clinical relevance considering that it may be helpful for the design of a radiotherapy treatment plan.

4.6 Validation and assessment of results

Diagnostics automation will only add one member to the medical team advising the physician. AI systems are highly specialized, must be verified in clinical settings, and are subject to medical practitioner supervision (Arslan et al., 2020). Stakeholders should collaborate with IT developers,

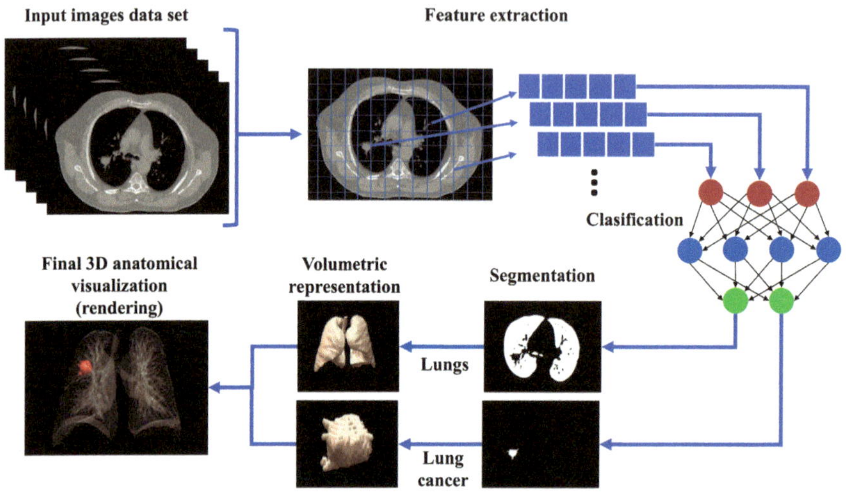

Figure 4.2. Example of a CT lung cancer segmentation and 3D visualization (rendering) process based on AI.

government agencies, and public organizations to ensure the correctness of AI algorithms, decrease unconscious bias, and guarantee patient safety (Huang et al., 2021). Clinicians in medical imaging are the gold standard for validating AI algorithms and the key to promoting cross-industry cooperation. Validation can give insight into the algorithm's robustness, reproducibility, generalizability regarding quantification, and diagnostic or prognostic precision. Table 4.7 shows the most-used metrics for validating image segmentation; depending on the relations between the metrics, their nature, and their definition, the metrics were grouped into five categories (Taha and Hanbury, 2015): overlap, volume, pair-counting, probabilistic, and spatial distance based.

4.7 Conclusion

In this chapter, an introduction to the AI concept was presented, alongside the description of its main subsets, such as ML and DL, in order to understand the potential of AI for healthcare, particularly for cancer detection. Clinical data sources (electronic health records, medical imaging, patient or disease registers, and patient-generated health data) for AI systems were also explored. In addition, a brief description of the main tasks involved in cancer detection through AI was covered, where the importance of the quality and quantity of data was evident for a feature extraction phase in order to have accurate information on an excellent classification or segmentation of cancer using different ML and DL approaches, as well as how the outcome of the AI

Table 4.7. Commonly used metrics to validate image segmentation results.

Category	Metric	Equation
Overlap	Dice (F1-measure)	$Dice = \frac{2TP}{2TP+FP+FN}$ *TP*: true positive, *TN*: true negative, *FP*: false positive, *FN*: false negative.
	Jaccard index	$J(A,B) = \frac{\mid A \cap B \mid}{\mid A \cup B \mid}$ *A* and *B*: segmentations.
	True positive rate (sensitivity, recall)	$TPR = \frac{TP}{TP+FN}$
	True negative rate (specificity)	$TNR = \frac{TN}{TN+FP}$
	False positive rate (1-specificity, fallout)	$FPR = \frac{FP}{TN+FP}$
	False negative rate (1-sensitivity)	$FNR = \frac{FN}{TP+FN}$
	Global consistency error	$GCE = \frac{1}{n}\left\{\frac{FN(FN+2TP)}{TP+FN} + \frac{FP(FP+2TN)}{TN+FP}, \frac{FP(FP+2TP)}{TP+FP} + \frac{FN(FN+2TN)}{TN+FN}\right\}$
Volume	Volumetric similarity	$VS = 1 - \frac{\mid FN-FP \mid}{2TP+FP+FN}$
Pair counting	Rand index	$a = \frac{1}{2}[TP(TP-1)+FP(FP-1)+TN(TN-1)+FN(FN-1)]$ $b = \frac{1}{2}[(TP+FN)^2+(TN+FP)^2-(TP^2+TN^2+FP^2+FN^2)]$ $c = \frac{1}{2}[(TP+FP)^2+(TN+FN)^2-(TP^2+TN^2+FP^2+FN^2)]$ $d = \frac{1}{2}\left[\frac{n(n-1)}{2}-(a+b+c)\right]$ $RI = \frac{a+b}{a+b+c+d}$
	Adjusted Rand index	$ARI = \frac{2(ad-bc)}{c^2+b^2+2ad+(a+d)(c+b)}$

Table 4.7 contd. ...

...Table 4.7 contd.

Category	Metric	Equation
Probabilistic	Interclass correlation	$$ICC = \frac{\sigma_S^2}{\sigma_S^2 + \sigma_\varepsilon^2}$$ σ_S: variance between the segmentations. σ_ε: variance between the points in the segmentations.
	Probabilistic distance	$$PDB(A,B) = \frac{\int \lvert P_A - P_B \rvert}{2\int P_{AB}}$$ P_A and P_B: probability distributions representing the segmentations. P_{AB}: pooled joint probability distribution.
	Cohens kappa	$$KAP = \frac{f_a - f_c}{N - f_c}$$ $$f_a = TP + TN$$ $$f_c = \frac{(TN+FN)(TN+FP)+(FP+TP)(FN+TP)}{N}$$ N: number of observations.
	Area under ROC curve	$$AUC = 1 - \frac{FPR + FNR}{2}$$
Spatial distance	Hausdorff distance	$$HD(A,B) = max(h(A,B), h(B,A))$$ $$h(A,B) = \lVert a - b \rVert$$
	Average distance	$$AVD(A,B) = max\ (d(A,B), d(B,A))$$ $$d(B,A) = \frac{1}{N}\sum_{a \in A} \lVert a - b \rVert$$
	Mahalanobis distance	$$MHD(x,y) = \sqrt{(x-y)^T S^{-1}(x-y)}$$ S^{-1}: inverse of the covariance matrix S. T: transpose matrix.

system should be visualized and validated. Finally, the information presented in this chapter suggests that with the advance of technology, the complexity and quantity of clinical data may lead to an increase of AI systems applied within the healthcare field.

Acronym/Abbreviation

AI	Artificial Intelligence
ANN	Artificial Neural Network
CNN	Convolutional Neural Network

CRN	Convolutional Residual Network
CV	Computer Vision
DL	Deep Learning
EHR	Electronic Health Records
FCN	Fully Convolutional Network
FE	Feature Extraction
FS	Feature Selection
ICA	Independent Component Analysis
LDA	Linear Discriminant Analysis
MI	Mutual Information
ML	Machine Learning
NLP	Natural Language Processing
PCA	Principal Components Analysis
PGHD	Patient-Generated Health Data
RF	Random Forest
RNN	Recurrent Neural Network
ROC	Receiver Operating Characteristic
SVM	Support Vector Machines

Reference list

Arslan, J., Benke, K.K. and Baird, P.N. (2020). Artificial Intelligence in medicine: What are the implications for the medical practitioner? *ACNEM Journal*, 39(2): 21–27.

Chi, Ethan Andrew, Gordon Chi, Cheuk To Tsui, Yan Jiang et al. (2021). Development and validation of an ArtificialIntelligence system to optimize clinician review of patient records. *JAMA Network Open*, 4(7). https://doi.org/10.1001/jamanetworkopen.2021.17391.

Felix, E.A. and Lee, S.P. (2019). Systematic literature review of preprocessing techniques for imbalanced data. *IET Software*, 13(6): 479–496. https://doi.org/10.1049/iet-sen.2018.5193.

HealthIT. (2018). What Are Patient-Generated Health Data? *Official Website of The Office of the National Coordinator for Health Information Technology*. January 19, 2018.

Hesamian, Mohammad Hesam, Jia Wenjing, He Xiangjiang and Kennedy Paul. (2019). Deep learning techniques for medical image segmentation: Achievements and challenges. *Journal of Digital Imaging*, 32(4): 582–596. https://doi.org/10.1007/s10278-019-00227-x.

Holzinger, Andreas, Benjamin Haibe-Kains and Igor Jurisica. (2019). Why imaging data alone is not enough: AI-based integration of imaging, omics, and clinical data. *European Journal of Nuclear Medicine and Molecular Imaging*, 46(13): 2722–30. https://doi.org/10.1007/s00259-019-04382-9.

Huang, Shigao, Jie Yang, Lijian Tan, Simon Fong et al. (2021). Current medical imaging and Artificial Intelligence and its future. *In Current and Future Application of Artificial Intelligence*, 2021: 59–81. https://doi.org/10.2174/97816810884191210101.

Jim, Heather S.L., Aasha I. Hoogland, Naomi C. Brownstein, Anna Barata, Adam P. Dicker et al. (2020). Innovations in research and clinical care using patient-generated health data. *CA: A Cancer Journal for Clinicians*, 70(3): 182–99. https://doi.org/10.3322/caac.21608.

Kaul, Vivek, Sarah Enslin and Seth Gross. (2020). History of artificial intelligence in medicine. *Gastrointestinal Endoscopy*, 92(4): 807–812.

Makhlouf, Yasmine, Manuel Salto-Tellez, Jacqueline James, Paul O'reilly and Perry Maxwell et al. (2022). General roadmap and core steps for the development of AI tools in digital pathology. *Diagnostics*, 12(5). https://doi.org/10.3390/diagnostics12051272.

National Cancer Institute. (2016). Uses of imaging. *Cancer Imaging Program. U.S. Department of Health and Human Services*. December 22, 2016.

Obaid, Omar Ibrahim, Mazin Abed Mohammed, Mohd Kanapi Abd Ghani, Salama Mostafa, Fahad Taha Al-Dhief et al. (2018). Evaluating the performance of machine learning techniques in the classification of Wisconsin Breast Cancer. *International Journal of Engineering & Technology*, 7(4.36): 160–166. https://doi.org/10.14419/ijet.v7i4.36.23737.

Taha, Abdel Aziz and Allan Hanbury. (2015). Metrics for evaluating 3D medical image segmentation: Analysis, selection, and tool. *BMC Medical Imaging*, 15(1). https://doi.org/10.1186/s12880-015-0068-x.

Yanase, Juri and Evangelos Triantaphyllou. (2019). A systematic survey of computer-aided diagnosis in medicine: Past and present developments. *Expert Systems with Applications*. Elsevier Ltd. https://doi.org/10.1016/j.eswa.2019.112821.

Zhang, Dejun, Lu Zou, Xiounghui Zhou and Fazi He. (2018). Integrating feature selection and feature extraction methods with deep learning to predict clinical outcome of breast cancer. *IEEE Access*, 6: 28936–28944.

Chapter 5

Hyperspectral Imaging for Cancer Applications

Ines A. Cruz-Guerrero,[1,*] *Raquel Leon*,[2] *Aldo Rodrigo Mejía-Rodríguez*,[1] *Daniel Ulises Campos-Delgado*,[3] *Himar Fabelo*,[2,4] *Samuel Ortega*[2,5] and *Gustavo M. Callico*[2]

Hyperspectral imaging (HSI) is a well-established technique in remote sensing that has been successfully translated into the medical field, especially for diagnosis and guided surgery. HS images are composed of spatial and spectral information merged together into a three-dimensional array. Hence, the spectral information makes it possible to discriminate, at pixel level, the materials present in the captured scene according to their optical properties. Furthermore, one key advantage is that HSI is a contactless, non-ionizing, non-labeled, and non-invasive technology. In the medical area, HSI has been proposed for the detection of various types of tumor tissues in cancer diagnosis. Thus, this chapter presents a review of HSI for detecting and evaluating various types of cancer. First, the main types of HSI technologies and instrumentation in the medical field

[1] School of Sciences, Autonomous University of San Luis Potosi, San Luis Potosí, S.L.P. 78295 México.
[2] Research Institute for Applied Microelectronics, University of Las Palmas de Gran Canaria, Las Palmas de Gran Canaria, Gran Canaria E35017 Spain.
[3] Institute for Optical Communication Research, Autonomous University of San Luis Potosi, San Luis Potosí, S.L.P. 78210 México.
[4] Canarian Foundation Canarian Institute for Health Research, Las Palmas de Gran Canaria, Gran Canaria 35019 Spain.
[5] Norwegian Institute of Food, Fisheries and Aquaculture Research, Breivika, Tromsø NO 9291 Norway.
Emails: slmartin@iuma.ulpgc.es; ducd@fciencias.uaslp.mx; aldo.mejia@uaslp.mx; hfabelo@iuma.ulpgc.es; sortega@iuma.ulpgc.es; gustavo@iuma.ulpgc.es
* Corresponding author: alejandro.guerrero@uaslp.mx

are described. The analysis algorithms based on spectral unmixing and artificial intelligence are discussed. Finally, the state of the art in applications of HSI for skin, brain, gastrointestinal, head & neck cancer and in histological samples for tumor detection are presented in this work.

5.1 Introduction

Hyperspectral (HS) imaging (HSI) is an emerging modality originated in the remote sensing field, that has expanded its applications to other research and industrial areas in the past years (Gu et al., 2021), such as food quality inspection (Vejarano et al., 2017), quality control of pharmaceutical products (Al Ktash et al., 2021), monitoring of marine ecosystems (Montes-Herrera et al., 2021), soil pollution monitoring (Lassalle et al., 2021), petrochemical industry (Arnold et al., 2022) or defense and security (Shimoni et al., 2019). HS images are structured by spatial and spectral information, conforming a three-dimensional array, also called a *HS cube* or *hypercube*. Each spatial pixel is related to a vector of light intensity values that belong to hundreds of different spectral wavelengths, also called *channels* or *bands*. This vector describes a continuous spectrum that is commonly named *spectral signature*. Unlike standard digital color cameras (see Figure 5.1B), which capture RGB (red, green, and blue) images with only three wavelengths, HS cameras are able to cover broadband spectral ranges employing different sensor types, such as (i) visual and near infrared (VNIR) between 400 and 1000 nm, (ii) near infrared (NIR) from 900 to 1700 nm, or (iii) short-wave infrared (SWIR) from 900 to 2500 nm (see Figure 5.1A) (Govender et al., 2009). The spectral signature allows the differentiation of the materials presented at pixel level in the captured scene based on their chemical composition, and resulting optical properties (Kamruzzaman and Sun, 2016).

HSI techniques have found a presence in various fields of study, one of the most relevant being the medical field due to their contactless, non-ionizing, non-labeled, and non-invasive nature. Some examples of HSI applications in the medical field are focused on measuring deoxyhemoglobin and oxyhemoglobin by employing isosbestic points of the hemoglobin absorption spectra in the region 510 to 590 nm, or absorbance in the oxygen-sensitive wavelength of 600 nm (Holmer et al., 2016). In addition, skin lesions can be identified using HSI systems by analyzing the spectral properties of the skin, which are affected by chromophores (Rey-Barroso et al., 2021). Another emerging application is the organ quality assessment during perfusion, in which HSIs were used to predict the tissue water index in kidneys, allowing analysis of tissue-related damage during *ex-vivo* preservation (Markgraf et al., 2020). On the other hand, Alzheimer's disease has been investigated by

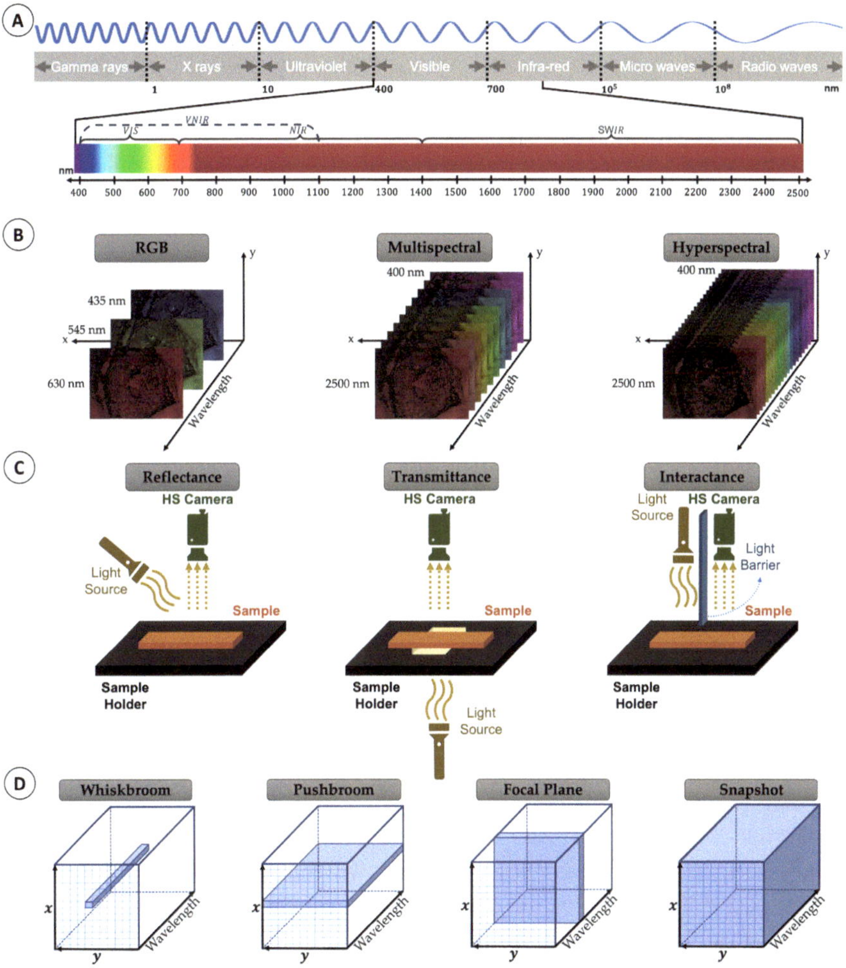

Figure 5.1. Hyperspectral imaging technique schematic: (A) Spectral resolution, (B) image modalities, (C) acquisition methodologies, and (D) scanning strategies.

analyzing amyloid-beta protein in the retina in combination with HSI systems, with no requirement of contrast agents (Lim et al., 2021). All these applications are possible because the optical characteristics of reflection, diffraction, and scattering of tissues change as the pathology progresses (Fei, 2020).

The applications of HSI in the healthcare environment have evolved significantly, to the point that in recent years this technology has been investigated and employed for the identification of various types of tumor tissues. For this reason, this chapter briefly describes the use of HSI to detect and evaluate diverse types of cancer. The overall contribution of this chapter is presented in five sections: introduction, HSI instrumentation, HSI analysis algorithms, HSI applications in cancer detection, and conclusions.

5.2 HSI instrumentation

In HSI, instrumentation is a crucial element to have reliable, efficient, and high-quality spectral data acquisition. Usually, a HSI platform consists of a HS camera, a light source, a computer with the acquisition software, and, in some instances, a motorized mobile station, which depends on the scanning mode employed by the HS camera (Fu and Chen, 2019). The most common acquisition modes are reflectance, transmittance, and interactance, whose configurations are depicted in Figure 5.1C.

The HS camera is the main component of the acquisition system, which consists of two main structures: spectrographs or spectrometers and a detector or array of photo-sensitive detectors (Fu and Chen, 2019). Spectrographs allow dispersion of the polychromatic incident light into beams with specific wavelengths, with three types of devices (Ma et al., 2019): monochromator, optical bandpass filter, and single-shot imager. The scattering devices focus the narrow wavelength light toward each of the detectors. In this sense, the photosensors most used in HSI are charge-coupled devices (CCD) and complementary metal-oxide semiconductors (CMOS) (Adão et al., 2017). The principal difference between these two sensors lies in the transmission scheme of the incoming signals. On the one hand, CCD sensors focus on measuring the luminous intensity, transferring the resulting multi-sensor signal to a digital/analog converter. On the other hand, CMOS sensors incorporate the photodetector and the digital/analog converter together, thus the information from each sensor is independent of the rest. Because of this difference, CMOS sensors are faster in measuring and capturing photons, but these sensors are susceptible to the presence of noise and are mostly affected by dark currents (Adão et al., 2017). This situation is compensated by CCD sensors, since by digitizing the signals outside the photodiode allows the inclusion of components with different characteristics that mitigate noise, dark current, and acquisition speed. In addition, CCDs have better sensitivity for visible and VNIR wavelengths, while CMOS have higher efficiency in the infrared range.

In general, HSI cameras are classified depending on the scanning method used to generate the hypercubes, with four main types of scanning: whiskbroom, pushbroom, focal plane, and snapshot (Adão et al., 2017). Whiskbroom or point-scanning cameras are characterized by capturing the spectral information of one pixel per acquisition-time (first column of Figure 5.1D); this means that to scan a particular region, it is necessary to have a mobile station that travels through the scanning area of the camera at each location in the X and Y dimensions, where the acquired hypercube is stored in a band-interleaved-by-pixel (BIP) format. Because of this, whiskbroom cameras require considerable time to acquire an image, so spatial resolution is often limited. Nevertheless, the main strength of these cameras is their high spectral resolution which permits to capture a large amount of information. Like the previous case, pushbroom or line-scanning cameras (second column of Figure 5.1D) acquire the complete spectra of several pixels continuously, i.e., the area of interest is scanned line by line until the entire image is captured, storing the hypercube in a band-interleaved-by-line (BIL) format. This scanning mode requires a mobile station to traverse the scan line through the area of interest, however motion artifacts may occur. Pushbroom camera provides high spatial and spectral resolution, and for this reason it is the most used today.

Focal plane and snapshot cameras (third and fourth column of Figure 5.1D) allow scanning of complete areas, capturing spatial and spectral information together. Focal plane cameras acquire a 2D monochromatic image at a given wavelength, i.e., each spectrum is captured independently until completing the hypercube, using a band sequential format (BSQ) as a form of storage. The main advantage of these cameras is that they can capture a single wavelength or several wavelengths, but they are susceptible to the presence of motion artifacts. As in the previous case, snapshot cameras acquire spatial and spectral information simultaneously, but unlike focal plane cameras, this type produces the hypercube in a single shot, which results in a reduction of the acquisition-time. However, snapshot cameras can capture a limited number of spectral bands, so the spectral resolution is lower than with other types.

In this sense, the light source is another crucial component of the HSI acquisition system, since light is the medium that provides information about the objects under study. Currently, halogen lamps are the most widely used because of their broad-spectrum, which is continuous, soft, and without sharp peaks (Fu and Chen, 2019). However, this type of illumination has certain disadvantages, such as a rise in the sample temperature, short lifetime, and non-constant spectral properties due to variations in temperature, voltage, and time of use. On the other hand, light-emitting diodes (LEDs) have begun to be used as light sources due to their long lifetime, fast response, compact size, low power consumption, and low heat generation (Fu and Chen, 2019). LEDs can produce broad and short spectra in the ultraviolet, visible, and infrared regions. However, they are not very efficient in dissipating heat, which reduces their lifetime and affects their spectrum. Finally, lasers are light sources with a narrow bandwidth, linearly directed, and used mainly in fluorescence and photoluminescence applications.

Once the raw HS image (I_0) is acquired, it is necessary to perform a preprocessing stage to eliminate the effects of temperature and illumination changes, and aging of the light source. This preprocessing is known as spectral calibration, so that the raw image is modified based on a dark (*D*) and white (*W*) reference images. The *D* reference is captured by closing the camera shutter, while *W* is obtained from an image of a highly reflective and uniform white surface. These two reference images are used to calculate the relative reflectance (*I*) (ElMasry and Sun, 2010) of each pixel by Equation 5.1.

$$I = 100 \times \frac{I_0 - D}{W - D} \quad (5.1)$$

The HS corrected image *I* can also be expressed in terms of absorbance (*A*) (ElMasry and Sun, 2010) by evaluating Equation 5.2.

$$A = -\log_{(10)}((I_0 - D)/(W - D)) \quad (5.2)$$

5.3 HSI analysis algorithms

Many studies in the literature have demonstrated the high potential of HSI for remote and nondestructive detection of the tissue chemical composition in different applications, obtaining promising results. However, a proper analysis of HS images is not an easy task, considering that the key idea relies on the analysis of spectral and spatial information presented in the HS image to identify the spectral signatures of the basic components, called endmembers, and their corresponding contributions or abundances (Gutierrez-Navarro et al., 2013).

In the literature, several approaches have been proposed to identify and classify endmembers present in a HS image, but the two main strategies are based on Spectral Unmixing (SU) and Machine Learning (ML) (Gutierrez-Navarro et al., 2013; Rico-Jimenez et al., 2016). First, SU allows identifying the different interactions of light by objects in the scene based on a mathematical mixing model, which determines the relationship between the endmembers and their abundances (Campos-Delgado et al., 2022). The simplest approach in SU assumes that within the HS image there are regions that possess pure materials of each class or linear combiations of them, meaning that a linear mixing model (LMM) might solve the problem; however, this approach is only suitable for simple cases. The light-tissue interactions may present scattering and nonuniform reflection patterns. Therefore, a nonlinear mixture model (NMM), that considers multiple reflections of photons, should be considered to pursue a SU for more realistic scenarios (Campos-Delgado et al., 2022; Gao et al., 2021).

NMM's can be divided into two main categories by their order: bilinear, and high-order mixing models. The generalized bilinear model (GBM) proposed by Fan et al., and the linear quadratic mixing model (LQM) are some of the most representative approaches of the first category (Fan et al., 2009). On the other hand, the p-linear model, the polynomial post-nonlinear model (PPNM), and the multilinear mixing model (MMM) are examples of higher-order ones (Heylen et al., 2014). These models possess a linear component and one or more nonlinear terms that depend on the assumed optical effects.

Proper analysis of HSI under a SU approach requires not only to estimate endmembers and their abundances, but also the approximation of the specific NMM parameters (Dobigeon et al., 2013). If prior studies of materials or tissue elements with their respective spectral signatures are available, a supervised strategy could be considered, where the endmembers are assumed known. However, in many real scenarios, including medical application of HSI, it is very difficult to have prior studies to define the endmembers' spectral information. Therefore, for the cases where endmembers are unknown, an unsupervised approach (also known as blind unmixing methodology) could be considered where the information has to be jointly estimated (Campos-Delgado et al., 2019). For more technical details on the different SU methodologies, the reader is referred to (Campos-Delgado et al., 2022; Dobigeon et al., 2013; Heylen et al., 2014).

The analysis of HSI by means of ML techniques allows the identification and classification of endmembers and their respective abundances through the recognition of features or patterns, mainly in a supervised manner. However, there are still open problems in this approach, as the necessity of large amounts of training data, the high-dimensionality of HS images, the difficulty to obtain labeled HS images, and the artifacts that might be present due to acquisition issues. HS image classification approaches might be grouped as supervised, unsupervised, and semisupervised methods according to the prior classification information for training. In the literature, several approaches have been used with HSI for clinical applications, such as spectral angle mapper (SAM), support vector machines (SVM), artificial neural networks (ANN), convolutional neuronal networks (CNN), spectral information divergence (SID), among others. For more specific details on the ML approaches for HSI classification, the reader is referred to (Lu et al., 2020; Lv and Wang, 2020; Qin et al., 2009).

5.4 HSI applications in cancer detection

As mentioned earlier, several applications of HSI have been proposed in the healthcare environment for *in-vivo* and *ex-vivo* cancer diagnosis during surgical procedures, pathological analysis, or early assessment in clinical routine practice. Next, we will describe some of these applications related to cancer diagnosis.

5.4.1 Skin cancer applications

The occurrence of skin cancer has increased around the world, being one of the most common kinds. This pathology includes malignant melanoma and non-melanoma skin cancer (NMSC), which considers basal cell carcinoma (BCC), squamous cell carcinoma (SCC) and other types with minor prevalence. Melanoma is the 17th most frequent cancer globally with 325,000 new cases in 2020, while NMSC is the 5th one with 1,200,000 new cases in 2020 (Sung et al., 2021). Among NMSC, BCC is the most frequent skin cancer with 80–85% of all cases, followed by SCC with 15–20% (Zambrano-Román et al., 2022). A summary of HSI systems employed in the literature is shown in Table 5.1.

Traditionally, skin cancer is detected during visual inspection by the naked eye and dermoscopy. A preliminary diagnosis is performed by evaluating the shape and color through the ABCDE (Asymmetry of the mole, Border irregularity, Color uniformity, Diameter and Evolving size) protocol (Tsao et al., 2015). Nowadays, different imaging techniques are employed in clinical practices: confocal microscopy, polarized imaging, three-dimensional topography, thermal imaging, multispectral (MS) and HS imaging (Rey-Barroso et al., 2021). In this sense, HS and MS images can capture information beyond the visible range related to distinct chromophores (Jolivot et al., 2013). Multispectral imaging (MSI) technology has been widely employed to identify skin cancer, with commercial systems such as MelaFind (Elbaum et al., 2001) developed for melanoma detection. This commercial device uses a snapshot imaging system to capture skin lesions. In 2015, an HSI system prototype was presented by Neittaanmäki-Perttu et al., to delimitate the contour of lentigo maligna and its progression to melanoma (Neittaanmäki-Perttu et al., 2015). The HS images were analyzed by employing a linear mixture model to identify the spectral responses of malignant lesions and normal skin to produce abundance maps that delineate the lesion borders,

Table 5.1. HSI systems used to identify skin cancer. B – Benign; BCC – Basal Cell Carcinoma; LM – Lentigo Maligna; LMM – Lentigo Maligna Melanoma; M – Malignant; MM – Malignant Melanoma; N – Nevus; SCC – Squamous Cell Carcinoma.

Publication	System	Wavelengths (nm)	Bands	Lesion
(Elbaum et al., 2001)	MSI - MelaFind	430–950	10	MM
(Neittaanmäki-Perttu et al., 2015)	HSI	500–900	76	LM, LMM
(Zherdeva et al., 2016)	HSI	450–750	61	MM, BCC
(Hosking et al., 2019)	HSI - mAID	350–950	21	MM, N
(Fabelo et al., 2019a)	HSI	450–950	125	B, M
(Courtenay et al., 2021)	HSI	398–995	270	BCC, SCC
(Lindholm et al., 2022)	HSI - SICSURFIS	475–975	33	B, M

obtaining a match of 94.7%. In 2019, this system was employed to delineate BCC in a pilot study (Salmivuori et al., 2019), where 75% accuracy was reached. In 2021, the same system was tested to distinguish between BCC and melanoma (Räsänen et al., 2021). In that work, a convolutional neural network classifier was employed to identify 26 pigmented lesions, obtaining a sensitivity of 100% and a specificity of 90%.

Zherdeva et al., proposed an HSI system to differentiate between different types of skin cancer utilizing the optical density of hemoglobin and melanin (Zherdeva et al., 2016). This work evaluated 45 skin lesions using 61 bands in the 450–750 nm range, yielding 84 and 87% in sensitivity and specificity metrics, respectively. Hosking et al., employed a Melanoma Advanced Imaging Dermatoscope (mAID) based on a non-polarized LED-driven HS camera (Hosking et al., 2019). The illumination system covered 21 distinct wavelengths in the spectral range between 350–950 nm. The study collected 70 HS images of skin lesions, performing a classification between nevus and melanoma with a sensitivity and specificity of 100% and 36%, respectively.

Fabelo et al., developed a system that captured HS images of 50 × 50 pixels and 125 spectral bands in the 450–950 nm range employing a cold light halogen device, and evaluating 49 HS images of skin lesions corresponding to 36 patients (Fabelo et al., 2019a). Later on, Leon et al., proposed a methodology that combined unsupervised and supervised techniques for the automated classification of pigmented skin lesions (Leon et al., 2020). In this study, excellent differentiation of malignant and benign tinted skin lesions was achieved with 87.5% sensitivity and 100% specificity.

In 2021, Courtenay et al., employed a pushbroom HS camera to discriminate unhealthy from healthy skin (Courtenay et al., 2021). The acquisition system consisted of a HS camera with two 60 W halogen light lamps on each side. The platform captured 270 spectral bands in the 398–995 nm range. A total of 60 patients were employed to perform robust statistical tests on 41 BCC and 19 SCC diagnosed cases. In a later work, the combination of a CNN with a SVM final activation layer was proposed to classify the same dataset, reaching up to 90% overall accuracy (Courtenay et al., 2022).

Recently, the SICSURFIS system was proposed as a hand-held HSI tool for complex skin surfaces (Lindholm et al., 2022). The system works in a spectral range of 475 to 975 nm capable of capturing thousands of spectral bands; but in this work only 33 were employed. A total of 42 skin lesions were evaluated to differentiate benign tumors from malignant pigmented and non-pigmented skin tumors. The classification and delineation method was proposed using a CNN employing spectral, spatial, and skin-surface models. The proposed approach had 87% sensitivity and 93% specificity for the different tissues evaluated.

5.4.2 Brain cancer applications

Primary brain and central nervous system cancers represent a significant cause of morbidity and mortality worldwide. The treatment consists of biopsy or aggressive surgical resection with postoperative radiation and chemotherapy (Patel et al., 2019). However, to achieve a successful resection that increases the probability of survival, precise delineation of the boundaries between tumor and normal tissue is necessary (D'Amico et al., 2017). During surgery different tools are often used to act as surgical guidance, examples of this are intraoperative image guided stereotactic neuronavigation, intraoperative magnetic resonance imaging, or fluorescent tumor markers like 5-aminolevulinic acid (Belykh et al., 2016).

In addition, HSI has emerged as a new intraoperative guidance tool (see Table 5.2). The initial efforts in HSI systems monitored brain oxygenation and hemodynamics in animals (Giannoni et al., 2020). Recently, Fabelo et al., designed a HS intraoperative system for the identification of human cancer tissue during in-vivo brain surgery (Fabelo et al., 2018). The system was based on two pushbroom HS cameras, an illumination source, and a scanning platform. One HS camera covered the VNIR spectral range between 400 and 1000 nm and could acquire 1004 pixels with 826 spectral bands. Another HS camera imaged the NIR range between 900 and 1700 nm, capturing 172 spectral bands and 320 spatial pixels. The illumination system was based on a quartz tungsten halogen (QTH) lamp of 150 W with a bandwidth emission in the range of 400 to 2200 nm. An optical fiber transmits the light from the lamp to a cold light emitter, isolating the high temperature produced by the QTH lamp from the brain surface. To provide the necessary movement to make the HS cubes, the HS cameras and the illumination system were coupled to a scanning platform. Employing this system, a HS human brain database was obtained from 22 patients (Fabelo et al., 2019c).

Several works employ this database (Fabelo et al., 2019c) to perform brain cancer classification and border delimitation. A hybrid approach mixing supervised and unsupervised ML method was produced to perform a spatio-spectral classification. The supervised stage was based on the SVM algorithm with a pixel-wise classification, where the resultant classification map is smoothed spatially by K-Nearest Neighbors (KNN) filtering on a representative

Table 5.2. HSI systems used to identify brain cancer.

Publication	System	Wavelengths (nm)	Bands
(Fabelo et al., 2018)	Pushbroom HSI	400–1000	826
(Urbanos et al., 2021)	Snapshot HSI	655–975	25
(Caredda et al., 2020)	Snapshot HSI	675–975	25

band of the HS cube. The output produced by this stage is combined using a majority voting algorithm with the unsupervised stage utilizing a Hierarchical K-Means strategy to obtain a segmentation map. The results prove the ability to accurately distinguish between normal tissue, tumor tissue, blood vessels, and background with an overall accuracy close to 100%. Apart from the traditional machine learning methods, deep learning approaches have been proposed to identify glioblastoma tumors (Fabelo et al., 2019b). The framework proposed was able to identify the parenchymal region, which corresponds to the principal surgical zone of the brain, employing a fully 2D CNN. In addition, blood vessels were identified using 2D-CNN. After that, the HS cube is classified by a 1-dimensional deep neural network (1D-DNN) generating a classification map with the four tissue classes. The blood vessels and parenchyma maps were joined into the 1D-DNN classification map. This framework was able to identify glioblastoma tumors obtaining a general mean accuracy of 80%. Another research employed a blind linear unmixing method to identify glioblastomas as a low computational time cost alternative (Cruz-Guerrero et al., 2020). This method was compared with a supervised SVM strategy, which required a high training time, achieving similar classification results but with a speedup factor of $\sim 429\times$. Recently, a method conjugating multiple ML models was proposed to use spectral and spatial information to identify glioblastomas (Hao et al., 2021). The main strength of the framework is the joint implementation of 1D-DNN and 2D-CNN architectures, which yield spectral and spectral-spatial HSI feature extraction and classification, respectively. These structures, in conjunction with fusion and optimization based on edge-preserving filtering and background class estimation by a fully CNN, enabled the proposed method to achieve an overall accuracy of 96.69% for a four-class classification, and 96.34% for glioblastoma identification.

Urbanos et al., presented a HS acquisition system to acquire and process HS images in a surgical environment (Urbanos et al., 2021). The system is based on a snapshot HS camera able to capture them in 25 bands in the range of 655 to 975 nm. The illumination system is based on 150 W halogen lamp linked to two fiber-optic connections to eliminate thermal exposure. In this study, an HS database was generated and composed by more than 50 images of different pathologies, and labeled into five different classes: normal tissue, tumor, dura mater, and venous and arterial blood vessels. Finally, 13 images with advanced stages of gliomas (Grade III and IV) were employed to train and evaluate the SVM, Random Forest (RF), and CNN classifiers achieving an overall accuracy between 60% and 95%.

In addition, HSI systems have been employed during neurosurgical procedures to monitor the oxygenated and deoxygenated hemoglobin level differences occuring in the brain (Caredda et al., 2020). The system was based on a HS camera able to capture 25 bands in the spectrum from 675 to 975 nm. Another HSI system employed the spectral range between 400 and 800 nm for monitoring intraoperative changes in brain surface hemodynamics to identify the postoperative cerebral hyperperfusion syndrome (Iwaki et al., 2021).

5.4.3 Gastrointestinal cancer applications

Gastrointestinal cancer includes stomach, liver, oesophagus, pancreas and colorectum, and this modality represents 26% of the global incidence and 35% of mortality in 2018 (Arnold et al., 2020). Endoscopic tools are employed to detect gastrointestinal cancers and other abnormalities. These tools include gastroscopy, colonoscopy, and wireless capsule endoscopy employing RGB cameras mainly (Du et al., 2019). To increase its versatility , HSI has been implemented in endoscopy to exploit the spectral potential (see Table 5.3). In 2018, Lin et al., developed a system named ICL SLHSI (Structured Light and Hyperspectral Imager), which employed a pushbroom HS camera covering the spectral range of 400–1000 nm with 270 spectral bands (Lin et al., 2018). Another work suggests a technique to distinguish early esophageal cancerous lesions through endoscopy and HSI endoscopies with a spectral range 350–800 nm (Wu et al., 2018). A study by Yoon et al., reported a line-scanning HSI endoscopy capable of capturing 100 spectral bands spanning the wavelength range of 680 to 730 nm (Yoon et al., 2019). The HSI system was employed to enhance polyp discrimination for detection and resection in seven patients undergoing routine colonoscopy screening. The KNN algorithm was employed as a classifier to differentiate patients with and without polyps (Yoon et al., 2021). Köhler et al., developed a HSI laparoscope able to capture 100 spectral bands ranging in wavelengths from 500 to 1000 nm, and the platform was tested with resected human tissue (Köhler et al., 2020).

Meanwhile, Sato et al., employed a pushbroom NIR HS camera in the range of 1000–2350 nm with 256 bands (Sato et al., 2020). During their study, 12 *ex vivo* gastrointestinal stromal tumors were imaged. The SVM

Table 5.3. HSI systems used to identify gastrointestinal cancer.

Publication	System	Wavelengths (nm)	Bands	Sample
(Lin et al., 2018)	Pushbroom HSI	400–1000	270	*In vivo*
(Wu et al., 2018)	Monochromator HSI	350–800	–	*In vivo*
(Yoon et al., 2019)	Line-scanning HSI	680–730	100	*In vivo*
(Köhler et al., 2020)	Pushbroom HSI	500–1000	100	*In vivo*
(Sato et al., 2020)	Pushbroom HSI	1000–2350	256	*Ex vivo*
(Jansen-Winkeln et al., 2019)	TIVITA HSI	500–1000	100	*Ex vivo*

algorithm was employed to predict normal and lesion regions, achieving a performance higher than 73% in all evaluated metrics. A Commercial HSI system, TIVITA® Tissue System (Diaspective Vision GmbH, Am Salzhaff, Germany) has been employed in different works. This system was able to capture images with a spatial dimension of 640 by 480 pixels and a spectral range of 500 to 1000 nm. This tool was evaluated for the determination of the resection margin during colorectal surgery in 24 patients (Jansen-Winkeln et al., 2019). Another study with the same commercial tool was deployed to detect colorectal carcinoma employing 54 patients (Jansen-winkeln et al., 2021). By using a neural network, tumor and healthy mucosa in colorectal carcinoma was classified with a sensitivity of 86% and a specificity of 95%.

5.4.4 Head and neck cancer applications

Head and neck cancer includes the oral cavity, nasopharynx, pharynx, and larynx. The diagnosis strategies which are quite diverse: (i) oral cavity tumors are often detected by patient self-identification, (ii) laryngeal tumors are diagnosed at an early stage by presenting voice changes or florid hoarseness, (iii) nasopharyngeal carcinoma can present hearing loss or cranial nerve palsies (Johnson et al., 2020). For this type of cancer, the applications of HSI are in an early development phase, see Table 5.4. Halicek et al., proposed a method to use HSI and convolutional neural networks to perform an optical biopsy of *ex vivo* head and neck cancer (Halicek et al., 2019). The acquisition structure employed a CRI Maestro imaging system (Perkin Elmer Inc., Waltham, Massachusetts), which is composed of a xenon white-light illumination lamp, a liquid crystal tunable filter, and a 16-bit CCD camera capable of capturing 91 bands with wavelengths ranging from 450 to 900 nm. Recently, Eggert et al., performed a prospective clinical observational study to classify the tissue into of laryngeal, hypopharyngeal and oropharyngeal mucosa into healthy and tumorous (Eggert et al., 2022). The HSI system was able to capture 30 spectral bands from 390 to 680 nm. In this work, 98 patients were examined due to suspicious lesions of the mucosal membrane before *in vivo* surgery. Deep learning methods were employed to achieve an average accuracy of 81%, a sensitivity of 83%, and a specificity of 79%.

Table 5.4. HSI systems used to identify head and neck cancer. LCTF – Liquid Crystal Tunable Filter.

Publication	System	Wavelengths (nm)	Bands	Sample
(Halicek et al., 2019)	LCTF HSI	450–900	91	*Ex vivo*
(Eggert et al., 2022)	Monochromator HSI	390–680	30	*In vivo*

5.4.5 Histological samples in cancer applications

Histological samples are examined by an expert physician using digital pathology to identify several diseases. The samples are digitalized employing microscopy, so partial or complete images are captured in higher magnification (Klein et al., 2021). HSI has been employed in different works for histological analysis using microscopy (Ortega et al., 2020b), see Table 5.5.

Ortega et al., proposed a process to correctly configure a pushbroom HS microscope to attain high-quality HS images (Samuel Ortega et al., 2019). The HS camera works in the spectral range from 400 to 1000 nm by capturing 826 spectral bands and 1004 spatial pixels. The HS camera was directly coupled to a conventional light microscope. However, the instrumentation restricted the adequate spectral range to approximately 400–800 nm. Employing this system, a dataset of 83 HS images was obtained for 5× and 10× magnifications by 13 pathology slides from biopsies of the human brain tissue resected during surgery to patients affected by Grade IV Glioblastoma tumor (Ortega et al., 2016). The HS images, captured using a 5× magnification, were classified using three different supervised classification algorithms: SVM, neural networks, and RF. Competitive results in the distinction between normal and tumorous tissue were obtained, achieving results above 80% accuracy (Samuel Ortega et al., 2018). In a recent work, the pushbroom HS microscope system was modified to remove the limitation of the effective spectral range (400–800 nm). To achieve this goal, the microscope was replaced, obtaining an effective spectral range of approximately 400–1000 nm. A new database was collected with 527 HS images, where 337 were non-tumor brain samples and 190 were diagnosed as glioblastoma (Ortega et al., 2020a). A CNN was employed to detect glioblastoma samples, reaching mean sensitivity and specificity values of 88% and 77%. In addition, the same HS system was employed for differentiating normal and tumor breast cancer cells. This way, 112 HS images were captured from histologic samples of human patients using 20× magnification. A deep learning neural network was employed achieving more than 89% of area under the curve for all the experiments.

In another work, Ma et al., developed an HS microscopic imaging system employing a SnapScan HS camera covering a spectral range of

Table 5.5. HSI systems employed to identify cancer using histological samples. LCTF – Liquid Crystal Tunable Filter.

Publication	System	Wavelengths (nm)	Bands	Cancer
(Samuel Ortega et al., 2019)	Pushbroom HSI	400–800	826	Brain
(Samuel Ortega et al., 2020)	Pushbroom HSI	400–1000	826	Brain, breast
(Ma et al., 2020)	SnapScan HSI	460–750	87	Head and neck
(Souza et al., 2021)	LCTF HSI	400–720	–	Animal model

460 to 750 nm with 87 spectral bands (Ma et al., 2020). A total of 15 histologic slides of larynx and hypopharynx from 15 head and neck cancer patients were collected at 40× magnification. The author proposed a nuclei segmentation strategy that relies on principle component analysis. After that, spectral-based SVM and patch-based CNN were used for nuclei classification. The average accuracies of the spectral-based SVM classification and CNN were 68% and 82.4%.

Finally, Souza et al., presented a system to acquire HS images using a liquid crystal tunable filter (LCTF) and a conventional microscope (Souza et al., 2021). The system can filter light in diverse wavelengths from 400 to 720 nm by polarization. This system was tested employing H&E-stained slides of rat skin treated with ALA-mediated photodynamic therapy. Four different algorithms were employed (KNN, SVM, and RF) with an accuracy between 96% and 98%.

5.5 Conclusions

In this chapter, the most recent advances in HSI for cancer applications were presented, with special emphasis on skin, brain, gastrointestinal, and head & neck cancers, as well as in the analysis of histological samples for the detection of tumor tissue. Moreover, the HS instrumentation commonly used for HS image acquisition and some of the processing methodologies in the literature were described. These technological advances demonstrate the potential of HS image analysis in the medical field; due to the advantages of its non-contact, non-ionizing, label-free, and minimally invasive nature. In fact, HSI techniques are also reaching other types of applications and are a field of great interest in research and industry. Some of the existing challenges in HSI are related to the compromise in the acquisition technologies among spectral bandwidth, spectral resolution, and capturing time. Furthermore, the light source must be carefully selected in the medical field to match the spectral bandwidth of the HS camera and the expected reflectance of the tissue, as well as, to consider its heat dissipation during experimentation.

Acronym/Abbreviation

1D-DNN	1-Dimensional Deep Neural Network
2D-CNN	2-Dimensional Convolutional Neural Network
ABCDE	Asymmetry of the mole, Border irregularity, Color uniformity, Diameter and Evolving size
ANN	Artificial Neural Networks
BCC	Basal Cell Carcinoma
BIL	Band-Interleaved-by-Line
BIP	Band-Interleaved-by-Pixel

BSQ	Band Sequential Format
CCD	Charge-Coupled Devices
CMOS	Complementary Metal-Oxide Semiconductors
CNN	Convolutional Neural Networks
GBM	Generalized Bilinear Model
HS	Hyperspectral
HSI	Hyperspectral Imaging
KNN	K-Nearest Neighbors
LCTF	Liquid Crystal Tunable Filter
LED	Light-Emitting Diode
LMM	Linear Mixing Model
LQMM	Linear Quadratic Mixing Model
mAID	Melanoma Advanced Imaging Dermatoscope
ML	Machine Learning
MMM	Multilinear Mixing Model
MS	Multispectral
MSI	Multispectral Imaging
NIR	Near Infrared
NMM	Nonlinear Mixture Model
NMSC	Non-Melanoma Skin Cancer
PPNM	Polynomial Post-Nonlinear Model
QTH	Quartz Tungsten Halogen
RF	Random Forest
RGB	Red, Green, and Blue
SAM	Spectral Angle Mapper
SCC	Squamous Cell Carcinoma
SID	Spectral Information Divergence
SLHSI	Structured Light and Hyperspectral Imager
SU	Spectral Unmixing
SVM	Support Vector Machines
SWIR	Short-Wave Infrared
VNIR	Visual and Near Infrared

Reference list

Adão, T., Hruška, J., Pádua, L., Bessa, J., Peres, E. et al. (2017). Hyperspectral imaging: A review on UAV-based sensors, data processing and applications for agriculture and forestry. *Remote Sensing*, 9(11). https://doi.org/10.3390/rs9111110.

Al Ktash, M., Stefanakis, M., Boldrini, B., Ostertag, E. and Brecht, M. (2021). Characterization of pharmaceutical tablets using UV hyperspectral imaging as a rapid in-line analysis tool. *Sensors*, 21(13): 4436. https://doi.org/10.3390/s21134436.

Arnold, M., Abnet, C.C., Neale, R.E., Vignat, J., Giovannucci, E.L. et al. (2020). Global burden of 5 major types of gastrointestinal cancer. *Gastroenterology*, 159(1): 335–349.e15. https://doi.org/10.1053/J.GASTRO.2020.02.068.

Arnold, T., De Biasio, M., Bereczki, T., Baumgart, M. and Horn, A. (2022). Development of inspection system for the detection and analysis of solid particles and oil droplets in process water of the petrochemical industry using hyperspectral imaging and fluorescence imaging. p. 37. *In*: Messinger, D.W. and Velez-Reyes, M. (eds.). *Algorithms, Technologies, and Applications for Multispectral and Hyperspectral Imaging XXVIII*. SPIE. https://doi.org/10.1117/12.2618760.

Belykh, E., Martirosyan, N.L., Yagmurlu, K., Miller, E.J., Eschbacher, J.M., et al. (2016). Intraoperative fluorescence imaging for personalized brain tumor resection: Current state and future directions. In *Frontiers in Surgery* (Vol. 3, p. 55). Frontiers Media S.A. https://doi.org/10.3389/fsurg.2016.00055.

Campos-Delgado, D.U., Gutierrez-Navarro, O., Rico-Jimenez, J.J., Duran-Sierra, E., Fabelo, H. et al. (2019). Extended blind end-member and abundance extraction for biomedical imaging applications. *IEEE Access*, 7: 178539–178552.

Campos-Delgado, D.U., Cruz-Guerrero, I.A., Mendoza-Chavarria, J.N., Mejia-Rodriguez, A.R., Ortega, S. et al. (2022). Nonlinear extended blind end-member and abundance extraction for hyperspectral images. *SSRN Electronic Journal*, 201: 108718. https://doi.org/10.2139/ssrn.4111839.

Caredda, C., Mahieu-Williame, L., Sablong, R., Sdika, M., Guyotat, J. et al. (2020). Optimal spectral combination of a hyperspectral camera for intraoperative hemodynamic and metabolic brain mapping. *Applied Sciences*, 10(15): 5158. https://doi.org/10.3390/APP10155158.

Courtenay, L.A., González-Aguilera, D., Lagüela, S., Del Pozo, S., Ruiz, C. et al. (2021). Hyperspectral imaging and robust statistics in non-melanoma skin cancer analysis. *Biomedical Optics Express*, 12(8): 5107–5127. https://doi.org/10.1364/BOE.428143.

Courtenay, L. A., González-Aguilera, D., Lagüela, S., Pozo, S. del, Ruiz-Mendez, C. et al. (2022). Deep convolutional neural support vector machines for the classification of basal cell carcinoma hyperspectral signatures. *Journal of Clinical Medicine*, 11(9): 2315. https://doi.org/10.3390/JCM11092315.

Cruz-Guerrero, I.A., Leon, R., Campos-Delgado, D.U., Ortega, S., Fabelo, H. et al. (2020). Classification of hyperspectral *in vivo* brain tissue based on linear unmixing. *Applied Sciences*, 10(16): 5686.

D'Amico, R.S., Englander, Z.K., Canoll, P. and Bruce, J.N. (2017). Extent of resection in glioma–A review of the cutting edge. In *World Neurosurgery* (Vol. 103, pp. 538–549). Elsevier Inc. https://doi.org/10.1016/j.wneu.2017.04.041.

Dobigeon, N., Tourneret, J.-Y., Richard, C., Bermudez, J.C.M., McLaughlin, S. et al. (2013). Nonlinear unmixing of hyperspectral images: Models and algorithms. *IEEE Signal Processing Magazine*, 31(1): 82–94.

Du, W., Rao, N., Liu, D., Jiang, H., Luo, C. et al. (2019). Review on the applications of deep learning in the analysis of gastrointestinal endoscopy images. *IEEE Access*, 7: 142053–142069. https://doi.org/10.1109/ACCESS.2019.2944676.

Eggert, D., Bengs, M., Westermann, S., Gessert, N., Gerstner, A.O.H. et al. (2022). *In vivo* detection of head and neck tumors by hyperspectral imaging combined with deep learning methods. *Journal of Biophotonics*, 15(3): e202100167. https://doi.org/10.1002/JBIO.202100167.

Elbaum, M., Kopf, A.W., Rabinovitz, H.S., Langley, R.G.B., Kamino, H. et al. (2001). Automatic differentiation of melanoma from melanocytic nevi with multispectral digital dermoscopy: A feasibility study. *Journal of the American Academy of Dermatology*, 44(2): 207–218. https://doi.org/10.1067/mjd.2001.110395.

Fabelo, H., Carretero, G., Almeida, P., Garcia, A., Hernandez, J.A.et al. (2019). Dermatologic hyperspectral imaging system for skin cancer diagnosis assistance. *2019 34th Conference on Design of Circuits and Integrated Systems, DCIS*, 2019: 5–10. https://doi.org/10.1109/DCIS201949030.2019.8959869.

Fabelo, H., Halicek, M., Ortega, S., Shahedi, M., Szolna, A. et al. (2019). Deep learning-based framework for *in vivo* identification of glioblastoma tumor using hyperspectral images of human brain. *Sensors (Switzerland)*, 19(4). https://doi.org/10.3390/s19040920.

Fabelo, H., Ortega, S., Lazcano, R., Madroñal, D.M., Callicó, G. et al. (2018). An intraoperative visualization system using hyperspectral imaging to aid in brain tumor delineation. *Sensors*, 18(2): 430. https://doi.org/10.3390/s18020430.

Fabelo, H., Ortega, S., Szolna, A., Bulters, D., Pineiro, J.F. et al. (2019). *In vivo* hyperspectral human brain image database for brain cancer detection. *IEEE Access*, 7: 39098–39116. https://doi.org/10.1109/ACCESS.2019.2904788.

Fan, W., Hu, B., Miller, J. and Li, M. (2009). Comparative study between a new nonlinear model and common linear model for analysing laboratory simulated-forest hyperspectral data. *International Journal of Remote Sensing*, 30(11): 2951–2962. https://doi.org/10.1080/01431160802558659.

Fei, B. (2020). Hyperspectral imaging in medical applications. In *Data Handling in Science and Technology* (Vol. 32, pp. 523–565). Elsevier.

Fu, X. and Chen, J. (2019). A review of hyperspectral imaging for chicken meat safety and quality evaluation: application, hardware, and software. In *Comprehensive Reviews in Food Science and Food Safety* (Vol. 18, Issue 2, pp. 535–547). Blackwell Publishing Inc. https://doi.org/10.1111/1541-4337.12428.

Gao, L., Wang, Z., Zhuang, L., Yu, H., Zhang, B. et al. (2021). Using low-rank representation of abundance maps and nonnegative tensor factorization for hyperspectral nonlinear unmixing. *IEEE Transactions on Geoscience and Remote Sensing*, 60: 1–17.

Giannoni, L., Lange, F. and Tachtsidis, I. (2020). Investigation of the quantification of hemoglobin and cytochrome-c-oxidase in the exposed cortex with near-infrared hyperspectral imaging: A simulation study. *Https://Doi.Org/10.1117/1.JBO.25.4.046001*, 25(4): 046001. https://doi.org/10.1117/1.JBO.25.4.046001.

Govender, M., Chetty, K. and Bulcock, H. (2009). A review of hyperspectral remote sensing and its application in vegetation and water resource studies. *Water {SA}*, 33(2). https://doi.org/10.4314/wsa.v33i2.49049.

Gu, Y., Liu, T., Gao, G., Ren, G., Ma, Y. et al. (2021). Multimodal hyperspectral remote sensing: An overview and perspective. *Science China Information Sciences*, 64(2): 1–24. https://doi.org/10.1007/S11432-020-3084-1.

Gutierrez-Navarro, O., Campos-Delgado, D.U., Arce-Santana, E.R., Mendez, M.O. and Jo, J.A. (2013). Blind end-member and abundance extraction for multispectral fluorescence lifetime imaging microscopy data. *IEEE Journal of Biomedical and Health Informatics*, 18(2): 606–617.

Halicek, M., Little, J.V., Wang, X., Chen, A.Y. and Fei, B. (2019). Optical biopsy of head and neck cancer using hyperspectral imaging and convolutional neural networks. *Https://Doi.Org/10.1117/1.JBO.24.3.036007*, 24(3): 036007. https://doi.org/10.1117/1.JBO.24.3.036007.

Hao, Q., Pei, Y., Zhou, R., Sun, B., Sun, J. et al. (2021). Fusing multiple deep models for *in vivo* human brain hyperspectral image classification to identify glioblastoma tumor. *IEEE Transactions on Instrumentation and Measurement*, 70. https://doi.org/10.1109/TIM.2021.3117634.

Heylen, R., Parente, M. and Gader, P. (2014). A review of nonlinear hyperspectral unmixing methods. *IEEE Journal of Selected Topics in Applied Earth Observations and Remote Sensing*, 7(6): 1844–1868. https://doi.org/10.1109/JSTARS.2014.2320576.

Holmer, A., Tetschke, F., Marotz, J., Malberg, H., Markgraf, W. et al. (2016). Oxygenation and perfusion monitoring with a hyperspectral camera system for chemical based tissue analysis of skin and organs. *Physiological Measurement*, 37(11): 2064. https://doi.org/10.1088/0967-3334/37/11/2064.

Hosking, A.M., Coakley, B.J., Chang, D., Talebi-Liasi, F., Lish, S. et al. (2019). Hyperspectral imaging in automated digital dermoscopy screening for melanoma. *Lasers in Surgery and Medicine*, 51(3): 214–222. https://doi.org/10.1002/LSM.23055.

Iwaki, K., Takagishi, S., Arimura, K., Murata, M., Chiba, T. et al. (2021). A novel hyperspectral imaging system for intraoperative prediction of cerebral hyperperfusion syndrome after superficial temporal artery-middle cerebral artery anastomosis in patients with moyamoya disease. *Cerebrovascular Diseases*, 50(2): 208–215. https://doi.org/10.1159/000513289.

Jansen-Winkeln, B., Holfert, N., Köhler, H., Moulla, Y., Takoh, J.P. et al. (2019). Determination of the transaction margin during colorectal resection with hyperspectral imaging (HSI). *International Journal of Colorectal Disease*, 34(4): 731–739. https://doi.org/10.1007/S00384-019-03250-0/FIGURES/6.

Jansen-Winkeln, B., Barberio, M., Chalopin, C., Schierle, K., Diana, M. et al. (2021). Feedforward artificial neural network-based colorectal cancer detection using hyperspectral imaging: A step towards automatic optical biopsy. *Cancers*, 13(5): 967. https://doi.org/10.3390/CANCERS13050967.

Johnson, D.E., Burtness, B., Leemans, C.R., Lui, V.W.Y., Bauman, J.E. et al. (2020). Head and neck squamous cell carcinoma. *Nature Reviews Disease Primers*, 6(1): 1–22. https://doi.org/10.1038/s41572-020-00224-3.

Jolivot, R., Benezeth, Y. and Marzani, F. (2013). Skin parameter map retrieval from a dedicated multispectral imaging system applied to dermatology/cosmetology. *International Journal of Biomedical Imaging*, 2013. https://doi.org/10.1155/2013/978289.

Kamruzzaman, M. and Sun, D.-W. (2016). Introduction to hyperspectral imaging technology. In *Computer Vision Technology for Food Quality Evaluation*, pp. 111–139. Elsevier. https://doi.org/10.1016/B978-0-12-802232-0.00005-0.

Klein, C., Zeng, Q., Arbaretaz, F., Devêvre, E., Calderaro, J. et al. (2021). Artificial intelligence for solid tumour diagnosis in digital pathology. *British Journal of Pharmacology*, 178(21): 4291–4315. https://doi.org/10.1111/BPH.15633.

Köhler, H., Kulcke, A., Maktabi, M., Moulla, Y., Jansen-Winkeln, B. et al. (2020). Laparoscopic system for simultaneous high-resolution video and rapid hyperspectral imaging in the visible and near-infrared spectral range. *Https://Doi.Org/10.1117/1.JBO.25.8.086004*, 25(8): 086004. https://doi.org/10.1117/1.JBO.25.8.086004.

Lassalle, G., Fabre, S., Credoz, A., Hédacq, R., Dubucq, D. et al. (2021). Mapping leaf metal content over industrial brownfields using airborne hyperspectral imaging and optimized vegetation indices. *Scientific Reports*, 11(1): 2. https://doi.org/10.1038/s41598-020-79439-z.

Leon, R., Martinez-Vega, B., Fabelo, H., Ortega, S., Melian, V. et al. (2020). Non-invasive skin cancer diagnosis using hyperspectral imaging for *in situ* clinical support. *Journal of Clinical Medicine*, 9(6). https://doi.org/10.3390/jcm9061662.

Lim, J.K.H., Li, Q.-X., Ryan, T., Bedggood, P., Metha, A. et al. (2021). Retinal hyperspectral imaging in the 5xFAD mouse model of Alzheimer's disease. *Scientific Reports*, 11(1): 6387. https://doi.org/10.1038/s41598-021-85554-2.

Lin, J., Clancy, N.T., Qi, J., Hu, Y., Tatla, T. et al. (2018). Dual-modality endoscopic probe for tissue surface shape reconstruction and hyperspectral imaging enabled by deep neural networks. *Medical Image Analysis*, 48: 162–176. https://doi.org/10.1016/J.MEDIA.2018.06.004.

Lindholm, V., Raita-Hakola, A.M., Annala, L., Salmivuori, M., Jeskanen, L. et al. (2022). Differentiating malignant from benign pigmented or non-pigmented skin Tumours—A pilot study on 3D hyperspectral imaging of complex skin surfaces and convolutional neural networks. *Journal of Clinical Medicine*, 11(7): 1914. https://doi.org/10.3390/JCM11071914.

Lu, B., Dao, P. D., Liu, J., He, Y. and Shang, J. (2020). Recent advances of hyperspectral imaging technology and applications in agriculture. *Remote Sensing*, 12(16): 1–44. https://doi.org/10.3390/RS12162659.

Lv, W. and Wang, X. (2020). Overview of hyperspectral image classification. *Journal of Sensors*, 2020.

Ma, J., Sun, D.W., Pu, H., Cheng, J.H. and Wei, Q. (2019). Advanced techniques for hyperspectral imaging in the food industry: Principles and recent applications. *Annual Review of Food Science and Technology*, 10: 197–220. https://doi.org/10.1146/annurev-food-032818-121155.

Ma, L., Halicek, M., Zhou, X., Dormer, J.D. and Fei, B. (2020). Hyperspectral microscopic imaging for automatic detection of head and neck squamous cell carcinoma using histologic image and machine learning. *Proceedings of SPIE—The International Society for Optical Engineering*, 11320: 31. https://doi.org/10.1117/12.2549369.

Markgraf, W., Lilienthal, J., Feistel, P., Thiele, C. and Malberg, H. (2020). Algorithm for mapping kidney tissue water content during normothermic machine perfusion using hyperspectral imaging. *Algorithms*, 13(11): 289. https://doi.org/10.3390/a13110289.

Montes-Herrera, J.C., Cimoli, E., Cummings, V., Hill, N., Lucieer, A. et al. (2021). Underwater hyperspectral imaging (UHI): A review of systems and applications for proximal seafloor ecosystem studies. *Remote Sensing*, 13(17): 3451. https://doi.org/10.3390/rs13173451.

Neittaanmäki-Perttu, N., Grönroos, M., Jeskanen, L., Pölönen, I., Ranki, A. et al. (2015). Delineating margins of lentigo maligna using a hyperspectral imaging system. *Acta Dermato-Venereologica*, 95(5): 549–552. https://doi.org/10.2340/00015555-2010/.

Ortega, S., Callico, G.M., Plaza, M.L., Camacho, R., Fabelo, H. et al. (2016). Hyperspectral database of pathological *in vitro* human brain samples to detect carcinogenic tissues. *Proceedings—International Symposium on Biomedical Imaging*, 2016 June. https://doi.org/10.1109/ISBI.2016.7493285.

Ortega, S., Fabelo, H., Camacho, R., de la Luz Plaza, M., Callicó, G.M. et al. (2018). Detecting brain tumor in pathological slides using hyperspectral imaging. *Biomedical Optics Express*, 9(2): 818. https://doi.org/10.1364/boe.9.000818.

Ortega, S., Guerra, R., Diaz, M., Fabelo, H., Lopez, S. et al. (2019). Hyperspectral push-broom microscope development and characterization. *IEEE Access*, 7: 122473–122491. https://doi.org/10.1109/ACCESS.2019.2937729.

Ortega, S., Halicek, M., Fabelo, H., Callico, G.M. and Fei, B. (2020). Hyperspectral and multispectral imaging in digital and computational pathology: A systematic review [Invited]. *Biomedical Optics Express*, 11(6): 3195. https://doi.org/10.1364/boe.386338.

Ortega, S., Halicek, M., Fabelo, H., Camacho, R., Plaza, M. de la L. et al. (2020). Hyperspectral imaging for the detection of glioblastoma tumor cells in H&E slides using convolutional neural networks. *Sensors*, 20(7): 1911.

Patel, A.P., Fisher, J.L., Nichols, E., Abd-Allah, F., Abdela, J. et al. (2019). Global, regional, and national burden of brain and other CNS cancer, 1990–2016: A systematic analysis for the Global Burden of Disease Study 2016. *The Lancet Neurology*, 18(4): 376–393. https://doi.org/10.1016/S1474-4422(18)30468-X.

Qin, J., Burks, T.F., Ritenour, M.A. and Bonn, W.G. (2009). Detection of citrus canker using hyperspectral reflectance imaging with spectral information divergence. *Journal of Food Engineering*, 93(2): 183–191.

Räsänen, J., Salmivuori, M., Pölönen, I., Grönroos, M. and Neittaanmäki, N. (2021). Hyperspectral imaging reveals spectral differences and can distinguish malignant melanoma from pigmented basal cell carcinomas: A pilot study. *Acta Dermato-Venereologica*, 101(2): adv00405–adv00405. https://doi.org/10.2340/00015555-3755.

Rey-Barroso, L., Peña-Gutiérrez, S., Yáñez, C., Burgos-Fernández, F.J., Vilaseca, M. et al. (2021). Optical technologies for the improvement of skin cancer diagnosis: A review. *Sensors*, 21(1): 252. https://doi.org/10.3390/s21010252.

Rico-Jimenez, J.J., Campos-Delgado, D.U., Villiger, M., Otsuka, K., Bouma, B.E. et al. (2016). Automatic classification of atherosclerotic plaques imaged with intravascular OCT. *Biomedical Optics Express*, 7(10): 4069–4085.

Salmivuori, M., Neittaanmäki, N., Pölönen, I., Jeskanen, L., Snellman, E. et al. (2019). Hyperspectral imaging system in the delineation of Ill-defined basal cell carcinomas: A pilot study. *Journal of the European Academy of Dermatology and Venereology*, 33(1): 71–78. https://doi.org/10.1111/JDV.15102.

Sato, D., Takamatsu, T., Umezawa, M., Kitagawa, Y., Maeda, K. et al. (2020). Distinction of surgically resected gastrointestinal stromal tumor by near-infrared hyperspectral imaging. *Scientific Reports*, 10(1): 1–9. https://doi.org/10.1038/s41598-020-79021-7.

Shimoni, M., Haelterman, R. and Perneel, C. (2019). Hypersectral imaging for military and security applications: Combining Myriad processing and sensing techniques. *IEEE Geoscience and Remote Sensing Magazine*, 7(2): 101–117. https://doi.org/10.1109/MGRS.2019.2902525.

Souza, M.M., Carvalho, F.A., Sverzut, E.F.V., Requena, M.B., Garcia, M.R. et al. (2021). Hyperspectral imaging system for tissue classification in h&e-stained histological slides. *2021 SBFoton International Optics and Photonics Conference (SBFoton IOPC)*, 1–4. https://doi.org/10.1109/SBFotonIOPC50774.2021.9461972.

Sung, H., Ferlay, J., Siegel, R.L., Laversanne, M., Soerjomataram, I. et al. (2021). Global Cancer Statistics 2020: GLOBOCAN estimates of incidence and mortality worldwide for 36 Cancers in 185 Countries. *CA: A Cancer Journal for Clinicians*, 71(3): 209–249. https://doi.org/10.3322/caac.21660.

Tsao, H., Olazagasti, J.M., Cordoro, K.M., Brewer, J.D., Taylor, S.C. et al. (2015). Early detection of melanoma: Reviewing the ABCDEs American academy of dermatology Ad Hoc task force for the ABCDEs of Melanoma. *Journal of the American Academy of Dermatology*, 72(4): 717–723. https://doi.org/10.1016/j.jaad.2015.01.025.

Urbanos, G., Martín, A., Vázquez, G., Villanueva, M., Villa, M. et al. (2021). Supervised machine learning methods and hyperspectral imaging techniques jointly applied for brain cancer classification. *Sensors*, 21(11): 3827. https://doi.org/10.3390/S21113827.

Vejarano, R., Siche, R. and Tesfaye, W. (2017). Evaluation of biological contaminants in foods by hyperspectral imaging: A review. *International Journal of Food Properties*, 20: 1–34. https://doi.org/10.1080/10942912.2017.1338729.

Wu, I.C., Syu, H.Y., Jen, C.P., Lu, M.Y., Chen, Y.T. et al. (2018). Early identification of esophageal squamous neoplasm by hyperspectral endoscopic imaging. *Scientific Reports*, 8(1): 1–10. https://doi.org/10.1038/s41598-018-32139-1.

Yoon, J., Joseph, J., Waterhouse, D.J., Borzy, C., Siemens, K. et al. (2019). A clinically translatable hyperspectral endoscopy (HySE) system for imaging the gastrointestinal tract. *Nature Communications*, 10(1): 1–13. https://doi.org/10.1038/s41467-019-09484-4.

Yoon, J., Joseph, J., Waterhouse, D.J., Luthman, A.S., Gordon, G.S.D. et al. (2021). First experience in clinical application of hyperspectral endoscopy for evaluation of colonic polyps. *Journal of Biophotonics*, 14(9): e202100078. https://doi.org/10.1002/JBIO.202100078.

Zambrano-Román, M., Padilla-Gutiérrez, J.R., Valle, Y., Muñoz-Valle, J.F. and Valdés-Alvarado, E. (2022). Non-Melanoma skin cancer: A genetic update and future perspectives. *Cancers*, 14(10): 2371. https://doi.org/10.3390/cancers14102371.

Zherdeva, L.A., Bratchenko, I.A., Myakinin, O.O., Moryatov, A.A., Kozlov, S.V. et al. (2016). *In vivo* hyperspectral imaging and differentiation of skin cancer. *In*: Luo, Q., Li, X., Gu, Y. and Tang, Y. (eds.). *Optics in Health Care and Biomedical Optics VII* (Vol. 10024, Issue November, p. 100244G). https://doi.org/10.1117/12.2246433.

CHAPTER 6

Oral Cancer Detection by Multi-Spectral Fluorescence Lifetime Imaging Microscopy (m-FLIM) and Linear Unmixing

Omar Gutiérrez Navarro,[1,*] *Aldo Rodrigo Mejía Rodríguez*,[2] *Daniel Ulises Campos Delgado*,[3] *Elvis de Jesús Durán Sierra*[4] and *Javier A Jo*[5]

Cancer patients' survival chances are highly dependent on early detection. The problem with oral cancer is that it is hard to distinguish it from benign oral lesions. This is complicated because patients rarely seek professional help until it is already too late. In this chapter, we will review the fundamentals and latest developments around Multi-Spectral Fluorescence Lifetime Imaging Microscopy, which is one the novel techniques employed in state-of-the-art research for early detection of oral carcinoma. We will start with a review of FLIM techniques and the particularities of m-FLIM. Next, we will talk about the optical instrumentation employed. The following sections

[1] Departament of Biomedical Engineering, Autonomous University of Aguascalientes, Ags. 20100, Mexico.
[2] School of Science, Autonomous University of San Luis Potosí, San Luis Potosí, S.L.P. 78295 México.
[3] School of Science/Institute for Optical Communication Research, Autonomous University of San Luis Potosí, San Luis Potosí, S.L.P. 78210 México.
[4] The University of Texas MD Anderson Cancer Center, Houston, TX 77030, USA.
[5] University of Oklahoma, Norman, OK 73019-0631, USA.
Emails: aldo.mejia@uaslp.mx; ducd@fciencias.uaslp.mx; eduran3@mdanderson.org; javierjo@ou.edu
* Corresponding author: omar.gutierrezn@edu.uaa.mx

will discuss standard m-FLIM data processing techniques such as deconvolution and lifetime estimation. The amount of data generated by m-FLIM is not easy to understand by non-experts. A way to generate intuitive results with a physical interpretation is spectral unmixing techniques. The final sections are dedicated to linear unmixing and the EBEA algorithm for the decomposition of m-FLIM data. The method is tested using *in vivo* samples. The data is decomposed to obtain characteristic profiles and their abundance maps. These results illustrate how to locate the contribution of endogenous fluorophores in a sample, which could be used to monitor carcinogen metabolism.

6.1 Introduction

According to (Siegel, 2022), the American Cancer Society expects around 54,000 new oral cancer cases to be reported this year. These numbers include cancer in the oral cavity and pharynx. The five-year survival rate for diagnosed individuals is 67% of all races. Early detection is crucial just like other cancers. The current standard method for oral cancer diagnosis is a thorough clinical examination by a dentist. If experts detect a suspicious lesion, then they recommend a biopsy. Early-stage treatment may require minor procedures to remove the localized tumor. When oral cancer is detected at a late stage it undermines the quality of life since it implies removal of a considerable portion of the face and neck regions. Even though the oral cavity is an easily accessible region for health experts, only 31% of the cases are detected in the early stages. However, early-stage detection is challenging even for experienced healthcare professionals (Cheng, 2015; Epstein, 2012; Cleveland, 2013). Therefore, early-stage detection relies on the dentist's ability to differentiate oral cancer and dysplasia characteristics from mucosal abnormalities due to benign conditions.

There are different diagnostic tools which have been proposed in the last 20 years and they can be grouped (Cheng, 2015) into one of the following base technologies: vital staining, brush cytology, salivary diagnostic tests, and optical imaging techniques. There is plenty of published research on the sensitivity and specificity of these adjuncts, but there is no data for early-stage detection of oral cancer and dysplasia. Since the number of cancer diagnoses appears to be rising, there is a critical need for reliable technologies capable of early oral cancer detection.

What is commonly known as oral cancer is a group of different malignant cancers where the most usual form is Squamous Cell Carcinoma (SCC). The usual progression of cancer starts with mutated cells which grow into hyperplasia, followed by dysplasia, severe carcinoma, and finally metastasis.

One of the problems with the clinical management of SCC is that its carcinogenesis does not follow a linear progression. Oral epithelial lesions might be a warning for the development of SCC, considering that clinical visual manifestations such as leukoplakia (white patches) or erythroplakia (red patches) may evolve into oral cancer (Chakraborty et al., 2019). Yet, they might skip some dysplastic stages, or they might not progress to cancer at all, which is the case for most oral epithelial lesions (Abati, 2020).

This is further complicated by a low early-detection rate, as screening for oral cancer usually involves identification of signs and symptoms of precancerous lesions at different sites of the oral environment (tongue, oral gingiva, lips among others). Current diagnostic strategies for oral cancer, however, can identify lesions only after a certain degree of malignancy progression. As a matter of fact, patients usually seek help once cancer is in the more advanced stages (Kaur, 2021). The diagnosis usually starts with a clinical oral examination. If the dentist finds a suspicious lesion, then she/he might use auxiliary techniques. A definitive diagnosis is only provided by histopathologic analysis. Both procedures are subjective, invasive and the results might take several days. The current state of the art in oral cancer diagnosis is focused on finding early sign biomarkers as fast as possible. Some techniques employed are the identification of biomarkers from biological fluids. Another example is exfoliative cytology, where cell samples are scrapped and evaluated for the detection of dysplasia. Optical imaging methods aim at finding abnormalities in the oral cavity in a minimally invasive fashion. Such is the case with Raman spectroscopy, optical coherence, and fluorescence imaging-based techniques (Chakraborty, 2019).

6.2 Non-invasive mFLIM techniques

The most extensively evaluated optical modality for early detection of oral cancer is endogenous fluorescence or autofluorescence imaging or spectroscopy. According to the literature, the natural fluorescence of the tissue is altered by key biomarkers of oral cancer and dysplasia (Georgakoudi, 2002; Pavlova, 2008). This disease is characterized by collagen degradation and epithelial thickening, which results in a lower intensity of autofluorescence emission from collagen. Diagnostic adjuncts based on autofluorescence imaging typically rely on this biomarker to identify suspicious oral lesions. Unfortunately, many benign conditions are also characterized by a loss in autofluorescence intensity of stromal collagen, and this fact fundamentally limits the specificity of any approach based solely on collagen autofluorescence (Shin, 2010; Roblyer, 2010; Sweeny, 2011).

The oral mucosa is covered by an epithelial layer that contains two metabolic coenzymes: the reduced form of nicotinamide adenine dinucleotide

(NADH) and flavin adenine dinucleotide (FAD). These coenzymes intervene in multiple metabolic activities including oxidative phosphorylation and glycolysis, plus, they are the dominant endogenous fluorophores in the region (Pavlova, 2008; Shin, 2010; Roblyer, 2010; Sweeny, 2011; Skala, 2007). A highly specific change induced by oral dysplasia and cancer is the increase in metabolic activity. The malignant transformation of cells is characterized by changes in the oral tissue autofluorescence response due to the presence of NADH and FAD. This change can be detected using tools to monitor variations in the metabolic activity of tissue. Such is the case of the optical "redox-ratio" which is the proportion of autofluorescence intensity from NADH (an electron donor) divided by the contribution of FAD (an electron acceptor). This quantifies the proportion of electron donors and acceptors in a cell. Usually, a lower redox ratio indicates higher metabolic activity, which is the standard case when dealing with cancerous cells. Changes in the fluorescence lifetime of the free and protein-bound states of NADH and FAD are also linked with increased metabolic activity in malignant cells in oral mucosa. Skala and Ramanujam et al., performed multiphoton FLIM imaging *in vivo* in the normal and dysplastic epithelium of a hamster model of oral cancer and reported a decrease in bound NADH lifetime with dysplasia attributed to a shift from oxidative phosphorylation to glycolysis, consistent with neoplastic metabolism (Skala, 2005). The same authors reported an increase in the lifetime of bound FAD in dysplasia, which is associated with a reduction in intracellular NAD+ expected with abnormal growth (Skala, 2007). In summary, this array of metabolic biomarkers can form the foundation of a predictor for detecting oral cancer and dysplasia with clinically relevant sensitivity and specificity.

Fluorescence techniques are a popular way to study biological samples due to the plethora of synthetic and endogenous fluorescent dyes available. The use of exogenous dyes is a complicated process which requires dealing with potential toxicity, non-specific binding to the object of study, and even interference from the same biochemical environment. A novel trend is to take advantage of the endogenous fluorophores which also allows for in-vivo monitoring in a non-invasive fashion. Common endogenous fluorophores present in the human body are collagen and elastin as well as several vitamins. Some important chromophores for biomedical applications also emit a natural fluorescent response, such as in the case of hemoglobin, melanin, bilirubin, keratin, and eumelanin to name a few. Plus, tryptophan, tyrosine, and phenylalanine are three well known amino acids which also happen to be endogenous fluorophores. Two known biomarkers of malignant growth in oral tissue are reduced-form NADH and FAD.

In Fluorescence imaging, lifetime is defined as the average time a fluorophore maintains an electronically excited state before emitting light. This characteristic is useful to identify a substance and characterize its environment. For this goal, an ultraviolet (UV) laser-pulse serves as an excitation source over the imaged area or field of view (FOV). The fluorescence decays are measured in every spatial position over multiple spectral bands. This information is arranged in a 3D array, as depicted in Figure 6.1. When this optical technique is applied at a microscopic level, it is called multi-spectral fluorescence lifetime imaging microscopy (m-FLIM). The measured fluorescence decays over all spectral bands are concatenated in a vector, which is sampled for discrete time recording and processing.

The process of oral carcinogenesis causes morphological, functional, and biochemical changes that modify the natural fluorescence response of the oral epithelial tissue and the underlying lamina propria (Kaur, 2021; Pavlova, 2008). Specifically, the levels of NADH and FAD within oral epithelial cells

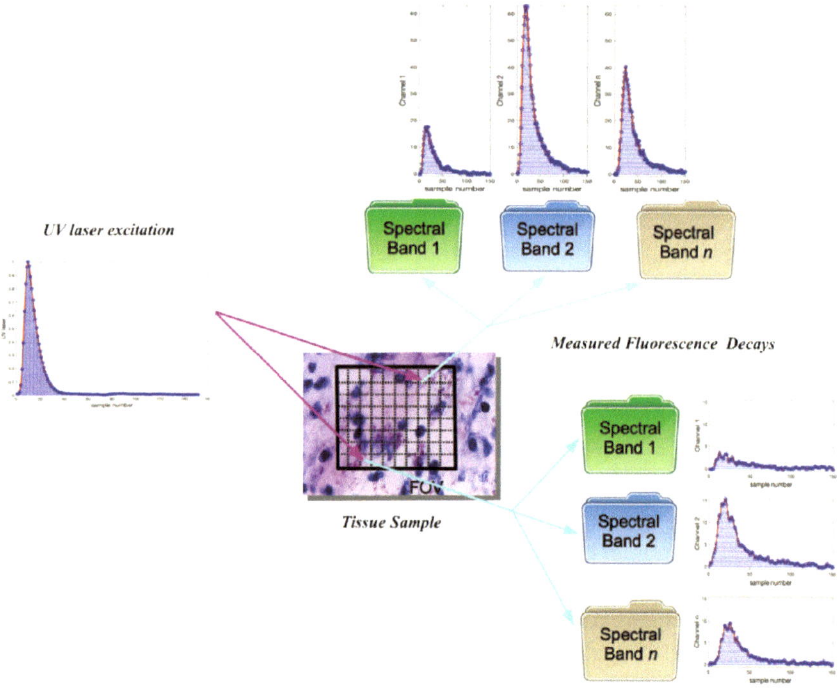

Figure 6.1. Schematic of the non-invasive m-FLIM Phenomenon. A tissue sample is excited using a UV laser and the fluorescence emission is recorded at multiple wavelengths and concatenated in a single vector. This operation is performed in multiple spatial positions generating a 3D array of data.

can change as oral cancer develops (Müller, 2003; Shah, 2014). Moreover, changes in the autofluorescence lifetimes of NADH and FAD can occur as oral cancer develops (Shah, 2017; Skala, 2007). Oral cancer usually leads to epithelial thickening and extracellular matrix remodeling in the lamina propria, resulting in decreased collagen autofluorescence (Pavlova, 2008; Pavlova, 2003). Therefore, imaging of collagen, FAD, and NADH autofluorescence responses can yield quantitative optical indicators of oral epithelial cancer through m-FLIM technology.

6.3 m-FLIM optical instrumentation

Fluorescence lifetime imaging microscopy (FLIM) has great potential for clinical applications considering that it offers a noninvasive approach to characterize *in vivo* biological tissue from a biochemical composition perspective. However, a clinically compatible FLIM system must meet the following requirements: (i) should be compact for easy handling and portability, (ii) should have a fast image acquisition rate, and (iii) should be able to acquire the fluorescence signal at multiple emission bands (m-FLIM image).

An example of an optical m-FLIM system used to non-invasively acquire clinical images of oral lesions from patients was developed by (Cheng, 2014). This system consisted of a handheld box fitted with a rigid endoscope that is placed in contact with the oral tissue. In this case, a 355 nm frequency-tripled Q-switched Nd:YAG laser (1 ns pulse width, ~ 1 μJ/pulse at the tissue) induces tissue autofluorescence through an excitation fiber and acquires the emission at 390 ± 20 nm, 452 ± 22.5 nm, and spectral bands above 500 nm, which were chosen to record the oral cavity fluorescence response originating from collagen, NADH, and FAD, respectively. The m-FLIM data is sampled with a circular field-of-view (FOV) of 10 mm diameter. The lateral resolution is close to 100 mm and the total acquisition time is less than 3 seconds. At each pixel, the time-resolved m-FLIM signal is acquired through a multichannel plate photomultiplier tube (25 ps transient time spread) followed by a preamplifier and a high-speed digitizer with a sampling rate of 6.25 giga-samples per second (temporal resolution of 160 ps). The total energy delivered by the system (< 2.8 mJ) has an order of magnitude lower than the maximum permissible exposure (MPE = 29.8 mJ) set by (ANSI, 2017). The reader is referred to (Cheng et al., 2014) to see a graphical representation of this m-FLIM endoscopic system with a more detailed description of its optical components.

6.4 m-FLIM data processing and fluorescence lifetime estimation

Fluorescence intensity on its own does not provide enough information to distinguish endogenous fluorophores. This is because they usually emit a broad non-localized response with high similarity between them. It is even more complicated when the task is to distinguish between healthy and malignant tissue. The goal of FLIM and m-FLIM systems is to provide complimentary information useful for such scenarios. These systems can monitor endogenous and exogenous fluorophores, and therefore monitor progression of diseases, treatments, and natural processes in tissue as well as interactions between cells. This is suitable for sensing molecular interactions, interaction between proteins with Föster Resonance Energy Transfer (FRET) as well as detecting biochemical parameters such as pH, and temperature. Several methods are employed to extract meaningful results in the state of the art (Suhling, 2015).

In this section we will review the common techniques used to process the output from both FLIM and m-FLIM systems. The simplest model assumes that the fluorescence lifetime decay or fluorescence impulse response (FIR) is a mono-exponential function of time t such as

$$h(t) = A_0 exp(-t/\tau) \tag{6.1}$$

where, τ represents the fluorescence lifetime A_0 and is the signal amplitude at time 0. However, a single pixel recorded by a m-FLIM instrument may record the fluorescence response from various sources. The measurement of the contribution of multiple fluorophores is best described as a multiexponential decay (Lakowicz, 2006) instead of a monoexponential decay (Equation 6.1). The raw time-domain FLIM data consists of fluorescence intensity time decay signals y(t). They are acquired at each emission spectral band at all positions. The measured fluorescence decay y(t) at each spatial location is the result of the convolution of the FIR $h(t)$ and the measured instrument response function (IRF) $u(t)$ as:

$$\mathrm{y(t)} = \int_0^t u(\sigma)h(t-\sigma)d\sigma = u(t) * h(t) \tag{6.2}$$

The main problem is how to estimate $h(t)$ and its properties such as the lifetime τ. Some authors assume the instrument response $u(t)$ is known and others estimate it experimentally (Straume, 2002). The problem on how to interpret the output of m-FLIM, and its variants, is how to model the fluorescence decays precisely and estimate the lifetime of the multiple components within each position and wavelength registered. The work by

(Datta, 2020) makes a great review of traditional methods. The curve fitting approach tries to estimate the parameters A_0, τ to fit a model. These models can be linear or non-linear and they are solved under several assumptions including a known u(t). This is in the case of methods such as least squares that provide a quick solution and are easy to implement. Other methods are fit to time and frequency domain FLIM systems; however, they are not time efficient.

A common approach is to estimate the sample FIR h(t) and the IRF u(t) employing deconvolution techniques. Temporal deconvolution of the raw m-FLIM data y(t) can be achieved through several methods reported in the literature. Some examples are the least squares deconvolution from (Zickus, 2020). There are optimization-based approaches which are popular since the Levenberg-Marquardt based method (Birch, 2002). Such is the case of the non-linear least squares and maximum likelihood approaches (Maus, 2001). A novel approach is presented in (Mannan, 2022), which works under the assumption of Poisson noise. This method is adapted from confocal microscopy and employs a regularized cost function to correct for the distortions in m-FLIM measurements due to optical convolution.

The most common and widely investigated temporal deconvolution method for m-FLIM data is the nonlinear least squares iterative deconvolution algorithm (Pande, 2010), in which the FIR h(t) is typically modeled as a bi-exponential decay function (Zhang, 2016):

$$h(t) = \alpha\, exp[-t/\tau_1] + (1-\alpha)\, exp[-t/\tau_2] \tag{6.3}$$

where τ_1 and τ_2 represent the lifetime of the fluorescence decay components. Meanwhile, α and $1-\alpha$ are the proportion of the fast and slow decay components, respectively.

Once the FIR h(t) (Equation 6.3) has been deconvolved from the measured decay y(t) (Equation 6.2), the spectral fluorescence intensity (I) can be computed at every position through numerical integration of the FIR:

$$I = \int_0^\infty h(t)dt \tag{6.4}$$

Additionally, the average fluorescence lifetime (τ_{avg}) at every spatial location and emission band can be estimated from the deconvolved FIR (Lakowicz, 2006):

$$\tau_{avg} = \frac{\int_0^\infty t\, h(t)dt}{\int_0^\infty h(t)dt} \tag{6.5}$$

Modern approaches make use of numerical optimization and deep learning techniques such as convolutional neural networks (Chen, 2022). These may provide a rapid solution; however, they require to be trained with enough representative data and are prone to overfitting.

These m-FLIM imaging features, namely multispectral fluorescence intensity (Equation 6.4) and average fluorescence lifetime (Equation 6.5), quantify autofluorescence biomarkers with great potential for the clinical diagnosis of malignant and premalignant oral lesions from patients. For instance, the studies performed by (Duran-Sierra, 2021; Duran-Sierra, 2020) demonstrated the potential of m-FLIM derived autofluorescence features for the automated discrimination of cancerous/pre-cancerous vs. healthy oral tissue. FLIM features have also shown promising results in the discrimination of benign vs. cancerous/pre-cancerous oral epithelial lesions as reported by (Caughlin, 2021; Duran-Sierra, 2022; Jo et al., 2018). Altogether, these studies provide evidence of the clinical applicability of FLIM and its relevance for the early detection of oral cancer and dysplasia.

6.5 Linear unmixing

For clinical use, the data generated by m-FLIM should be processed fast and easy to understand by non-experts. One of the methods employed to reach this goal is to assume that m-FLIM decays follow a linear mixture formulation. In this way, we assume that there are K m-FLIM measurements available. Each one is expressed as L-dimensional vectors $z_k \in R^L$ with $k \in [1, K]$. Then, the set of measurements $Z = \{z_1, \dots , z_K\}$ is normalized in a sum-to-one fashion. Therefore $y_k = \frac{1}{\mathbf{1}_L^T z_k} z_k \; \forall k \in [1, K]$, where $\mathbf{1}_L$ denotes a L-dimensional column vector with unitary entries. Given the set of signals $Y = \{y_1, \dots , y_K\}$, we assume a linear mixture model of N-th order such that:

$$y_k = \sum_{n=1}^{N} \alpha_{k,n} \, p_n + v_k = \boldsymbol{P}\boldsymbol{\alpha}_k + \boldsymbol{v}_{k'} \tag{6.6}$$

where $\boldsymbol{p}_n \in R^L$ is the n-th end-member $\forall n \in [1, N]$ ($\boldsymbol{p}_n \geq 0$ and $\mathbf{1}_L^{\mathrm{T}} \boldsymbol{p}_n = 1$), $\boldsymbol{P} = [\boldsymbol{p}_1 \dots \boldsymbol{p}_N] \in R^{L \times N}$ while $\alpha_{k,n} \geq 0$ represents its contribution or abundance in the k-th measurement and $\boldsymbol{\alpha}_k = [\alpha_{k,1} \dots \alpha_{k,N}]^T \in R^N$. The abundances in every k-th spatial location are normalized to sum-to-one, i.e., $\mathbf{1}_N^{\mathrm{T}} \boldsymbol{\alpha}_k = 1$. The vector $\boldsymbol{v}_k$ stands for independently and identically distributed Gaussian noise with zero mean and finite covariance matrix. The estimation problem of end-members and their abundances in each m-FLIM measurement (Equation 6.7) departs from the extended blind end-member and abundance extraction (EBEAE) procedure from (Campos-Delgado, 2019). The problem can be described as a cost function of the form,

$$\min_{\{\alpha_k\}_{k=1}^{K},P} \frac{1}{2K}\sum_{k=1}^{K}\frac{\| \boldsymbol{y}_k - \boldsymbol{P}\boldsymbol{\alpha}_k \|^2}{\| y_k \|^2} + \frac{\rho}{2\vartheta}\sum_{n=1}^{N-1}\sum_{j=n+1}^{N} \| \boldsymbol{p}_n - \boldsymbol{p}_j \|^2 - \frac{\mu}{2K}\frac{\lambda_{min}(\boldsymbol{P}^T\boldsymbol{P})}{\| y_k \|^2}\sum_{k=1}^{K} \| \boldsymbol{\alpha}_k \|^2 \tag{6.7}$$

with ρ and μ as regularization weights, $\vartheta \triangleq 1+\cdots+(N-1)$ for all $N \geq 2$, and $\|\cdot\|$ denotes the Euclidean norm. The optimization problem in (Equation 6.7) can be solved with two decoupled subproblems by a coordinate descent scheme (CDS) and assuming an initial end-members matrix $\boldsymbol{P}^0$: (i) abundances estimation $\{\boldsymbol{\alpha}_k\}_{k=1}^{K}$ over all spatial locations, and (ii) end-members extraction P over the whole dataset. These sub-problems are solved in an alternating scheme. That is, the abundances at all spatial locations $\{\boldsymbol{\alpha}_k\}_{k=1}^{K}$ are estimated while the end-members matrix $\boldsymbol{P}$ is considered constant; in the next step matrix $\boldsymbol{P}$ is calculated while the abundances $\{\boldsymbol{\alpha}_k\}_{k=1}^{K}$ are fixed. This process is performed until convergence of the overall estimation error or the maximum number of iterations is reached. For further detail of EBEAE, the reader is referred to (Campos-Delgado, 2019).

6.6 EBEAE analysis of m-FLIM datasets for oral cancer detection

In this section, we illustrate the use of m-FLIM for oral cancer detection. To identify cancerous tissue in the oral cavity, a suspicious lesion is imaged using the handheld m-FLIM endoscope mentioned in Section 6.3. The m-FLIM images included in the analysis presented in this section were obtained from the study performed by (Duran-Sierra, 2020), in which *in vivo* clinical m-FLIM images of cancerous and pre-cancerous oral lesions were acquired from patients. First, an experienced oral pathologist spotted the suspicious oral lesion from the patient during a standard clinical oral examination. Second, non-invasive m-FLIM imaging was performed. Only two m-FLIM images were acquired per subject. The first one is obtained from the suspicious site (lesion sample). A second image for control purposes is extracted from a normal contralateral site (contrast sample) within the mouth cavity. To validate the results, a biopsy was taken for histopathological analysis.

A first subject was imaged with a lesion in the tongue, labeled as Dataset 1. This sample was diagnosed as squamous cell carcinoma (SCC) according to the histopathological evaluation. The data had dimensions of $160 \times 160 \times 1125$, where the first two represent the number of spatial locations, and the last one the recorded m-FLIM response. The EBEAE method was employed to analyze the lesion and contrast samples simultaneously. The hyper-parameter values employed were $N = 3$, $\rho = 0.8$, $\mu = 0.2$. The hyper-parameters were tuned manually to obtain the best separability in the abundance maps. The results obtained for the lesion and contrast samples are shown in Figure 6.2. The abundance maps are shown in the top and bottom

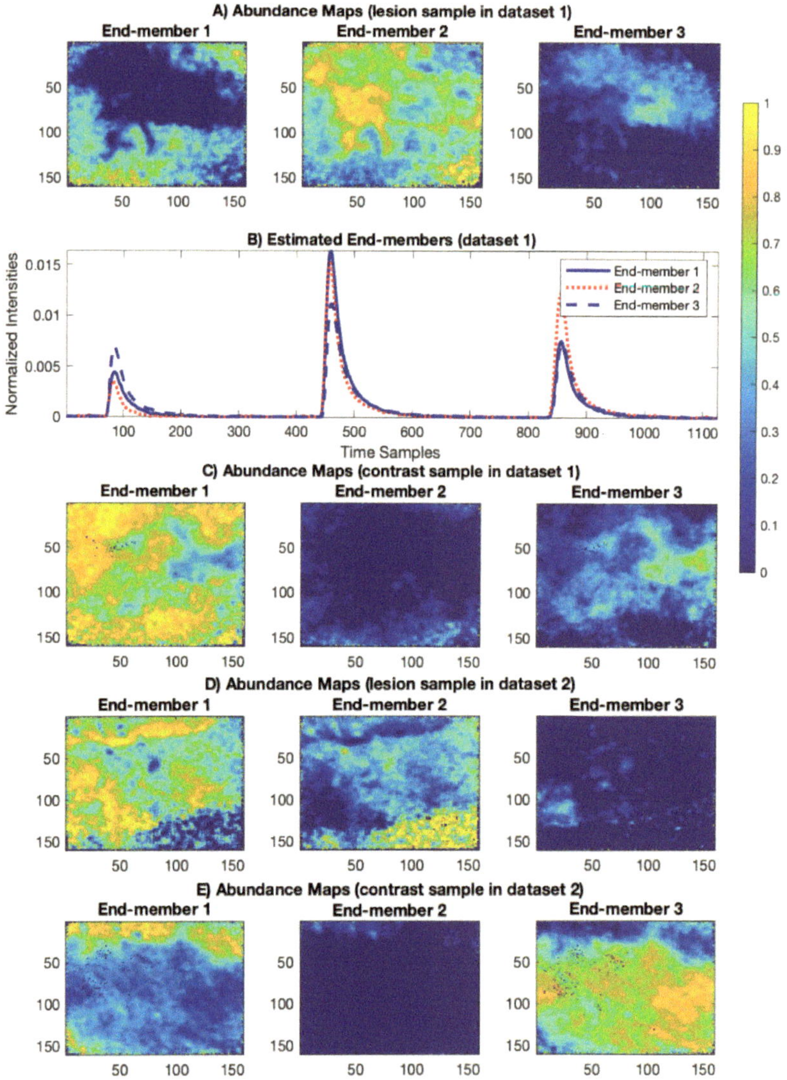

Figure 6.2. The results provided by EBEAE for m-FLIM data: For dataset 1, abundance maps in the interval [0,1] for three end-members: (A) lesion and (C) contrast samples, and (B) estimated end-members. For dataset 2, abundance maps for (D) lesion and (E) contrast samples.

images. The resulting end-members are displayed in the middle plot. These results show the contrasting characterizations for the lesion and control samples. In this case, it is possible to observe that end-member 2 is present only in the SCC (lesion) sample. Meanwhile end-members 1 and 3 are mostly present in the control sample. We can observe that the estimated end-members (see Figure 6.2B) have characteristic peak responses for each spectral band. Thus, it can be confirmed that the EBEAE algorithm detected three different fluorophores in both samples, which is backed up by the dissimilar abundance maps and the characteristic peaks in the estimated multi-spectral time-profiles. The EBEAE algorithm was also evaluated with a higher model order $N = 4$, however, one of the abundance maps was trivial which indicates a low contribution from the corresponding end-member and therefore does not contribute to explaining the data.

The results estimated for Dataset 2 using N = 3 end-members are shown in Figure 6.2(D) and (E). These abundance maps show the results obtained for the second tongue sample. Consider that Figure 6.2(D) shows that end-member 2 is present in the lesion sample only. Meanwhile, Figure 6.2(E) shows that the contrast samples include mostly contributions from end-members 1 and 3. Consequently, end-member 2 is a distinctive fluorophore of the SCC samples in Figure 6.2 for datasets 1 and 2. These results show that the application of EBEAE estimates a quantitative interpretation of the m-FLIM database which extracts characteristic features useful to pinpoint SCC lesions.

6.7 Conclusions

Current diagnosis of SCC is performed by clinical examination followed by a biopsy of any suspicious lesion. This intrusive procedure discourages early detection as patients usually elude it. The goal of modern screening tools is to assess the diagnosis as minimally invasive as possible. To do this, they look for biomarkers that can characterize the carcinogenesis process. Such is the case of FAD and NADH, which are autofluorescent molecules that provide a markerless way to monitor the biochemical information from a sample.

There are several m-FLIM technologies available which are able to monitor SCC fluorescent biomarkers, such as TCSPC, time gating and pulse sampling techniques. The m-FLIM devices record the fluorescence decay in multiple positions in multiple wavelength channels simultaneously. This information is helpful to identify fluorophores with similar responses and detect variations due to their chemometric environment. Frequency-resolved m-FLIM is suitable for real-time screening although it requires high photon counts when sampling.

Theoretically, m-FLIM can pinpoint the position in a sample which contains biomarkers associated with SCC. However, the interpretation of the raw data output is an open problem. The estimation of the fluorescence lifetime is performed using techniques such as curve fitting and phasor analysis. The main disadvantages of m-FLIM are inherent to fluorescence measurements. It can provide great insight into the biochemical composition of a sample but provides no structural information. These measurements are only superficial since light is only able to penetrate a few millimeters without being absorbed.

In this chapter, we recapitulate an approach based on the linear mixture model which is performed once the raw data has been deconvoluted to separate the lifetime decay from the system response. The solution to the linear unmixing model allows us to identify and quantify the endogenous fluorophores present. Such is the case of the scenario presented in Figure 6.2. These representative results are easy to understand and do not require great expertise in the numerical methods employed. In addition, they highlight the utility of m-FLIM as a diagnostic tool with clinical applicability for oral cancer screening. Ongoing research efforts are focused on the use of m-FLIM as a diagnostic aid for early detection of SCC as well as to study carcinogen metabolism.

Acronym/Abbreviation

CDS	Coordinate Descent Scheme
EBEAE	Extended Blind End-member Extraction
FAD	Flavin Adenine Dinucleotide
FIR	Fluorescence Impulse Response
FLIM	Fluorescence Lifetime Imaging Microscopy
FOV	Field of View
FRET	Föster Resonance Energy Transfer
IRF	Instrument Response Function
m-FLIM	multispectral Fluorescence Lifetime Imaging Microscopy
MPE	Maximum Permissible Exposure
NADH	Nicotinamide Adenine Dinucleotide
SCC	Squamous Cell Carcinoma
TCSPC	Time Correlated Single Photon Counting
UV	Ultraviolet

Reference list

Abati, S., Bramati, C., Bondi, S., Lissoni, A., Trimarchi, M. et al. (2020). Oral cancer and precancer: A narrative review on the relevance of early diagnosis. *International Journal of Environmental Research and Public Health*, 17(24): 9160. doi: 10.3390/ijerph17249160.

Birch, D.J. and Imhof, R.E. (2002). Time-domain fluorescence spectroscopy using time-correlated single-photon counting. *In Topics in Fluorescence Spectroscopy*, pp. 1–95. Springer, Boston, MA.

Campos-Delgado, D.U., Gutierrez-Navarro, O., Rico-Jimenez, J.J., Duran-Sierra, E., Fabelo, H. et al. (2019). Extended blind end-member and abundance extraction for biomedical imaging applications. *IEEE Access*, 7: 178539–178552.

Caughlin, K., Duran-Sierra, E., Cheng, S., Cuenca, R., Ahmed, B. et al. (2021, November). End-to-End neural network for feature extraction and cancer diagnosis of *in vivo* fluorescence lifetime images of oral lesions. *In 2021 43rd Annual International Conference of the IEEE Engineering in Medicine & Biology Society (EMBC)*, pp. 3894–3897. IEEE.

Chakraborty, D., Natarajan, C. and Mukherjee, A. (2019). Advances in oral cancer detection. *Advances in Clinical Chemistry*, 91: 181–200.

Chen, Y.I., Chang, Y.J., Liao, S.C., Nguyen, T.D., Yang, et al. (2022). Generative adversarial network enables rapid and robust fluorescence lifetime image analysis in live cells. *Communications Biology*, 5(1): 1–11.

Cheng, S., Cuenca, R.M., Liu, B., Malik, B.H., Jabbour, et al. (2014). Handheld multispectral fluorescence lifetime imaging system for *in vivo* applications. *Biomedical Optics Express*, 5(3): 921–931.

Cheng, Y.S., Rees, T. and Wright, J. (2015). Updates regarding diagnostic adjuncts for oral squamous cell carcinoma. *Texas Dental Journal*, 132(8): 538–549.

Cleveland, J.L. and Robison, V.A. (2013). Clinical oral examinations may not be predictive of dysplasia or oral squamous cell carcinoma. *Journal of Evidence Based Dental Practice*, 13(4): 151–154.

Datta, R., Heaster, T.M., Sharick, J.T., Gillette, A.A., Skala, M.C. et al. (2020). Fluorescence lifetime imaging microscopy: Fundamentals and advances in instrumentation, analysis, and applications. *Journal of Biomedical Optics*, 25(7): 071203.

Duran-Sierra, E., Cheng, S., Cuenca-Martinez, R., Malik, B., Maitland, K.C. et al. (2020). Clinical label-free biochemical and metabolic fluorescence lifetime endoscopic imaging of precancerous and cancerous oral lesions. *Oral Oncology*, 105: 104635.

Duran-Sierra, E., Cheng, S., Cuenca, R., Ahmed, B., Ji, J., Yakovlev et al. (2021). Machine-learning assisted discrimination of precancerous and cancerous from healthy oral tissue based on multispectral autofluorescence lifetime imaging endoscopy. *Cancers*, 13(19): 4751.

Duran-Sierra, E., Cheng, S., Cuenca, R., Ahmed, B., Ji, J., Yakovlev et al. (2022). Clinical label-free endoscopic imaging of biochemical and metabolic autofluorescence biomarkers of benign, precancerous, and cancerous oral lesions. *Biomedical Optics Express*, 13(7): 3685–3698.

Epstein, J.B., Güneri, P., Boyacioglu, H. and Abt, E. (2012). The limitations of the clinical oral examination in detecting dysplastic oral lesions and oral squamous cell carcinoma. *The Journal of the American Dental Association*, 143(12): 1332–1342.

Georgakoudi, I., Jacobson, B.C., Muller, M.G., Sheets, E.E., Badizadegan, K. et al. (2002). NAD (P) H and collagen as *in vivo* quantitative fluorescent biomarkers of epithelial precancerous changes. *Cancer Research*, 62(3): 682–687.

Jo, J.A., Cheng, S., Cuenca-Martinez, R., Duran-Sierra, E., Malik, B. et al. (2018, July). Endogenous fluorescence lifetime imaging (FLIM) endoscopy for early detection of oral cancer and dysplasia. In *2018 40th Annual International Conference of the IEEE Engineering in Medicine and Biology Society (EMBC)*, pp. 3009–3012. IEEE.

Kaur, J., Srivastava, R. and Borse, V. (2021). Recent advances in point-of-care diagnostics for oral cancer. *Biosensors and Bioelectronics*, 178: 112995. doi: 10.1016/j.bios.2021.112995.

Lakowicz, J.R. (ed.). (2006). *Principles of Fluorescence Spectroscopy*. Boston, MA: springer US.

Mannam, V., Yuan, X. and Howard, S. (2022, March). Deconvolution of fluorescence lifetime imaging microscopy (FLIM). In *Multiphoton Microscopy in the Biomedical Sciences XXII* (Vol. 11965, pp. 38–43). SPIE.

Maus, M., Cotlet, M., Hofkens, J., Gensch, T., De Schryver, F.C. et al. (2001). An experimental comparison of the maximum likelihood estimation and nonlinear least-squares fluorescence lifetime analysis of single molecules. *Analytical Chemistry*, 73(9): 2078–2086.

Müller, M.G., Valdez, T.A., Georgakoudi, I., Backman, V., Fuentes, C. et al. (2003). Spectroscopic detection and evaluation of morphologic and biochemical changes in early human oral carcinoma. *Cancer: Interdisciplinary International Journal of the American Cancer Society*, 97(7): 1681–1692.

Pande, P. and Jo, J.A. (2010). Automated analysis of fluorescence lifetime imaging microscopy (FLIM) data based on the Laguerre deconvolution method. *IEEE Transactions on Biomedical Engineering*, 58(1): 172–181.

Pavlova, I., Sokolov, K., Drezek, R., Malpica, A., Follen, M. et al. (2003). Microanatomical and biochemical origins of normal and precancerous cervical autofluorescence using laser-scanning fluorescence confocal microscopy. *Photochemistry and Photobiology*, 77(5): 550–555.

Pavlova, I., Williams, M., El-Naggar, A., Richards-Kortum, R., Gillenwater, A. et al. (2008). Understanding the biological basis of autofluorescence imaging for oral cancer detection: High-resolution fluorescence microscopy in viable tissue. *Clinical Cancer Research*, 14(8): 2396–2404.

Roblyer, D., Kurachi, C., Stepanek, V., Schwarz, R.A., Williams, M.D. et al. (2010). Comparison of multispectral wide-field optical imaging modalities to maximize image contrast for objective discrimination of oral neoplasia. *Journal of Biomedical Optics*, 15: 066017–066017.

Safe Use of Lasers, ANSI Z136.1–2007, L.I.o. America, Editor. 2007, American National Standards Institute.

Shah, A.T., Demory Beckler, M., Walsh, A.J., Jones, W.P., Pohlmann, P.R. et al. (2014). Optical metabolic imaging of treatment response in human head and neck squamous cell carcinoma. *PloS One*, 9(3): e90746.

Shah, A.T., Heaster, T.M. and Skala, M.C. (2017). Metabolic imaging of head and neck cancer organoids. *PloS One*, 12(1): e0170415.

Shin, D., Vigneswaran, N., Gillenwater, A. and Richards-Kortum, R. (2010). Advances in fluorescence imaging techniques to detect oral cancer and its precursors. *Future Oncology*, 6: 1143–1154.

Siegel, R.L., Miller, K.D., Fuchs, H.E. and Jemal, A. (2022). Cancer statistics, 2022. *CA: A Cancer Journal for Clinicians*, 72(1): 7–33. https://doi.org/10.3322/caac.21708.

Skala, M.C., Squirrell, J.M., Vrotsos, K.M., Eickhoff, J.C., Gendron-Fitzpatrick, A., Eliceiri, K.W. and Ramanujam, N. (2005). Multiphoton microscopy of endogenous fluorescence differentiates normal, precancerous, and cancerous squamous epithelial tissues. *Cancer Research*, 65(4): 1180–1186.

Skala, M.C., Riching, K.M., Gendron-Fitzpatrick, A., Eickhoff, J., Eliceiri, K.W. et al. (2007). *In vivo* multiphoton microscopy of NADH and FAD redox states, fluorescence lifetimes, and cellular morphology in precancerous epithelia. *Proceedings of the National Academy of Sciences*, 104(49): 19494–19499.

Straume, M., Frasier-Cadoret, S.G. and Johnson, M.L. (2002). Least-squares analysis of fluorescence data. In *Topics in Fluorescence Spectroscopy* (pp. 177–240). Springer, Boston, MA.

Sweeny, L., Dean, N.R., Magnuson, J.S., Carroll, W.R., Clemons, L. et al. (2011). Assessment of tissue autofluorescence and reflectance for oral cavity cancer screening. *Otolaryngology —Head and Neck Surgery*, 956–960.

Suhling, K., Hirvonen, L.M., Levitt, J.A., Chung, P.H., Tregidgo et al. (2015). Fluorescence lifetime imaging (FLIM): Basic concepts and some recent developments. *Medical Photonics*, 27: 3–40.

Zhang, Y., Chen, Y. and Li, D.D.U. (2016). Optimizing Laguerre expansion based deconvolution methods for analysing bi-exponential fluorescence lifetime images. *Optics Express*, 24(13): 13894–13905.

Zickus, V., Wu, M.L., Morimoto, K., Kapitany, V., Fatima et al. (2020). Fluorescence lifetime imaging with a megapixel SPAD camera and neural network lifetime estimation. *Scientific Reports*, 10(1): 1–10.

CHAPTER 7

Thermotherapies based on Microwaves (MW) and Radiofrequencies (RF)

Citlalli J. Trujillo-Romero[1,*] and *Rocío Ortega Palacios*[2]

In the last 30 years, research related to the application of heat as an alternative to treat cancer has shown its effectiveness. Treatments are divided according to the attained temperature and application time as cryotherapy, diathermy, hyperthermia, and thermal ablation. Physical principles of thermotherapies that use electromagnetic energy at radio and microwave frequencies, such as hyperthermia and thermal ablation, are described. The process to go from the applicator design to its use in clinical applications is described. Before the clinical application, the applicators must be validated by using tissue-mimicking phantoms; therefore, tissue properties play a significant role in the absorption of EM energy that produces the temperature increase and the therapeutic effect. For clinical applications, tools like treatment planning based on computational modeling, treatment quality protocols, and guidelines are considered; these are quality indicators that allow the improvement of the thermotherapy applications. All these aspects are addressed in the present chapter.

[1] Division of Medical Engineering Research, National Institute of Rehabilitation LGII. Calz. Mexico Xochimilco No. 289, Col. Arenal de Guadalupe, Mexico City, 14389, Mexico.

[2] Universidad Politécnica de Pachuca, Ingeniería Biomédica. Carretera Ciudad Sahagún-Pachuca Km. 20, Ex-Hacienda de Santa Bárbara, 43830 Zempoala, Hgo.

Email: rortega@upp.edu.mx

* Corresponding author: cjtrujillo@inr.gob.mx

7.1 Introduction

Thermal therapies are based on the transfer of thermal energy into the body. These therapies can be either increasing or decreasing the body's normal temperature. One of the medical applications of low temperatures is cryotherapy ablation, using temperatures below –20°C for the destruction of cancer cells. High temperature applications are divided into three distinct protocols for raising temperature diathermy (low temperature hyperthermia), hyperthermia (moderate temperature hyperthermia) and thermic ablation (high temperature hyperthermia). Low and moderate temperature hyperthermia (41°C–45°C) causes physiological changes in the tissue such as increased blood perfusion, vascular permeability, pH, among others, these transient effects are reversible, so it is recommended that these temperatures are maintained for a long time or in conjunction with conventional treatments such as radiotherapy or chemotherapy. On the other hand, it has been found that thermal ablation (60°C–100°C) can produce vascular stasis, protein denaturation, cell coagulation and cell necrosis, these effects are irreversible and cause cell death almost immediately. Several applicators based on Radiofrequencies (RF) and Microwaves (MW) as well as different tools have been developed to improve the quality of thermal treatments and be adopted as regular treatments. Although thermal therapies have been successfully used to treat different kinds of tumors, several aspects such as patient specific treatment planning, applicator design and more are still under development. This chapter provides an overview of the most important aspects on thermal therapies based on RF and MW.

7.2 Thermotherapies classification

For the last few years, thermotherapies using either minimum or non-invasive techniques have been used as an alternative to treat cancer and repair musculoskeletal injuries. Thermotherapies are classified according to the tissue temperatures attained and application time. Temperatures go from –50°C to 100°C and its classification can be as follow:

a) Cryotherapy: This is a minimally invasive therapy; the tumor must reach freezing temperatures (below –20°C) to destroy the cancer cells by the formation of intracellular ice crystals. Liquid nitrogen or high-pressure argon circulates throughout a cryoprobe producing temperatures around –190°C, due to the Joule-Thomson effect (Sprenkle et al., 2010).
b) Diathermy *(low-temperature hyperthermia)*: It is mainly used to treat orthopedic injuries (musculoskeletal pathologies) by heating the tissue at temperatures around 41°C from 30 to 60 minutes. It is produced by the interaction between electromagnetic waves and tissue (Laufer et al.,

2012). It reduces pain and increases the range of motion by increasing metabolic functions and blood flow.

c) Hyperthermia *(moderate temperature hyperthermia)*: The therapeutic effect occurs around 41°C–45°C; however, it is a function of temperature and exposition time. Healthy tissue can hold temperatures up to 44°C for approximately 1 hour (Habash et al., 2006). At these temperatures, protein denaturation induces tumor cell death. Hyperthermia produces an adjuvant effect that increases the effectiveness of radiotherapy and chemotherapy because tissue cells become more susceptible to them.

d) Thermal ablation *(high-temperature hyperthermia)*: In this case, very high temperatures (60°C–100°C) must be reached and kept for short periods to produce almost instant cell death (coagulation). Temperatures around 50°C–55°C produce cellular damage if the exposition time is increased to approximately 4 min–6 min. The lower the temperature attained, the longer the exposure time and vice-versa. The tumor can be completely eradicated if it is adequately heated (Riadh et al., 2007).

7.2.1 Physical principles of Microwaves (MW) and Radiofrequency (RF)

Recently, several options to treat cancer based on different physical principles have been developed. Worldwide thermotherapy systems based on Electromagnetic (EM) radiation have been developed. RF and MW are the most used heating principles to produce thermotherapy. Both interact in different ways with the human body. Thermal therapies based on RF (frequencies lower than 30 MHz) generate a local tissue temperature increase due to an ionic movement induced by the applied EM field (Joule heating). Therapies based on MW (frequencies >100 MHz) generate a tissue temperature increment due to the mechanical friction between the polar water molecules induced by the applied time-varying electric field. Figure 7.1 describes the section of the EM spectrum that includes RF and MW and describes the frequencies most used to apply thermal therapies.

Radiofrequency (RF)

Energy used to apply heat to a tumor includes radio frequencies from 100 kHz to 150 MHz. Moreover, to heat big tumors in depth, RF fields among 10–120 MHz are used. Thermal heating based on radiofrequency is due to the alternating electrical currents that oscillate in frequencies around 200 kHz–1.2 MHz. The configuration to apply RF treatment is a closed-loop circuit, connecting a RF generator to a large ground pad electrode (low-ohmic neutral electrode) located at the skin surface of the patient, and a

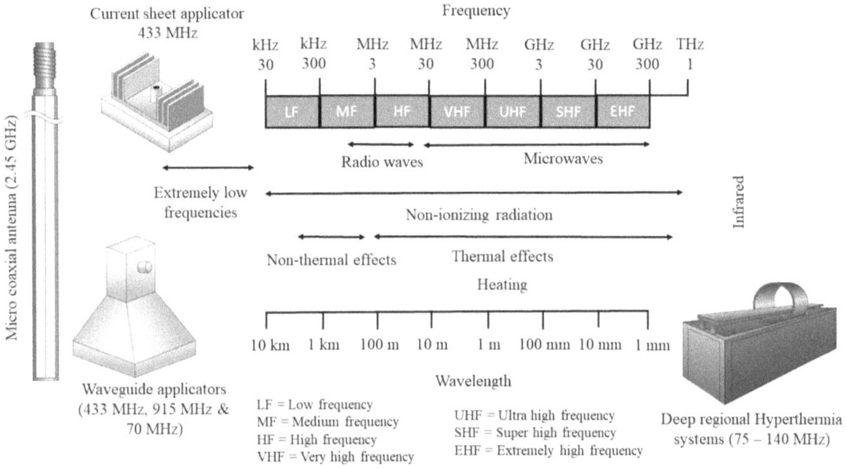

Figure 7.1. Section of the EM spectrum that includes RF and MW, including examples of the most used frequencies to apply thermotherapies either by RF or MW.

needle electrode inserted over the tumor, in series (Buscarini et al., 2005). The ground pad electrode and the needle electrode are active, and the patient performs as a resistor. Therefore, an alternating electric field is generated in the body tissues. Once the RF energy is delivered, both electrical fields and high-density currents are generated around the needle electrode in the tumor; consequently, a high-temperature increase results. The heat generation is focused on the needle electrode because of the high difference between its surface area and the surface of the ground pad electrodes; therefore, the final size of the tumor damage will always depend on tissue electrical conductivity and heat diffusion.

Microwaves (MW)

Thermal therapies based on Microwave irradiation are some of the most promising techniques to treat cancer. The temperature increase produced by microwaves helps to either reduce or destroy the treated tumors. The most used frequencies are 434, 915, and 2450 MHz. In this case, tissue water content plays an important role because the water molecules are polar, i.e., their electric charges are not symmetric. In water molecules, the hydrogen side has a positive charge, while the oxygen side has a negative charge. Moreover, EM radiation also has an electrical charge that flips from positive to negative. Consequently, the heat produced by an applicator is due to the water molecules' agitation at the surrounding tissue; this agitation produces friction between molecules and therefore the required heat to produce cellular death.

7.2.2 RF and MW applicators

All over the world, several thermotherapy devices had been developed and tested in the past. Nevertheless, different aspects concerning applicators, treatment protocols, mathematical models and treatment planning still require improvement. Applicators are classified as non-invasive and minimally invasive. The main challenges are treating deep-seated tumors by using non-invasive applicators and the design of new applicators to treat specific kinds of tumors.

Non-invasive applicators

Hyperthermia treatments using MW external applicators, to treat deep-seated tumors, are difficult due to the penetration depth of the waves at these frequencies being just about some millimeters, i.e., the attenuation of the EM fields is higher. The penetration depth of waves can be increased by using RFs; however, it is not possible to control the size of the affected area. Several heating techniques have been developed to solve the drawbacks of EM hyperthermia: local, regional, whole body, and magnetic fluid as described in Table 7.3, reported in Section 7.2.2.

Minimally invasive applicators

Minimally invasive applicators generate a uniform temperature distribution; moreover, the heating can be easily focused on the tumor. However, their main disadvantage is the necessity to be inserted into the tumor; therefore, these applicators are not feasible for many tumor locations. The most common applicators are the radiating microwave antennas; they can be used by themselves or as an antenna array. A typical implanted antenna is a coaxial dipole antenna built in a thin semi-rigid micro coaxial cable.

Mimic tissue phantoms for applicator evaluation

The evaluation of the electromagnetic interaction of RF/MW systems with the human body in a quantitative way is extremely important in the development of new devices for thermotherapy. RF and MW applicators must be validated; moreover, quality assurance (QA) studies must also be implemented. These validation tests should be performed as many times as required, under controlled conditions; this is required for the validation of all the possible case scenarios and assures the safety of these devices. Therefore, it is not possible to implement such a validation test in human beings; evaluation in humans not only can produce serious ethical issues, but also a variety of uncertainties in the validation process (cardiovascular vibration, respiratory movements and more). Nevertheless, to perform these evaluations under controlled and repeatable conditions, tissue-mimicking phantoms are used.

Phantoms are designed to emulate human tissue/organ properties; they are used either to validate new RF and MW applicators before being used in clinical applications or QA. They are used to investigate the interaction of RF/MW fields with biological tissues; therefore, the improvement of phantoms that mimic tissue properties accurately and reliably is extremely important. Phantoms are found in different compositions, shapes, and sizes; moreover, they can be in solid, semi-solid, and liquid states. Precision in the emulated tissue properties, lifetime, and the possibility to obtain a human shape are some of the features that can be emulated. The main advantage of phantoms is that their dielectric and thermal properties can be fit during the fabrication process.

Three necessities are covered by the development of tissue-mimicking phantoms: (1) Emulate a patient under treatment, (2) Maintenance of medical equipment, and (3) Research and design new devices. In these cases, phantoms are of great interest because such evaluations can be done under controlled, repeatable, and standardized conditions. In thermotherapy, the next two levels of equivalence between biological tissue and its phantom are the most important.

(1) **Evaluation of electromagnetic power deposition:** Dielectric properties of body tissues such as relative permittivity (ε_r) and electrical conductivity (σ) are strongly related to the interaction of RF/MW fields and tissues; therefore, they must be emulated. ε_r indicates how easily a material exposed to an EM field can be polarized, while the σ represents the electrical losses in a material due to the currents driven by an EM field. No heat transfer with the environment must occur; therefore, materials with low thermal conductivity and low heat capacity are recommended.

(2) **Evaluation of temperature distributions:** Tissue thermal properties, such as thermal capacity and conductivity must be emulated. Tissue thermal capacity represents its capacity to absorb heat when it is heated by an RF/MW applicator. Thermal conductivity refers to the ability of the tissue to conduct the heat produced during thermotherapy. Properties such as blood flow can be partially reproduced by a more complex phantom.

Tissue-mimicking phantoms must satisfy the following requirements:

Reproducibility and Reliability in its preparation: Its properties must be homogeneous and reproducible, while the preparation method must be reliable and repeatable. At the end of any experiment, its properties must go back to their initial values.

Stability: The phantom properties must not change over a large period; its lifetime must be long enough to reproduce the experiments as many times as necessary.

Toxicity: Although the components used to produce a phantom could be toxic, the phantom itself must not be toxic.

To develop phantoms, several materials with dielectric properties like those from the tissue to be emulated are used; however, phantoms are typically water-based materials. Table 7.1 describes different kinds of phantoms used to emulate tissue at several frequencies.

On the other hand, Table 7.2 describes different materials used in the tissue-mimicking phantoms' preparation.

7.3 Clinical applications

7.3.1 Requirements for the clinical application of thermal therapies

RF/MW Generation Systems

To implement the thermotherapies as a clinical application, and to give an effective treatment in a safe, controlled, and repeatable way, a common set-up composed by the applicator, the RF/MW, thermometry, and impedance matching systems, as described in Figure 7.2, is required.

***RF/MW generation system*:** It supplies the power to the applicator; composed of:

***Signal generator*:** Used to produce a time-dependent signal at the work frequency and amplitude. Most used frequencies are those corresponding to the ISM band (Industrial, Scientific, and Medical) such as 95 MHz, 443.5 MHz, and 2.45 GHz.

***Power amplifier*:** Used to amplify the signal power transmitted to the RF/MW applicator. By using the power amplifier, it is possible to work with high input power, even more than 100 W. Input power depends on the applicator type, treatment, type, and location of the tumor.

***Wattmeter*:** Used to monitor the incident and reflected power in real-time during the treatment; it helps to quantify the amount of power transmitted to the tissue and the power loss that can damage the equipment and cause burns to the patient.

***Thermometry System*:** Used to register the temperature in the tumor and healthy tissue around it. Temperature sensors are based on fiber optics to avoid electromagnetic interference and distortion of the applicator EM pattern. It requires a computer to record and visualize the temperature profiles during the treatment.

Table 7.1. Summary of different tissue-mimicking phantoms used for thermotherapy.

Tissue-mimicking phantom/ type	**Frequency (MHz)**	**Materials**	**Application**	**σ (S/m)**	**ε_r**	**Heat capacity (kg/K)**
Muscle-Phantom/ TX-150 Gel (Chou et al., 1984)	2, 450, 915, 750, 433, 300, 200	TX-150 Polyethylene NaCl Water	RF/MW studies.	2.17–1.06	47.4–56.7	–
Muscle-Phantom/ Agar-Gel (Ito et al., 2001)	430	Agar NaCl NaN_3 Polyethylene TX-151 Water	RF ablation.	1.41	53	–
Brain-Phantom/ Agar-Gel (Ito, 2001)	900	Agar NaCl NaN_3 Polyethylene TX-151 Water	MW ablation.	0.85	43	–
Muscle-Phantom/ Acrylamide-Gel (Surowiec et al., 1992)	500–3000	Acrylamide MBAA TEMED (N,N,NN-Tetramethylethylenedi) AMPS NaCl Water	RF/MW ablation.	–	63	–
Lung-Phantom/ Acrylamide-Gel (Surowiec, 1992)	500–3000	Acrylamide MBAA TEMED AMPS NaCl Water	RF/MW ablation.	–	38	–
Fat-Phantom/ Oil-gelatin (Kavousi et al., 2019)	13.56	Oil Gelatin Water Surfactant Formaldehyde	RF Hyperthermia.	0.03	11.8	2300

Table 7.1 contd. ...

...Table 7.1 contd.

Tissue-mimicking phantom/ type	**Frequency (MHz)**	**Materials**	**Application**	**σ (S/m)**	**ε_r**	**Heat capacity (kg/K)**
Gland-Phantom/ Oil-gelatin (Kavousi, 2019)	13.56	Oil Gelatin Water Surfactant Formaldehyde	RF Hyperthermia.	0.63	138.4	3670.5
Tumor-Phantom/ Oil-gelatin (Kavousi, 2019)	13.56	Oil Gelatin Water Salt Surfactant Formaldehyde	RF Hyperthermia.	0.71	142.5	3670.5
Fat-Phantom/ gelatin (Dobšíček Trefná et al., 2021)	100–3000	Water Oil Lecthin Gelatin CNC DAC	MW Hyperthermia.	–	7–14.5	3218
Tumor-Breast-Phantom/ Agarose (Ortega-Palacios et al., 2020)	2450	NaCl Ethanol Agarose Water	MW ablation.	2.88	55.88	–
Healthy-Breast-Phantom/ Agarose (Ortega-Palacios, 2020)	2450	Detergent Oil Agarose Water	MW ablation.	0.13	4.44	–
Healthy cortical and spongy bone, muscle and fat tissue (Ramírez-Guzmán et al., 2021)	2450	Corn oil Distilled water Wheat flour Dextrose Ethanol NaCl Agarose Neutral soap	MW ablation.	11.4 18.5 52.7 10.8	0.39 0.80 1.74 0.26	–

Table 7.2. Different materials used to develop tissue-mimicking phantoms used for thermotherapy.

Material	**Function**
Water	The main component of semi-solid and liquid phantoms. Used to reach the ε_r.
Polyacrylamide (C_3H_5NO)	To emulate tissue electrical parameters; by combining it with other materials, several tissues can be mimicked.
TEMED (tetra-metyl-ethylenediamine)	Used as catalyzer agent for heat-sensitive phantoms.
Sodium chloride (NaCl), sucrose	Increase ε_r and increase σ.
Agar/agarose	Solidifying agent. Modified ε_r and σ. Helps to obtain different phantom shapes. Provides stability at high temperatures and mechanical strength.
Oil	To emulate tissues with low water content.
Polyoxyethylene-sorbitanmonolaurate (Tween 20)	To produce stable oil-in-water emulsions.
Bactericide	To prevent microbial growth. Increase lifetime.
Polyvinyl Chloride (PVC)	To exponentially decrease σ and linearly decrease ε_r (Chen et al., 2020)
Graphite powder	To adjust σ and ε_r
Carbon fiber	To build semi-solid phantoms.
Glycerol, ethanol	Decrease ε_r
Polyethylene powder and aluminum powder	To modify ε_r. To emulate high-water-content tissues.
TX-150/ TX-151 (super stuff)	Gelling agent.
Sodium azide (NaN_3)	Increase preservation and slightly increase σ.
Ethylene glycol	Decreased ε_r.
N,N'-methylene-bis-acrylamide (MBAA)	Coagulant agent (Dabbagh et al., 2014).
ammonium persulfate (AMPS)	To reduce σ (Dabbagh, 2014).
Formaldehyde	Cross-linking agent.
Cellulose Nanocrystals (CNC) and Dialdehyde CNC (DAC)	Modifies viscosity, reinforces thermal integrity, and provides strength.

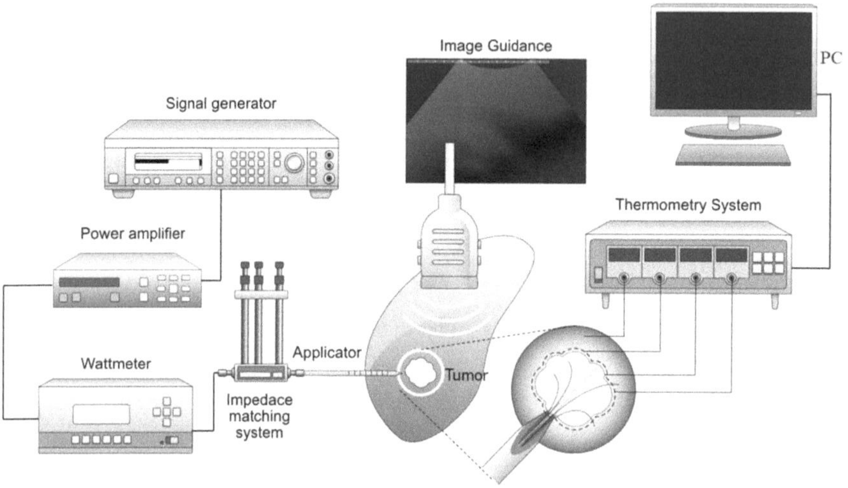

Figure 7.2. Common clinical set-up used in thermotherapies.

Impedance Matching System: Used to prevent damage to the equipment or the patient. It matches the impedance between the RF/MW system and the applicator, allowing maximum power transmission to the treated region.

Image Guidance: Used to locate tumor and applicator position, usually Ultrasound (US), Computed Tomography (CT), or Magnetic Resonance Imaging (MRI). Image guidance helps to plan, perform, and evaluate ablation treatment.

7.3.2 Main application and features according to the body region

Hyperthermia is classified as local, regional, whole-body, and magnetic fluid hyperthermia. Moreover, local hyperthermia is divided into external, interstitial, and endocavitary. Regional hyperthermia is classified as deep, regional perfusion, and continuous hyperthermic peritoneal perfusion. Table 7.3 presents the main clinical applications of thermotherapies according to the heating approach and tumor location, including frequency, type of applicator, and Clinical Protocol (hyperthermia and thermal ablation).

7.3.3 Treatment quality and clinical studies

Treatment quality helps to work out technology, performance, and organization problems during thermotherapy treatments. It includes measurements (thermal-dose effect) to control and apply therapies in quality assurance for patient safety. Quality indicators have been implemented to enable action plans to continue and improve the quality of these treatments. Some

Table 7.3. Main clinical applications of thermotherapies according to the heating approach and tumor location.

Heating approach	**Tumor location**	**Frequency/applicator type**	**Body region/ clinical application**	**Clinical protocol**
Hyperthermia: It is used to generate temperatures around 40–44°C for treatment times of 30–60 min.				
Local hyperthermia	Very high levels of heat are delivered in small areas of cancer cells or tumors (5–6 cm) located either superficially (≤ 4 cm) or in a body cavity like the rectum or oesophagus.			
Superficial (RF/MW)	Superficial tumors (< 4 cm deep).	Superficial and intraluminal applicators can be used to apply microwaves or radio waves. 434-MHz/waveguide applicator (de Bruijne et al., 2007). 433 MHz/HYPERcollar applicator (Rijnen et al., 2015). 8.2 MHz Thermatron applicator (Huilgol et al., 2010).	Breast tumors, head & neck cancers.	QA guidelines for the application of superficial hyperthermia (Dobšíček Trefná et al., 2017)
Intracavital/ Intraluminal	Tumors within or near body cavities.	434 MHz/Patch antennas (Rajendran et al., 2021). Endotract electrodes are inserted into the lumens of the human body to heat the tumor.	Gastrointestinal tumors (esophagus, rectum). Gynecological tumors (vagina, cervix, and uterus), Genitourinary tumors (prostate, bladder). Pulmonary tumors (trachea, bronchus) (Habash, 2006).	Locoregional deep hyperthermia of pelvic tumors (Fatehi et al., 2007).
Interstitial	Used to treat tumors deep within the body.	Arrays of needle-shaped applicators implanted directly or applicators inserted within catheters that are implanted into the target volume are used (Dobšíček Trefná et al., 2019a). Most common applicator: BSD-500 – Pyrexar (915 MHz) (*Pyrexar Medical Inc. BSD-500 Features*, n.d.).	Brain, breast, head and neck, cervix, prostate, and bladder.	QA guidelines for the application of interstitial hyperthermia (Dobšíček Trefná et al., 2019b)

Table 7.3 contd. ...

Heating approach	Tumor location	Frequency/applicator type	Body region/ clinical application	Clinical protocol
Regional hyperthermia	Low levels of heat to treat larger regions are used (organs, limbs, or hollow spaces inside the body) (Bruggmoser, 2012; Bruggmoser et al., 2011).			
Abdominal/ Pelvic/Limbs	Deep-seated tumors. Locally advanced tumors.	Applicators usually contain coherent annular phased arrays placed around the patient (Behrouzkia et al., 2016). Either MW or RF energy has been used. Sigma-60 applicator (Canters et al., 2013).	Cervical, rectal, bladder, ovarian, and prostate cancer. Soft tissue sarcomas. Peritoneal carcinomatosis.	Guidelines for the application of regional deep hyperthermia by Atzelsberg research group of the German Cancer Society (Bruggmoser, 2012, 2011)
Whole body hyperthermia	The body temperature is raised.			
	Disseminated tumors/ metastasis.	Hot blankets, immersion in a warm water bath, inductive loops, thermal chambers, and radiant heat with ultraviolet radiation are used to warm the patient body.	Malignant melanoma Recurrent soft tissue sarcomas Ovarian, colorectal, adenocarcinoma, and lung cancer (Lassche et al., 2019).	Clinical trials of whole-body hyperthermia (Jia and Liu, 2010)
Ablation: it is used to generate temperatures > 60°C for short periods of treatment time.				
Radiofrequency (RF)	Tumors with a defined volume located either where surgery cannot be done (liver) or if the preservation of the organ function is desired (prostate and uterus).	Needle electrodes are used to deliver a rapidly alternating current. Monopolar system: a needle with an electrode is located in the tumor and the counter electrode is located over the skin (large adhesive electrode). Bipolar systems: both electrodes are in the needle. Frequencies between 200 kHz and 1200 kHz are used (Ziegle et al., 2018).	Tumors in the liver, kidney, lung, rectum, breast, prostate, and musculoskeletal system.	American Association for the Study of Liver Diseases guidelines for Hepatocellular carcinoma (Reinhardt et al., 2017)

Table 7.3 contd. ...

...Table 7.3 contd.

Heating approach	Tumor location	Frequency/applicator type	Body region/ clinical application	Clinical protocol
Microwaves (MW)	Tumors with a defined volume located either where surgery cannot be done or if the preservation of the organ function is desired.	Antennas A microwave antenna is inserted directly into the tumor. In this case, an EM microwave is emitted from the antenna (Ziegle, 2018) (Trujillo-Romero et al., 2021).	Tumors in the liver, kidney, lung, rectum, breast, prostate, and musculoskeletal system.	Clinical guidelines for Microwave ablation of bone tumors in extremities (Zheng et al., 2020)

guidelines must be considered to evaluate the risk or benefit of alternative therapy, like hyperthermia or ablation. The QA objective is to ensure detailed documentation about the thermotherapy treatment. The European Society for Hyperthermic Oncology (ESHO)(www.esho.info/) and the American Association of Physicists in Medicine (AAPM) (www.aapm.org) had made a big contribution to the medical and scientific community by defining different guidelines to effectively provide thermotherapy treatments. QA guidelines are necessary to achieve uniform temperature application in clinical trials, facilitate the correct assessment if a tumor can be heated with the specific device or not, and, therefore, a determination if a patient is eligible to participate in a clinical trial (Dobšíček Trefná, 2017). The protocols were approved by the regional ethical association, i.e., treatment evaluation for superficial hyperthermia at the Erasmus University Medical Center (de Bruijne et al., 2011), and some of them are registered at clinicaltrials.gov. The main aspects to be evaluated are population, anesthesia, image guidance, equipment characterization, treatment planning, treatment, safety aspects, treatment documentation, and instrumentation for quality assurance (Lachenmayer et al., 2019).

The most important factor in ablation treatment is the volume of injured tissue achieved to validate the efficiency of the procedure. Therefore, many researchers have focused their efforts on developing a tool to determine the dimensions of the damaged tissue. Chen ct al., established a mathematical model that allows the determination of the number of ablation spheres to ensure that the tumor was covered in its entirety including the safety zone (Chen et al., 2004).

Different intervention procedures are implemented according to the application and the therapy validation. Lachenmayer (Lachenmayer, 2019) presented a protocol consisting of four main steps: (1) planning stage, microwave ablation probe direction planning, and simulation: once the tumor was located by CT guidance, the ablation area was simulated as a function of the applied energy and time. (2) navigation stage: the pointing device of the microwave antenna ensures stability and guides the antenna insertion, then the insertion antenna length was calculated using the computer. (3) probe validation and ablation stage: the antenna position is verified by CT imaging, and the treatment is applied. (4) ablation site validation: safety margin was determined by superposition of pre and post-CT images, if necessary, planned another ablation treatment. Table 7.3 shows the treatment quality and QA guidelines used for different clinical applications.

7.4 Computational modeling and treatment planning

The main goal of computational modeling is to predict the applicator performance by obtaining power deposition and temperature distributions reached in the tumor and healthy surrounding tissues. Thermal therapy computational modeling can be approached in different ways depending on our necessities (3 and 2 dimensions, 2D-axisymmetric). Computational modeling is mainly used for the evaluation of dosimetry, i.e., the accurate measurement of doses (EM and thermal). EM dosimetry is the calculation of the EM radiation absorbed by the tissues. Dosimetry requires the measurement/calculation of either induced current density or specific absorption rate (SAR), and the patterns in phantoms, animals, and humans, exposed to EM fields (Habash et al., 2007). Dosimetry can be used to establish guidelines to evaluate the performance of the applicators under the same clinical conditions, to predict the treatment conditions for a specific patient as well as to validate thermal models.

7.4.1 Electromagnetic models (MW and RF)

There are several ways to model the thermal therapies outcome by using partial differential equations (PDE). To predict electric fields, Specific Absorption Rate (SAR) and temperature distribution models based on Maxwell and Bioheat equations must be solved by using numerical methods such as the Finite Element Method (FEM) and Finite-Difference methods (FDM). In these computational models, parameters such as tissue properties, applicator type, boundary conditions, input power, treatment time and mesh (domain discretization) must be defined to obtain an accurate prediction of the applicator performance.

Normally the Specific Absorption Rate (SAR) is computed using EM computational models; it describes the EM field placed by a unit of mass in the tissue (W/kg). The SAR produced by the applicator is defined by Equation (7.1) (Trujillo-Romero et al. 2021)

$$SAR = \frac{\sigma}{2\rho} |E|^2, \tag{7.1}$$

where σ (S/m) and ρ (kg/m³) are tissue conductivity and density, respectively, and *E* is the electric field generated by the applicator. Power deposition is frequently calculated in a steady state. A current source (J_{imp}) that excited the applicator and irradiated an EM field modified by the applicator body can be considered to develop the FEM model and predict the applicator's outcome. Maxwell's equations (Equations 7.2–7.5), govern the applicator performance (Jin et al., 2009).

$$\nabla \times E = -j\omega \overleftrightarrow{\mu} \cdot H - M_{imp} \tag{7.2}$$

$$\nabla \times H = -j\omega \overleftrightarrow{\varepsilon} \cdot E - J_{imp} \tag{7.3}$$

$$\nabla \cdot (\overleftrightarrow{\varepsilon} \cdot E) = -\frac{1}{j\omega} \nabla \cdot J_{imp} \tag{7.4}$$

$$\nabla \cdot (\overleftrightarrow{\mu} \cdot H) = -\frac{1}{j\omega} \nabla \cdot M_{imp} \tag{7.5}$$

where *E* and *H* are the electric and magnetic field intensities, respectively, M_{imp} is the magnetic current density, $\overleftrightarrow{\varepsilon}$ is the permittivity, and $\overleftrightarrow{\mu}$ is the permeability of the irradiated tissues. Therefore, to estimate the effect of the EM fields irradiated to the tissue by the antenna, Maxwell's equations must be solved. Nevertheless, tissue dielectric properties and boundary conditions must be known (Prakash, 2010).

Equation (7.6) describes the boundary condition and must be applied to solve the above equations (Jin, 2009).

$$n \times E = 0 \quad r \in S_{PEC} \tag{7.6}$$

where S_{PEC} describes the surface of the body applicator as a Perfect Electric Conductor (PEC). If the EM propagation is limited to a finite space (S_0), a new boundary condition is considered. It must be transparent, then the EM field can travel without any reflection. To emulate this behavior, Equation (7.7) describes the first-order absorbing boundary condition, which is used.

$$\hat{n} \times \nabla \times (EH) + jk_0 \hat{n} \times \hat{n} \times (EH) \approx 0 \quad r \in S_0 \tag{7.7}$$

where $\hat{\boldsymbol{n}}$ is the unit vector normal to $S_{0,}$ pointed towards outer space. S_0 must be fixed minimally at half-wavelength from the applicator to get a feasible model.

Moreover, the surfaces must be designed with a low reflecting condition, i.e., the boundary does not disturb the EM field distribution. Therefore, power deposition in tissue can be predicted by using the Finite Element Method (FEM) or the Finite-Difference Time-Domain method (FDTD). Moreover, the integral form of Maxwell equations can be used to obtain the power deposition by using the conjugate gradient fast Fourier transform.

7.4.2 Thermal models

Due to the inaccuracies of thermal tissue properties, the main problem to be solved in thermal models is the accuracy of the predicted temperature distributions. Thermal tissue properties not only vary between tissues, but also among patients, time, and temperature. To implement an accurate model, very complex models must be used; however, this increases the necessity of computational resources as well as the solution time required to get results. Therefore, a continuum model has been widely used to implement thermal models. The Bioheat equation has been commonly used to predict temperature distributions (Pennes, 1948). This equation is used to model the impact of perfusion as an isotropic heat sink, as described by Equation (7.8).

$$\rho C\frac{\partial T}{\partial t} - \nabla \cdot (k\nabla T) + \rho Q + \rho SAR - \rho_b C_b \rho W_b (T - T_b) \tag{7.8}$$

where C (J/kg/K) is the heat capacity, ρ (kg/m^3) density, k (W/m/K) thermal conductivity, ρ_b (kg/m^3) blood density, C_b (J/kg/K) blood heat capacity, W_b (kg/m^3/s) blood perfusion, T_b (K) blood temperature, Q (W/kg) rate of heat generated by metabolism and SAR (W/kg) the specific absorption rate. As can be observed, SAR is used to predict the temperature distributions generated by the applicators. Normally, to reduce complexity SAR and temperature distributions are independently calculated. The Bioheat equation has been validated to model a temperature pattern in tissue with a healthy microvasculature and blood flowing throughout vessels with orientations distributed isotropically. However, for blood vessels with a diameter higher than 0.2 mm the effects of the direction of blood flow must be considered (Paulides et al., 2013). Although the Bioheat equation has been the most used, throughout the years, several modifications/new equations have been presented to represent the blood vessels in the human body in a better way. Consequently, more accurate predictions of temperature distributions in the human body due to an external heat source can be obtained. Thermal models are divided into continuous models where the perfusion is described by a continuum flow; while the discrete vasculature models the blood vessels discretely to predict local inhomogeneities in temperature. Table 7.4 summarizes the different equations proposed to generate the thermal models.

Table 7.4. Main thermal models developed by different research groups.

Model	Description
Pennes (Pennes, 1948)	Models perfusion as an isotropic heat sink. The energy exchange among blood vessels and tissue is across the capillary walls (low blood velocity).
Mitchell (Mitchell et al., 1968)	Includes the effect of countercurrent heat exchange between an artery and an adjacent vein. Heat conduction is not considered.
Chen-Holmes (Chen et al., 1980)	Evaluate the thermal equilibration length of individual vessels. It is related to vessel dimensions, local blood velocity, and heat transfer coefficient. Thermal equilibrium of blood with tissue occurs in vessels (0.2–05 mm diameters) instead of in capillaries.
Weinbaum-Jiji-Lemons (Weinbaum et al., 1984)	The heat transfer produced by the blood perfusion is related to an incomplete countercurrent heat exchange among a pair of arteries and veins.
Mooibroek-Lagendijk (Mooibroek et al., 1991)	Calculates the time-dependent temperature distributions in inhomogeneous vascularized tissue. Simulates a vessel tree using elementary cubic nodes.
Crezee-Lagendijk (Crezee et al., 1992)	Large vessels are modeled; moreover, vessel size and velocity, and direction of blood flow are considered.
Kotte-Leeuwen-Lagendijk (Kotte et al., 1999)	The discrete Vasculature model (DIVA) describes a genuine vessel tree, with the possibility to create vessels, shifting, rotating, and independence of tissue voxel resolution. It considers the directional heat flow in the tissue.
Hirata (Hirata et al., 2008)	Convective and radiative heat losses are included in the heat transfer coefficient.

7.4.3 Treatment planning

Treatment planning is a powerful tool that helps to predict the treatment parameters to maximize its quality, i.e., by using it, treatment parameters for a specific patient, such as type, number, and position of the applicator, input power and treatment time can be set. This tool is based on computational modeling; during the last year, several advances in electromagnetic, and thermal simulation techniques had been reached; therefore, the accuracy of the treatment planning systems has been improved. Since the European Society for Hyperthermic Oncology (ESHO) includes treatment planning in the QA guidelines for hyperthermia treatments, it has been playing an important role (Bruggmoser et al., 2012). The flow chart of a thermotherapy treatment system is depicted in Figure 7.3 and described below.

Anatomical model: The body region to be treated must be identified; afterward, the 3D human model is generated. Different 3D human models that can be used to perform treatment planning have been developed (Christ et al., 2010). However, to improve the accuracy of the treatment, a patient-specific

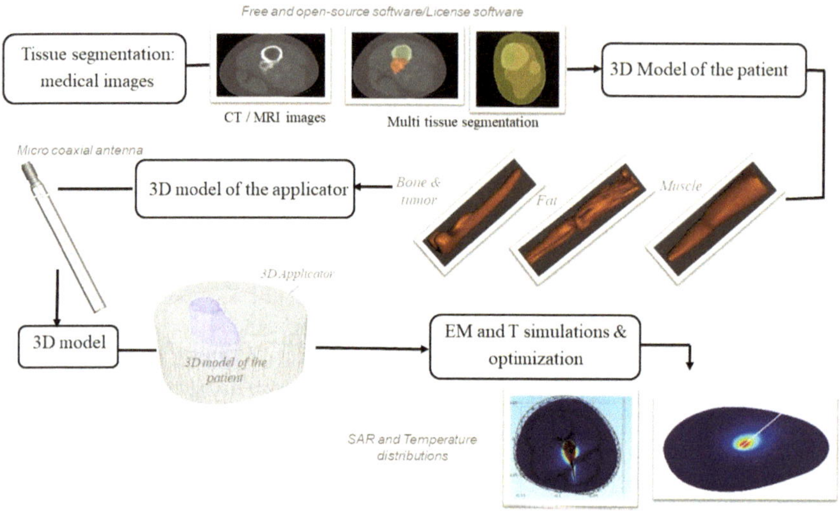

Figure 7.3. Flowchart to perform treatment planning for thermotherapies.

model must be generated by using medical images, either magnetic resonance images (MRI) or computed tomography (Trujillo-Romero et al., 2014). The first step consists of tissue segmentation. This can be done by using a free license or commercial software (Flores Cuautle et al., 2021). However, the implementation of time-efficient and easy-to-use protocols to develop these models is still a challenge; mainly because of the requirements of the clinical application (Paulides, 2013). Therefore, since the tissue segmentation is usually manual, the generation of the patient 3D model is highly time-consuming. Recently, advances in machine learning have been helping in the improvement of the techniques to generate semi-automatic/automatic tissue segmentation (Kashyap et al., 2022).

3D model of the applicator: A 3D model of the applicator that works as a heat external source is generated. The 3D model must match the real dimensions of the applicator but most importantly, the numerical model must represent the performance of the applicator; therefore, adequate discretization must be implemented.

Electromagnetic simulations (power absorption): Several methods based on numerical analysis had been developed; therefore, multi-physics phenomena such as thermotherapies can be solved (Kok et al., 2015). Once the 3D patient model and the applicator are created, both are combined by emulating the position of the patient and the antenna location as in the real treatment to calculate the EM field distribution. Moreover, parameters such as work frequency, dielectric tissue properties, and boundary conditions must be defined. Recent developments in computational sciences allow the generation of more accurate and realistic 3D models. The numerical solution is approached by solving Maxwell's equations as described in Section 7.3.1.

Thermal simulations: 3D temperature distributions are calculated by considering the output of the EM simulations as input. Thermal simulations are governed by a heat transfer equation that includes the external heat source (RF/MW applicator), as described in Section 7.3.2. Other parameters such as thermal conductivity, perfusion and metabolism are considered. Table 7.4 describes different thermal models developed to improve the thermal models. As described in Chapter 11, thermal tissue properties play an important role in the achievement of accurate thermal simulations, mainly because of the uncertainties of tissue properties and their heterogeneity. Therefore, getting an accurate thermal model in treatment planning is still a challenge.

Optimization of electromagnetic and thermal simulations: Different optimization techniques have been used to improve the heating over the tumor and to avoid hot spots over undesired regions. The reported SAR optimization process is based on techniques/algorithms such as the eigenvalue (Kotte et al, 1999) and the particle swarm optimization (PSO) (Cappiello et al., 2017) (Elkayal et al., 2021). EM optimization is helpful due to its direct correlation with the obtained temperature distributions. The goal of thermal distribution optimization is to maximize the tumor temperature, while the surrounding healthy tissue keeps its normal temperature. To reduce complexity and computational time, the optimization of the temperature distributions is based on the Bioheat equation.

7.5 Conclusion

Research in radiofrequency and microwave thermal therapies has shown beneficial results in early-stage cancer patients, with fewer side effects than conventional treatments. These therapies require validation, planning, and clinical protocols according to the organ and region to be treated in patients. Different tools such as tissue emulators and computational models can be used to determine the effectiveness of treatment in terms of temperature distribution, but there is still much work to be done to improve validation and create tools that allow healthcare experts to plan, simulate and optimize treatments to ensure quality treatment and patient safety. Around the world, many research groups are working to solve these issues as well as in the implementation of quality indicators to verify that tumor tissue has been eradicated.

Acronym/Abbreviation

AAMP	American Association of Physicists in Medicine
AMPS	Ammonium Persulfate
CNC	Cellulose Nanocrystals
CT	Computed Tomography
DAC	Dialdehyde CNC
DIVA	Discrete Vasculature Model
E	Electric Field
EM	Electromagnetic
ESHO	The European Society for Hyperthermic Oncology
FDM	Finite-Difference Methods
FDTD	Finite-Difference Time-Domain
FEM	Finite Element Method
H	Magnetic Field
ISM	Industrial, Scientific and Medical
MBAA	N,N'-methylene-bis-acrylamide
MRI	Magnetic Resonance Imaging
MW	Microwaves
NaCl	Sodium chloride
NaN_3	Sodium azide
QA	Quality Assurance
RF	Radiofrequency
SAR	Specific Absorption Rate
PDE	Partial Differential Equations
PEC	Perfect Electric Conductor
PSO	Particle Swarm Optimization
TEMED	Tetra-metyl-ethylenediamine
US	Ultrasound
ε_r	Relative Permittivity
σ	Electrical Conductivity
ρ	Density

Reference list

Behrouzkia, Z., Joveini, Z., Keshavarzi, B., Eyvazzadeh, N., Aghdam, R.Z. et al. (2016). Hyperthermia: How can it be used? *Oman Medical Journal*, 31(2): 89–97. https://doi.org/10.5001/omj.2016.19.

Bruggmoser, G. (2012). Some aspects of quality management in deep regional hyperthermia. *International Journal of Hyperthermia*, 28(6): 562–569. https://doi.org/10.3109/02656736.2012.714049.

Bruggmoser, G., Bauchowitz, S., Canters, R., Crezee, H., Ehmann, M. et al. (2011). Quality assurance for clinical studies in regional deep hyperthermia. *Strahlentherapie Und Onkologie*, 187(10): 605–610. https://doi.org/10.1007/s00066-011-1145-x.

Bruggmoser, G., Bauchowitz, S., Canters, R., Crezee, H., Ehmann, M. et al. (2012). Leitlinie für die klinische Applikation, die Dokumentation und die Analyse klinischer Studien bei der regionalen Tiefenhyperthermie: Qualitätsmanagement bei der regionalen Tiefenhyperthermie. *Strahlentherapie Und Onkologie*, 188(SUPPL. 2): 198–211. https://doi.org/10.1007/s00066-012-0176-2.

Buscarini, E., Savoia, A., Brambilla, G., Menozzi, F., Reduzzi, L. et al. (2005). Radiofrequency thermal ablation of liver tumors. *European Radiology*, 15(5): 884–894. https://doi.org/10.1007/s00330-005-2652-x.

Canters, R.aM., Paulides, M.M., Franckena, M., Mens, J.W., van Rhoon, G.C. et al. (2013). Benefit of replacing the Sigma-60 by the Sigma-Eye applicator. A Monte Carlo-based uncertainty analysis. *Strahlentherapie Und Onkologie: Organ Der Deutschen Röntgengesellschaft ... [et Al]*, 189(1): 74–80. https://doi.org/10.1007/s00066-012-0241-x.

Cappiello, G., McGinley, B., Elahi, M.A., Drizdal, T., Paulides, M.M. et al. (2017). Differential evolution optimization of the SAR distribution for head and neck hyperthermia. *IEEE Transactions on Biomedical Engineering*, 64(8): 1875–1885. https://doi.org/10.1109/TBME.2016.2627941.

Chen, M.H., Yang, W., Yan, K., Zou, M.W., Solbiati, L. et al. (2004). Large liver tumors: Protocol for radiofrequency ablation and its clinical application in 110 patients—Mathematic model, overlapping mode, and electrode placement process. *Radiology*, 232(1): 260–271. https://doi.org/10.1148/radiol.2321030821.

Chen, M.M. and Holmes, K.R. (1980). Microvascular contribution in tissue heat transfer. *Annals of the New York Academy of Sciences*, 335(1): 137–150. https://doi.org/10.1111/J.1749-6632.1980.TB50742.X.

Chen, W.J., Wang, Q. and Kim, C.Y. (2020). Gel pahntom models for radiofrequency and microwave ablation of the liver. *Dig. Dis. Interv.*, 4(3): 303–310. https://doi.org/10.1055/s-0040-1716737.Gel.

Chou, C.-K., Chen, G.-W., Guy, A.W. and Luk, K.H. (1984). Formulas for preparing phantom muscle tissue at various radiofrequencies. *Bioelectromagnetics*, 5(4): 435–441. https://doi.org/10.1002/bem.2250050408.

Christ, A., Kainz, W., Hahn, E.G., Honegger, K., Zefferer, M. et al. (2010). The virtual family-development of surface-based anatomical models of two adults and two children for dosimetric simulations. *Physics in Medicine and Biology*, 55(2). https://doi.org/10.1088/0031-9155/55/2/N01.

Crezee, J. and Lagendijk, J.J.W. (1992). Temperature uniformity during hyperthermia: The impact of large vessels. *Physics in Medicine and Biology*, 37(6): 1321–1337. https://doi.org/10.1088/0031-9155/37/6/009.

Dabbagh, A., Abdullah, B.J.J., Ramasindarum, C. and Abu Kasim, N.H. (2014). Tissue-mimicking gel phantoms for thermal therapy studies. *Ultrasonic Imaging*, 36(4): 291–316. https://doi.org/10.1177/0161734614526372.

de Bruijne, M., Van der Zee, J., Ameziane, A. and Van Rhoon, G.C. (2011). Quality control of superficial hyperthermia by treatment evaluation. *International Journal of Hyperthermia*, 27(3): 199–213. https://doi.org/10.3109/02656736.2010.525226.

de Bruijne, M., Wielheesen, D.H.M., van der Zee, J., Chavannes, N., van Rhoon, G.C. et al. (2007). Benefits of superficial hyperthermia treatment planning: Five case studies. *International Journal of Hyperthermia*, *23*(5), 417–429. https://doi.org/10.1080/02656730701502077.

Dobšíček Trefná, H., Crezee, J., Schmidt, M., Marder, D., Lamprecht, U. et al. (2017). Quality assurance guidelines for superficial hyperthermia clinical trials. *Strahlentherapie Und Onkologie*, 193(5): 351–366. https://doi.org/10.1007/s00066-017-1106-0.

Dobšíček Trefná, H., Llacer Navarro, S., Lorentzon, F., Nypelö, T., Ström, A. et al. (2021). Fat tissue equivalent phantoms for microwave applications by reinforcing gelatin with nanocellulose. *Biomedical Physics and Engineering Express*, 7(6). https://doi.org/10.1088/2057-1976/ac2634.

Dobšíček Trefná, H., Schmidt, M., van Rhoon, G.C., Kok, H.P., Gordeyev, S.S. et al. (2019a). Quality assurance guidelines for interstitial hyperthermia. *International Journal of Hyperthermia*, 36(1): 277–294. https://doi.org/10.1080/02656736.2018.1564155.

Dobšíček Trefná, H., Schmidt, M., van Rhoon, G.C., Kok, H.P., Gordeyev, S.S. et al. (2019b). Quality assurance guidelines for interstitial hyperthermia. *International Journal of Hyperthermia*, 36(1): 277–294. https://doi.org/10.1080/02656736.2018.1564155.

Elkayal, H.A. and Ismail, N.E. (2021). Efficient focusing of microwave hyperthermia for small deep-seated breast tumors treatment using particle swarm optimization. *Computer Methods in Biomechanics and Biomedical Engineering*, 24(9): 985–994. https://doi.org/10.1080/10255842.2020.1863379.

Fatehi, D., van der Zee, J., Notenboom, A. and van Rhoon, G.C. (2007). Comparison of intratumor and intraluminal temperatures during locoregional deep hyperthermia of pelvic tumors. *Strahlentherapie Und Onkologie: Organ Der Deutschen Röntgengesellschaft ... [et Al]*, 183(9): 479–486. https://doi.org/10.1007/s00066-007-1768-0.

Flores Cuautle, J., de, J.A., Martínez Valdez, R., Rodríguez Carmona, E.A., Posada Gomez, R., Trujillo Romero, C.J. et al. (2021). A computational evaluation of the temperature distribution generated by thermal splints designed to treat knee pain. *Journal of Thermal Biology*, 97: 102868. https://doi.org/10.1016/j.jtherbio.2021.102868.

Habash, R.W.Y., Bansal, R., Krewski, D. and Alhafid, H.T. (2006). Thermal therapy, part 2: Hyperthermia techniques. *Critical Reviews in Biomedical Engineering*, 34(6): 491–542. https://doi.org/4644b8e8318701ed,20ba3f104d3b2090 [pii].

Habash, R.W.Y., Bansal, R., Krewski, D. and Alhafid, H.T. (2007). Thermal therapy, Part IV: Electromagnetic and thermal dosimetry. *Critical Reviews in Biomedical Engineering*, 35(1-2): 123–182. https://doi.org/10.1615/CritRevBiomedEng.v35.i1-2.30.

Hirata, A., Asano, T. and Fujiwara, O. (2008). FDTD analysis of body-core temperature elevation in children and adults for whole-body exposure. *Physics in Medicine and Biology*, 53(18): 5223–5238. https://doi.org/10.1088/0031-9155/53/18/025.

Huilgol, N.G., Gupta, S. and Sridhar, C.R. (2010). Hyperthermia with radiation in the treatment of locally advanced head and neck cancer: A report of randomized trial. *Journal of Cancer Research and Therapeutics*, 6(4): 492–496. https://doi.org/10.4103/0973-1482.77101.

Ito, K., Furuya, K., Okano, Y. and Hamada, L. (2001). Development and characteristics of a biological tissue-equivalent phantom for microwaves. *Electronics and Communications in Japan, Part I: Communications (English Translation of Denshi Tsushin Gakkai Ronbunshi)*, 84(4): 67–77. https://doi.org/10.1002/1520-6424(200104)84:4<67::AID-ECJA8>3.0.CO;2-D.

Jia, D. and Liu, J. (2010). Current devices for hyperthermia therapy. *Expert Review of Medical Devices*, 7(3): 407–423.

Jin, J.-M. and Riley, D.J. (2009). *Finite Element Analysis of Antennas and Arrays*. Jhon Wiley & Sons, Inc.

Kashyap, A.K.K.P.S.A., Babu, E.S., Sathyamurthy, A., Ram, T.S., Arunachalam, K. et al. (2022). *Automated 3D Patient Model Generation using ML Technique for Hyperthermia Treatment Planning*, 1–3. https://doi.org/10.1109/TENSYMP54529.2022.9864509.

Kavousi, M., Saadatmand, E. and Alam, N.R. (2019). Physical parameters measurement of breast equivalent phantom for clinical studies in radiofrequency hyperthermia. *Frontiers in Biomedical Technologies*, 6(1): 28–34. https://doi.org/10.18502/fbt.v6i1.1100.

Kok, H.P., Wust, P., Stauffer, P.R., Bardati, F., van Rhoon, G.C. et al. (2015). Current state of the art of regional hyperthermia treatment planning: A review. *Radiation Oncology*, 10(1). https://doi.org/10.1186/s13014-015-0503-8.

Kotte, A., Van Leeuwen, G. and Lagendijk, J. (1999). Modeling the thermal impact of a discrete vessel tree. *Physics in Medicine and Biology*, 44: 57–74. https://doi.org/10.1088/0031-9155/41/5/004.

Lachenmayer, A., Tinguely, P., Maurer, M.H., Frehner, L., Knöpfli, M. et al. (2019). Stereotactic image-guided microwave ablation of hepatocellular carcinoma using a computer-assisted navigation system. *Liver International*, 39(10): 1975–1985. https://doi.org/10.1111/liv.14187.

Lassche, G., Crezee, J. and Van Herpen, C.M.L. (2019). Whole-body hyperthermia in combination with systemic therapy in advanced solid malignancies. *Critical Reviews in Oncology/Hematology*, 139: 67–74. https://doi.org/10.1016/J.CRITREVONC.2019.04.023.

Laufer, Y. and Dar, G. (2012). Effectiveness of thermal and athermal short-wave diathermy for the management of knee osteoarthritis: A systematic review and meta-analysis. *Osteoarthritis and Cartilage*, 20(9): 957–966. https://doi.org/10.1016/J.JOCA.2012.05.005.

Mitchell, J.W. and Myers, G.E. (1968). An analytical model of the counter-current heat exchange phenomena. *Biophysical Journal*, 8(8): 897–911. https://doi.org/10.1016/S0006-3495(68)86527-0.

Mooibroek, J. and Lagendijk, J.J.W. (1991). A fast and simple algorithm for the calculation of convective heat transfer by large vessels in three-dimensional inhomogeneous tissues. *IEEE Transactions on Biomedical Engineering*, 38(5): 490–501. https://doi.org/10.1109/10.81569.

Ortega-Palacios, R., Trujillo-Romero, C.J., Cepeda-Rubio, M.F.J., Leija, L., Hernández, A.V. et al. (2020). Heat transfer study in breast tumor phantom during microwave ablation: Modeling and experimental results for three different antennas. *Electronics (Switzerland)*, 9(3). https://doi.org/10.3390/electronics9030535.

Paulides, M.M., Stauffer, P.R., Neufeld, E., Maccarini, P.F., Kyriakou, A. et al. (2013). Simulation techniques in hyperthermia treatment planning. *International Journal of Hyperthermia*, 29(4): 346–357. https://doi.org/10.3109/02656736.2013.790092.

Pennes, H.H. (1948). Analysis of tissue and arterial blood temperatures in the resting human forearm. *Journal of Applied Physiology*, 85: 5–34. https://doi.org/9714612.

Prakash, P. (2010). Theoretical modeling for hepatic microwave ablation. *Open Biomed. Eng.*, 4: 27–38.

Pyrexar Medical Inc. BSD-500 Features. (n.d.). http://www.pyrexar.com/hyperthermia/bsd-500.

Rajendran, T. and Arunachalam, K. (2021). Microwave intracavitary applicator using 434 MHz conformal patch antennas for hyperthermia treatment of gynaecological cancers. *IET Microwaves, Antennas and Propagation*, 15(9): 1117–1126. https://doi.org/10.1049/mia2.12122.

Ramírez-Guzmán, T.J., Trujillo-Romero, C.J., Martínez-Valdez, R., Leija-Salas, L., Vera-Hernández, A. et al. (2021). Thermal evaluation of a micro-coaxial antenna set to treat bone tumors: design, parametric FEM modeling and evaluation in multilayer phantom and *ex vivo* porcine tissue. *Electronics*, 10(18): 2289. https://doi.org/10.3390/ELECTRONICS10182289.

Reinhardt, M., Brandmaier, P., Seider, D., Kolesnik, M., Jenniskens, S. et al. (2017). A prospective development study of software-guided radio-frequency ablation of primary and secondary liver tumors: Clinical intervention modelling, planning and proof for ablation cancer treatment (ClinicIMPPACT). *Contemporary Clinical Trials Communications*, 8(March): 25–32. https://doi.org/10.1016/j.conctc.2017.08.004.

Riadh, W.Y.H., Rajeev, B., Daniel, K. and Hafid, T.A. (2007). Thermal therapy, Part III: Ablation techniques. *Critical Reviews in Biomedical Engineering*, 35(1-2): 37–121. https://doi.org/5d41342c414d85d4,373c95d45dc0d376 [pii].

Rijnen, Z., Togni, P., Roskam, R., Van De Geer, S.G., Goossens, R.H.M. et al. (2015). Quality and comfort in head and neck hyperthermia: A redesign according to clinical experience and simulation studies. *International Journal of Hyperthermia*, 31(8): 823–830. https://doi.org/10.3109/02656736.2015.1076893.

Sprenkle, P.C., Mirabile, G., Durak, E., Edelstein, A., Gupta, M. et al. (2010). The effect of argon gas pressure on ice ball size and rate of formation. *Journal of Endourology*, 24(9): 1503–1507. https://doi.org/10.1089/end.2009.0587.

Surowiec, A., Shrivastava, P.N., Astrahan, M. and Petrovich, Z. (1992). Utilization of a multilayer polyacrylamide phantom for evaluation of hyperthermia applicators. *International Journal of Hyperthermia*, 8(6): 795–807. https://doi.org/10.3109/02656739209005027.

Trujillo-Romero, C.J., Leija-Salas, L., Vera-Hernández, A., Rico-Martínez, G., Gutiérrez-Martínez, J. et al. (2021). Double slot antenna for microwave thermal ablation to treat bone tumors: Modeling and experimental evaluation. *Electronics*, 10(7:761. https://doi.org/10.3390/electronics10070761.

Trujillo-Romero, C.J., Paulides, M.M., Drizdal, T. and van Rhoon, G.C. (2014). Impact of silicone and metal port-a-cath implants on superficial hyperthermia treatment quality. *International Journal of Hyperthermia*, 6736(September): 1–8. https://doi.org/10.3109/02656736.2014.985748.

Weinbaum, S., Jiji, L.M. and Lemons, D.E. (1984). Theory and experiment for the effect of vascular microstructure on surface tissue heat transfer—Part I: Anatomical foundation and model conceptualization. *Journal of Biomechanical Engineering*, 106(4): 321–330. https://doi.org/10.1115/1.3138501.

Zheng, K., Yu, X., Hu, Y., Zhang, Y., Wang, Z. et al. (2020). Clinical guideline for microwave ablation of bone tumors in extremities. *Orthopaedic Surgery*, 12(4): 1036–1044. https://doi.org/10.1111/os.12749.

Ziegle, J., Audigier, C., Krug, J., Ali, G., Kim, Y. et al. (2018). RF-ablation pattern shaping employing switching channels of dual bipolar needle electrodes: *ex vivo* results. *International Journal of Computer Assisted Radiology and Surgery*, 13(6): 905–916. https://doi.org/10.1007/s11548-018-1769-8.

Chapter 8

Thermotherapies based on Ultrasound

Raquel Martínez-Valdez[1,*] and *Juan Carlos García-López*[2]

Nowadays, with the increased evolution of technology, ultrasonic clinical usage has reemerged and new treatments based on thermal therapies have been proposed to have a different approach to cancer tumors located in soft tissues, bones, and brain disorders. Some proposals are being evaluated in tissue-mimicking materials, *ex vivo*, or may be under clinical trial testing; in a better case scenario, thermal therapy devices are already approved, commercialized and used in hospital facilities to treat malignant tumors, such as in the case of prostate cancer. In order to cause tumor necrosis, ultrasonic sources generate a high acoustic intensity that interacts with the target tissue producing a temperature increase of over 56°C that provokes irreversible cell damage due to thermal ablation. The ultrasonic pattern depends on the transducer geometry which is determined by the specific clinical application; the penetration depth is subject to the source operating frequency and ultrasonic attenuation of the beam while propagating through different tissue layers. Thermal ablation ultrasound-based devices can be classified as extracorporeal, intracavitary and interstitial due to the ultrasonic energy delivery fashion. This chapter includes a brief explanation of the physical

[1] Biomedical Engineering Program, Polytechnic University of Chiapas. Carretera Tuxtla Gutiérrez – Portillo Zaragoza km 21+500, Col. Las Brisas, Suchiapa, Chiapas, 29150, Mexico.
[2] 9 Once Distribuidora S. A. de C.V. 47 pte., Puebla City, Puebla, 72534, Mexico.
Email: juancarlosg737@gmail.com
* Corresponding author: rmartinez@ib.upchiapas.edu.mx

principles of ultrasound, ultrasonic propagation modeling, tissue-mimicking materials used in sources' validations, and current clinical applications.

8.1 Introduction

Thermal ablation based on ultrasonic energy takes advantage of the mechanical waves interaction with biological tissue to locally increase the temperature within the body over 56°C. This characteristic places ultrasonic thermal ablation as a promising alternative technique in the treatment of benign and malignant tumors. Treatment delivery is non-invasive, and in accordance with the target region, the ultrasonic device may be extracorporeal, intracavitary or interstitial. The design of the source strongly depends on the clinical applications; therefore, the geometry, operating frequency, number of piezo elements and electrical driving systems must be carefully established and tested. Furthermore, materials to fabricate the source need to be cautiously selected subject to the imaging guiding system, for example, ultrasound or magnetic resonance. This chapter presents an overall approach of ultrasound ablation from the physics generalities, computerized propagation modeling, validation in tissue-mimicking materials, type of devices and its applications, and finally to combining ultrasound with other therapies.

8.2 Physical principles of ultrasound

Ultrasound is defined as mechanical waves above the human audible frequency range (20 Hz–20 kHz). The acoustic vibrations are produced by a material sensitive to the inverse piezoelectric effect, which is commonly polarized in the thickness-extensional vibration mode to generate longitudinal wave propagation through media (Hutchins and Hayward, 1990). In clinical applications, the ultrasonic typical frequency range goes from 1 MHz to 10 MHz, and it can be used for diagnosis, physiotherapy, and thermal ablation. In accordance with the delivered acoustic intensity, clinical ultrasounds can be classified as low, mild and high intensity, respectively (Gutierrez et al., 2012; ter Haar, 1999).

8.2.1 Ultrasonic sources

Ultrasonic transducers produce mechanical waves at different points along the piezo element surface (Vives, 2008) or due to its transient response (Hutchins and Hayward, 1990). Piezo element geometry, dimensions and operating frequency play an important role on the acoustic pattern and the penetration depth (Hutchins and Hayward, 1990; ter Haar and Coussios,

2007a). The acoustic pattern is the result of the destructive and non-destructive interference between the generated waves; therefore, the ultrasonic sources are designed to produce a specific acoustic pattern defined by the clinical application such as collimated, focused, or diverging beams. Plane transducers produce a collimated beam formed by the Fresnel and the Fraunhofer zones. The Fresnel zone presents pressure variations until reaching the transducer's natural focus; then, at the Fraunhofer zone, the beam shows more uniformity as the mechanical wave travels away from the source in the z direction (Hutchins and Hayward, 1990). Focused beams can be obtained by concave transducers, acoustic lenses attached to plane transducers, phase controlled arrays and multiple piston transducers placed in a concave surface (Izadifar et al., 2020; Gail ter Haar, 2016). However, concave sources require higher electrical driving power, while array transducers need more complex driving systems (Diederich and Hynynen, 1999). Convex sources have been developed to reach difficult abdominal regions (Hill et al., 2004) during diagnosis. Figure 8.1 depicts the geometries, axial pressure distribution and acoustic pattern for a plane and concave and linear array transducers.

In thermal ablation applications, different devices and piezo element configurations are used to treat cancer in the breast, liver, spleen, kidney, uterine fibroids, benign prostatic hyperplasia and prostate cancer (Ren et al., 2007; ter Haar and Coussios, 2007a; Gail ter Haar, 1999). When the treatment region is large, it is suggested to wait from 30–60 s during exposures to avoid cumulative heat in healthy tissue (Diederich and Hynynen, 1999). Ultrasonic-based devices for thermal ablation therapy design depend on the clinical application; therefore, energy delivery may be extracorporeal, intracavitary or interstitial.

Extracorporeal devices used for tumor treatment in liver, kidney, breast cancer, uterine fibroids, spleen, bone metastasis (Hwang and Crum, 2009) and brain disorders (Almquist et al., 2016) consist of electronically controlled phase arrays of 208 to 251 elements (Hwang and Crum, 2009). This allows creating complex acoustic beams at different penetration depths, from 1 cm to 16.9 cm, in order to attain targets within the body without damaging surrounding tissues (Diederich and Hynynen, 1999). Extracorporeal devices realize high speed sweeps to radiate the entire tumor in a sequential fashion to guarantee the delivery of the same thermal dose to the target (Hokland et al., 2006). Some devices are magnetic resonance imaging compatible or ultrasonic guided, with operating frequencies between 0.9 MHz to 1.3 MHz. Water cooling systems are necessary to act as a propagating coupling medium and to regulate the temperature in the array-patient interface (Diederich and Hynynen, 1999).

Intracavitary devices are designed to reach regions near a body cavity wall, such as prostate cancer treatment via the rectum or the urethra (Chopra

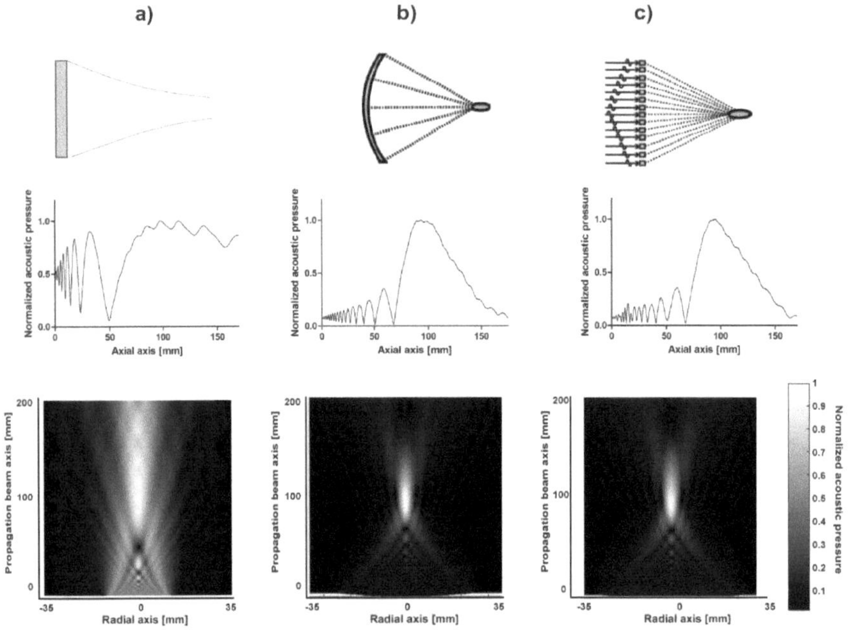

Figure 8.1. Example of typical ultrasonic sources, geometry, pressure distribution along the propagation axis and acoustic pattern; all three transducers have an operating frequency of 1 MHz and focal distance of 100 cm. The plane transducer has a radius of 12.25 mm, and the concave element and linear array have a radius of 25 mm. (a) Plane transducer, pressure distribution and acoustic pattern present pressure variations close to the transducer surface (Fresnel zone), afterwards, the acoustic pattern has more uniformity (Fraunhofer zone). (b) Concave transducer, pressure distribution and acoustic pattern show that the acoustic energy is only concentrated at the geometrical focus of the transducer. (c) Linear phase-controlled array shows that both pressure distribution and acoustic pattern present pressure variations near the array surface and a sharper focal zone.

et al., 2010; ter Haar and Coussios, 2007b). Intracavitary probes, transrectal and transvaginal, consist of phase-controlled arrays in order to control focus location and penetration depth which are mounted in a cylindrical plastic support (Diederich and Hynynen, 1999). Transrectal devices contain both therapy and imaging transducers in the same probe with operating frequencies of 3–4 MHz and 4–7.5 MHz, respectively, and reach penetration depths from 3 cm to 4 cm (Illing and Chapman, 2007). A transurethral devices consists of a multi-element planar transducer mounted on a rigid MRI compatible material with individual frequency, power and phase control, that is rotated in order to radiate the entire prostate gland (Chopra et al., 2010). The probe has a water cooling system to reduce the temperature in the transducer's surface that, also, acts like a coupling medium; and an endorectal cooling device to avoid thermal damage in the rectum (Chopra et al., 2010). Transvaginal devices proposed for cervical cancer comprise a linear array of tubular transducers that are integrated in a endocervical catheter to deliver the energy; the operating frequency of the applicator is 6.5–8 MHz (Wootton et al., 2011).

Thermal ablation and hyperthermia clinical devices must be validated in accordance with national or international standards prior to merchandising them. Some regulatory organizations are the European Committee (CE), the Food and Drug Administration (FDA), the State Food and Drug Administration (SFDA), among others. These clinical devices are categorized under class II and class III of the FDA regulatory norms, and must be standardized in order to guarantee that thermal therapies are safe for clinical application (Shaw and Ter Haar, 2006).

8.2.2 Acoustic propagation modeling

Ultrasonic propagation modeling is a helpful resource to analyze the acoustic pattern of a specific transducer and its interaction with the propagation media. Moreover, as described in Chapter 11, this interaction produces thermal effects in tissues that are a function of frequency and temperature-dependence of acoustic properties, absorbed acoustic energy and exposure time, among others. Computational modeling by the finite element method (FEM) allows engineers and researchers to obtain a numerical solution of a particular physic or multi-physics phenomena whether it is a transitory or a stationary analysis. The study and selected geometrical dimension, 2D or 3D, depends on the parameters and variables to be analyzed, saving time and computational resources.

A steady-state solution of the acoustic pattern produced by a transducer can be obtained by realizing a stationary study; besides, if the multi-physics problem is simplified and the geometry is symmetrical, then, the computational model can be reduced to a 2D axisymmetric model. Furthermore, the independent geometric variables are in the radial r and axial z coordinates; while, the dependent variable is the azimuthal coordinate φ. The governing equation for a time-harmonic model with axial symmetry in a homogeneous lossless medium is described by Equation 8.1,

$$\nabla \cdot \left(\left(-\frac{1}{\rho_0} \right) \nabla p - \frac{p}{\rho_0} \left(\frac{\omega^2}{c^2} \right) \right) = 0 \tag{8.1}$$

where p is the acoustic pressure (Pa), ρ_0 is the density of the propagation medium (kg/m^3), c is the speed of sound in the medium, and ω is the angular frequency.

By simplifying Equation 8.1, it results in Equation 8.2,

$$\nabla^2 p + k^2 p = 0 \tag{8.2}$$

where k is the wave number $\left(k = \frac{\omega}{c} = \frac{2\pi}{\lambda}, and, \lambda \text{ is the wavelength}\right)$. Boundary conditions considered in stationary studies are (Duraiswami, 2006) as follows,

* Soft wall (Dirichlet boundary condition): This condition describes a low acoustic impedance in the surface (transducer) compared with the impedance of the load (medium). Moreover, the pressure in the surfaces is not different to that of the source, however, the normal derivative of the pressure is unknown ($p = 0$) (Delannoy et al., 1979).

* Rigid wall (Neumman boundary condition): It is used when the material of the transducer surface has a higher acoustic impedance compared to that of the medium. In other words, the acoustic pressure of the surface does not contribute to the acoustic pattern formation, i.e., the normal derivative of the pressure in the boundary fades (Delannoy et al., 1979), and the normal acceleration is null as described in Equation 8.3,

$$\frac{dp}{dn} = 0 \quad or \quad n \cdot \left(\frac{1}{\rho_0} \nabla p\right) = 0 \tag{8.3}$$

* Impedance boundary (Robin boundary condition): it is used to model finite acoustic impedances in a boundary; it is a generalized form of the rigid wall and the soft wall that allows setting intermediate impedance points described by Equation 8.4,

$$n \cdot \left(\frac{1}{\rho_0} \nabla p\right) + \frac{1}{Z}\left(\frac{dp}{dt}\right) = 0 \tag{8.4}$$

where $Z = \rho_0 c$ (acoustic impedance, MRayls), and it reduces the reflection of the acoustic waves into other boundaries.

* Pressure boundary: this condition establishes a particular acoustic pressure in the boundary; in other words, $p = p_0$.

* Axial symmetry: it is used only in the symmetry axis of the geometry, where the boundary $r = 0$.

The acoustic properties of the media are included in the computational model as well and the transducer pressure can be obtained from the RMS voltage (V_{rms}) data acquired by the hydrophone sensitivity $\left(S = \frac{V}{MPa}\right)$ during the acoustic characterization of the sound source. In order to do so, the free-field spatial-peak temporal-average intensity (*Ispta*) must be calculated (Martínez-Valdez et al., 2016) with Equation 8.5,

$$I_{spta} = \frac{1}{\rho c}\left(\frac{V_{rms}}{S}\right)^2 \quad (8.5)$$

Thermal heating due to ultrasound in a medium can be model by means of the bio-heat equation (Pennes, 1948), described by Equation 8.6

$$\rho C \frac{\partial T}{\partial t} + \nabla \cdot (-k\nabla T) = \rho_b C_b \omega_b (T_b - T) + Q_{met} + Q_{ext} \quad (8.6)$$

where C is the tissue heat capacity ($kg \cdot m^2/s^2/K$), k is the tissue thermal conductivity $W/(m \cdot K)$, ρ_b is blood density (kg/m^3), C_b is blood heat capacity ($kg \cdot m^2/s^2/K$), ω_b is blood perfusion rate ($kg/s/m^3$), T_b is blood temperature (K), Q_{met} metabolism heat (W/m^2), and Q_{ext} is the heat produced by an external source.

When modeling in homogeneous media, heat sources due to blood and metabolism can be neglected, then, Equation 8.6 is reduced to Equation 8.7

$$\rho C \frac{\partial T}{\partial t} + \nabla \cdot (-k\nabla T) = Q_{ext} \quad (8.7)$$

where Q_{ext} determines the relation between ultrasound and tissue heating that depends on the acoustic intensity (I) and the absorption coefficient α, $Q_{ext} = 2\alpha I$ (Moros et al., 1993).

During thermal ablation, target temperatures may reach values near 100° that could be due to exposure time to ultrasonic energy or due to non-linear propagation of high pressure acoustic waves (Ginter, 2000). In order to obtain more realistic ultrasonic patterns and thermal distributions of thermal ablation models due to focused ultrasound (FUS), high intensity focused ultrasound (HIFU) or continuous sonication non-linear propagation effects should be considered (Jenne et al., 2012). Then, the time harmonic Helmholtz equation, Equation 8.1 should be replaced by the Westervelt equation (Lee et al., 2008), the Khokhlov-Zabolotskaya-Kuznetsov (KZK) equation (Fan et al., 2009; Jenne et al., 2012) or other approximations. Furthermore, tissue thermal properties and speed of sound temperature-dependence should be included (Bamber and Hill, 1979; Goss, 1978; Guntur et al., 2013); ultrasound attenuation temperature and frequency-dependence (Bamber and Hill, 1979; Goss, 1978), blood perfusion rate and metabolism heat should be included as well (Jenne et al., 2012). Nevertheless, thermal properties' temperature-dependence and acoustic properties' temperatures and frequency dependence must be widely studied for different tissues before including those parameters in a model and, consequently, computational resources must be increased.

8.2.3 Acoustic field characterization

Acoustic field measurement or characterization of an ultrasonic transducer is realized punctually in a finite space. A high bandwidth piezoelectric element is attached to the tip of a needle hydrophone; this element senses the transducer acoustic waves and converts them into electrical signals to be processed for calculating the pressure. The hydrophone is connected to a preamplifier that compensates the low sensitivity and the high electrical impedance of the detector (Hill et al., 2004). The transducer is placed on a fixed base, while the hydrophone is placed on a base controlled by a high resolution position system, positioned at 1–2 mm from the transducer emitting surface. Both the transducer and the hydrophone are immersed in a tank full of bidistilled degassed water; the hydrophone is displaced in three orthogonal planes to reconstruct its acoustic pattern (Vera et al., 2008). During scanning, the transducer is excited by a low power electrical signal comprised of a 10–15 cycles burst at the source operating frequency, the frequency repetition rate is compared to the burst duration to avoid waves interference.

HIFU fields are measured at low power, i.e., in a linear propagation fashion to avoid hydrophone damage due to high acoustic intensity. Usually, the measured acoustic pressure is linearly extrapolated to a higher acoustic power (Shaw and Ter Haar, 2006). Other methods to measure the acoustic pattern scan a thermocouple embedded in an absorbing material (ter Haar and Coussios, 2007b), Schlieren imaging (ter Haar and Coussios, 2007b; Zhou et al., 2006) or fiber-optic hydrophones (Zhou et al., 2006).

8.2.4 Tissue mimicking-material for ultrasonic source validation

The study of physical phenomena related to the ultrasound application in the human body needs to use materials with similar acoustic properties of the medium where it will propagate (ter Haar and Coussios, 2007b). These kinds of materials are called either tissue mimicking or phantoms. Ideally, an ultrasonic phantom has all the acoustic properties of the imitated structure such as speed of sound, acoustic impedance, absorption, attenuation coefficient, scattering and morphology (Menikou et al., 2015; ter Haar and Coussios, 2007b; Zell et al., 2007).

The ultrasonic phantoms are used during equipment and transducers' calibration either for imaging or therapy ultrasound (Arturo Vera et al., 2008) to determine thermal fields, mechanics properties of the tissues and assessment protocols guaranteeing its reliability and reproducibility. When a transducer is characterized to obtain an acoustic distribution and performance, all experimental measurements are made over phantoms (Ambrogio et al., 2022).

Degassed water is considered to be a general purpose phantom, because most of the structures in the human body are composed of a large quantity of water (Zell et al., 2007). Besides, phantoms can be fabricated from materials with acoustic properties similar to that of water such as hydrogels or water-swellable polymers (Choi et al., 2013; Zell et al., 2007). These materials have a jelly-like structure and, generally, other substances can be added to obtain the closest acoustic properties of the structure for simulation (Kaczmarek et al., 2019; McGarry et al., 2020). In soft tissues, we can find some materials like agar, silicone, polyvinyl alcohol gel (PVA), polyacrylamide gel (PAA), polyvinyl chloride (PVC) and polyvinyl chloride plastisol (PVCP) (Cortela et al., 2015; Zell et al., 2007). Furthermore, in recent years there has been an increasing interest to create rigid structure tissues as bone phantoms. Usually, these tissue mimicking material proposals are made with polymeric materials like resins, materials compatible with 3D printing materials, and present a high density (Clarke et al., 1994; McGarry et al., 2020; Menikou et al., 2015). Based on the latest features, some resins can be modified and any metal can be added (Clarke et al., 1994; McGarry et al., 2020; Wydra and Maev, 2013). Finally, for low density tissues such as lungs, materials such as polyurethane have proved to be effective. The acoustic properties of different human body tissues compared with some materials used to fabricate phantomsreported by different authors (Clarke et al., 1994; Cortela et al., 2015; De Matheo et al., 2017; Dodd et al., 2006; IT'IS Foundation, 2022; Zell et al., 2007), are shown in the Tables 8.1 and 8.2, respectively.

The ultrasonic systems specifically developed for hyperthermia treatments must guarantee the power deposition, desirable temperature distribution and SAR (specific absorption rate) distribution inside the tissue or the phantoms (Ambrogio et al., 2022; Arturo Vera et al., 2008). In order to obtain thermal profiles and distributions due to ultrasound hyperthermia, the phantoms are heated in a short time, and the temperature increase response is measured by an external device. These devices can be ultrasound imaging, fiber optic sensors, thermistors, MRI, tomography, and others , therefore, the phantom must be compatible with them (Vera et al., 2008). Thermal mapping measurement with non-invasive techniques such as MRI may not be accessible or expensive; meanwhile, less expensive methods like thermocouples or thermistors are considered minimally invasive techniques, may damage the phantom and can interfere hence modify the acoustic pattern. Currently, there has been a special interest in developing phantoms with thermochromic materials. A thermochromic phantom is made with materials that change their color in a particular temperature range, and allows visualizing the thermal fields and heating patterns during HIFU energy delivery (Eranki et al., 2019). These phantoms are usually reusable, but they have a low temperature range and loss

Table 8.1. Acoustic properties of different human body tissues.

Tissue	Density [kg/m³]	Speed of sound [m/s]	Acoustic impedance [MRayls]
Fat	911	1440	1.3118
Muscle	1090	1588.4	1.73135
Breast	911	1440.2	1.3120
Liver	1079	1585.7	1.71097
Kidney	1066	1554.3	1.6568
Brain	1046	1546.3	1.61742
Prostate	1045	1559.5	1.62967
Cortical bone	1908	3514	6.70471
Cancellous bone	1178	2117.5	2.49441

Table 8.2. Acoustic properties of materials used to fabricate ultrasonic phantoms.

Material	Density [kg/m³]	Speed of Sound [m/s]	Acoustic Impedance [MRayls]
Water	1000	1482	1.482
Agar	1040 ± 110	1500 ± 30	1.56
Agar with Nanomaterials	1005.5 ± 18.5	1494.145 ± 15.855	1.50236
Silicone	1070 ± 30	1030 ± 60	1.1021
PVA	1100 ± 50	1570 ± 20	1.727
Agar with Al_2O_3	1053.5	1549.8	1.63271
Agar with evaporated milk	1060	1470	1.5582
PAA	1024	1546 ± 30	1.58310
PVCP	953, 972 ± 24	1501.5 ± 0.7, 1331.8 ± 0.3	1.36210
Resin	940–1170	1844–3118	2.617455
3D	1040	2230–2270	2.34
Acrylic	1185	2685	3.181725

of transparency; therefore, the measurement must be made quickly to obtain an accurate measure.

8.3 Clinical applications

8.3.1 Requirements for the clinical applications

High intensity focused ultrasound has become an emerging oncological, palliative and rehabilitation therapy because it is non-ionizing (Izadifar et al., 2020; ter Haar and Coussios, 2007b). This technique is more effective when the target tissue (tumor) is at an early stage and when the mass is either superficial or surrounded only by soft tissue. The use of HIFU is convenient because heating in situ does not depend directly on the tissues' thermal conduction, therefore, the dissipation effect (skin) is minor (Elhelf et al., 2018). However, in order to deliver the ultrasonic energy to the target, it is important to determine a window to assure correct acoustic propagation and avoid hard tissue (long bones or ribs) that produces energy reflection and loss (Gélat et al., 2014). On the other hand, tumors located in the lungs would have a high attenuation and reflection due to the large difference in acoustic impedances (liquid-air), so HIFU ablation will not be effective enough (Elhelf et al., 2018; Mikhak and Pedersen, 2002).

In order to propose a protocol aimed either for oncological or palliative treatment, it is necessary to define the kind of tumor, its location and size. Also, the anthropometric and biochemical characteristics of the patient should be considered, as well as the disease stage (Choe et al., 2019); to be aware of the tumor stage is required when HIFU is used as palliative therapy. Before a patient becomes a candidate for HIFU therapy, the tumor must be located using imaging techniques (X-ray, tomography or MRI) and confirmed by means of a biopsy (Barkin, 2011). Taking into account the above general requirements, it will be possible to determine the exposure and/or dose during the treatment (ter Haar and Coussios, 2007b). The biological effects will depend on the different ways in which the procedure will be performed; exposure times, frequency, pulse (mode, time and amplitude), intensity, etc. (ter Haar and Coussios, 2007b; Gail ter Haar, 2016).

In the therapy process for HIFU energy delivery, the patient must be under general anaesthesia to reduce movements and to allow the accurate positioning of the ultrasonic beam and avoid damage to healthy tissues. The total treatment duration can vary from a few minutes up to some hours; this also will depend on the type of tumor and its size, as well as the temperature reached on the surrounding tissues (Barkin, 2011). Sometimes, the treatment may be accompanied by minimally invasive surgery to avoid some structures and deliver the energy as close as possible to the tumor. Some secondary effects have been reported after an application of HIFU treatment which involve moderate discomfort, "minimal" skin burn and irritation in the area; nevertheless, these symptoms usually disappear in a few days or with medical treatment (Brenin, 2019).

Clinical devices for HIFU therapy are usually of two types: extracorporeal and interstitial or intracavitary. Their main components consist of a transducer, a signal generator, an amplifier, a coupling circuit, a power meter, and, generally, an image monitoring device (Izadifar et al., 2020; ter Haar and Coussios, 2007b; Gail ter Haar, 2016). Additionally, the transducer in these devices can be joined to a positioning system (stereotaxic) which allows it to perform different sonication points, covering a larger volume or surface and reducing treatment time. The shape of the transducer will depend on the application surface and, therefore, in situ power will depend on this (Izadifar et al., 2020; ter Haar and Coussios, 2007b; ter Haar, 2016). In addition, to recording the temperature on the application surface, HIFU therapy systems are composed of imaging systems, which can be either magnetic resonance (MRgHIFU) or ultrasound imaging (USgHIFU) (Izadifar et al., 2020; ter Haar and Coussios, 2007b). The imaging guidance systems must comprise a real-time automatic feedback control with robustness from patient-to-patient, normal tissue safety, reduced treatment and thermal dose monitoring (Arora et al., 2005). A thermal dose is used to quantify the relationship between treatment efficacy and the target temperature as a function of time (Arora et al., 2005), and is an indicator for the medical specialist to determine if the target has received the sufficient ultrasonic energy. The specific clinical locations for getting a HIFU therapy will depend on the monitoring system used by the HIFU ablation device; less confined outpatient clinics are used for USgHIFU and MRI rooms for those using MRgHIFU (Zhang and Wong, 2020).

8.3.2 Extracorporeal applications

Due to HIFU advantages, tumors located in tissues such as breast, liver, kidney, lung, prostate, uterus and even some bones are considered as candidates for oncological thermal ablation (Elhelf et al., 2018; Izadifar et al., 2020; Maloney and Hwang, 2015). During treatment planning over these tissues, an acoustic window on the skin may be required to guarantee maximum energy delivery. In order to do so, coupling materials must also be used to obtain skin-like acoustic impedance (Martínez-Valdez et al., 2016). For MRgHIFU, it is necessary to ensure that the materials used to build the system are compatible with MRI to avoid accidents and distortion by magnetic fields that introduce artifacts (Gail ter Haar, 2016). When the therapy is guided by ultrasound imaging, an image transducer is used to monitor, guide and visualize the generated thermal fields (Ebbini and Ter Haar, 2015).

Primary and secondary liver tumors such as hepatocellular carcinomas and hepatoblastoma (Elhelf et al., 2018; Izadifar et al., 2020) have been treated with HIFU; frequently, patients who are not candidates for surgical resection with the presence of a degree of cirrhosis. Some studies have shown an ablation lesion higher than 70% after the treatment; the outcomes have an effectiveness rate between 50–80% for studies with up to 275 patients (Illing et al., 2005; Ji et al., 2020). Despite the large number of clinical and preclinical studies, obtaining an acoustic window has become the main problem of the application in this tissue (Illing et al., 2005; Maloney and Hwang, 2015).

Also, due to the energy path through the ribs, heating can be generated in insignificant tissues that would result in low acoustic intensity at the target site (tumor).

HIFU treatment for breast tumors is one of the FDA approved procedures due to the type of tissue and superficial location. Treatment has been performed with MRI and US imaging for benign and malignant tumors ranging in size from 0.5 cm to 10 cm (Brenin, 2019; Peek and Wu, 2018). Some authors have reported an efficacy of more than 80% in the effectiveness of treatment after analyzing coagulation necrosis, patient survival and non-recurrence of tumor; they also observed tumor necrosis and destruction of the cell membrane (Napoli et al., 2013; Peek and Wu, 2018). Treatment duration for breast tumor can be in a range of 40 minutes to 4 hours. Despite all, one of the risks for this application is the skin burns that can generate fat necrosis and mild to moderate discomfort (Brenin, 2019; Peek and Wu, 2018).

In the case of kidney tumors, some pilot tests have been carried out in conjunction with laparoscopic techniques, they have found effectiveness of more than 90% avoiding tumor recurrence and survival of the patients in a period up to 5 years (Izadifar et al., 2020). The biggest problems when performing a totally extracorporeal therapy are due to breathing induced movement of the tissue and loss of acoustic energy by the ribs through their pathway. In uterine fibroids, clinical studies have evaluated a total use of extracorporeal devices, as an attempt to find satisfactory results in necrosis, tumor coagulation and symptom reduction (Ren et al., 2007). Many HIFU treatments are carried out by recurrence, for myomectomy or due to a potential loss of the uterus, which prevents a future pregnancy. Since pancreatic tumors, are detected at stage IV, patients having them have a low life expectancy (Sofuni et al., 2021; Stanislavova et al., 2021). HIFU treatments with extracorporeal devices are usually palliative for unresectable tumors. Efficiency is more prolific in combination with chemotherapy, increasing the average life expectancy of patients. In addition, the decreasing tumor volume and local pain , allows relief of symptoms (Marinova et al., 2019; Sofuni et al., 2021). Finally, in recent years, the use of HIFU has been proposed as a curative and palliative treatment of primary and secondary bone tumors, both benign and malignant (Chen et al., 2010; Napoli et al., 2013; Rodrigues et al., 2015). It has taken main interest in cortical tumors, soft bone tumors and metastasis bone tumors, in which in some studies have observed decreased pain and bone tumor recurrence (Napoli et al., 2013; Rodrigues et al., 2015).

8.3.3 Intracavitary and interstitial applications

When benign or malignant tumors grow deep within the body or in unreachable body regions extracorporeal devices, intracavitary and interstitial applicators are developed. Interstitial methods are considered minimally invasive because the heating sources are directly in contact with the tumor (Diederich and Hynynen, 1999). Interstitial acoustic energy delivery is usually combined with radiotherapy to improve hyperthermic treatment of tumors; the devices may be catheter-cooled, direct-couple applicators or acoustic waveguide antennas (Diederich and Hynynen, 1999). Intracavitary probes allow ultrasonic waves to reach targets located near a natural body cavity or lumen (Zubair et al., 2021); the most studied clinical application is the treatment of benign or malignant prostate tumors. Commercially available devices for treating the prostate are Ablatherm® Fusion (EDAP-TMS, France) and Sonablate® (Sonablate Corp, Charlotte, NC, USA) both with transrectal ultrasound guidance, and Exablate® Prostate (Insightec, Israel) and TULSA-PRO® (Profound Medical Corp., Mississauga, Ontario, Canada) with MRI guidance (Jenne et al., 2012).

Cervical cancer thermal treatment has been explored via endocervical or interstitial probes along with brachytherapy, and a combination of endocervical and interstitial devices to improve and reach larger heating volumes (Wootton et al., 2011; Wootton et al., 2011). Recently, Zubair et al. (Zubair et al., 2021) proposed an endoluminal ultrasound deployable applicator to treat pancreas cancer through the oesophagus or even through laparoscopic surgery. This device consists of cylindrical ultrasonic transducers within a dual layer of conical balloon-based reflectors to reflect and redirect the acoustic energy in a focused beam or an annular pattern that produces a collimated beam into the treatment area. Intracavitary and interstitial devices are constrained to cavity and catheter dimensions (Diederich and Hynynen, 1999; Zubair et al., 2021). Moreover, the source of energy or heating depends on the type of cancer and its location (Burchardt and Roszak, 2018), as well as size and shape of the tumor.

8.3.4 Combining therapies and clinical studies

HIFU therapy has taken place as a new alternative for cancer treatment. As monotherapy, one of the most significant effects of HIFU application is a reduction in tumor volume (Dahan et al., 2022). To improve its performance, currently, there are some approaches to combine multiple cancer treatment modalities with HIFU to obtain mostly positive outcome rates. For example, HIFU has been used with other types of procedures such as radiofrequency therapy, chemotherapy, immunotherapy and local surgery (Borasi et al., 2014). In addition, HIFU has demonstrated potential as an adjuvant in the administration of energy and drugs by radiotherapy; the advantages that might be obtained when these two mentioned techniques are combined are described below (Borasi et al., 2013):

- Better penetration rates and energy focus by therapeutic behavior in some soft tissues.
- Efficacy of HIFU as a second procedure on solid tumors avoiding adverse effects due to ionizing radiation.
- The combined use of the two therapies can enhance the improvement process due to cell inactivation by ionizing radiation.
- Cavitation generated by HIFU as sonodynamic therapy (SDT) can enhance the thermal effect and it can activate systemic chemotherapy drug delivery.

When HIFU is combined with immunotherapies, the results have been focused on an antitumor immune response (Dahan et al., 2022). The latter was reported in the SDT application with the HiPorfin sensitizer, which has shown an antitumor response in liver cells and a decrease in anti-inflammatories. Furthermore, the results by immunohistochemistry analysis of primary tumors showed induction of cell death by apoptosis and necrosis. For example, for breast cancer, the effectiveness of drugs as systemic therapies before surgery in metastatic tumors is investigated on the antibody drug trastuzumab emtansine (T-DM1) (von Minckwitz et al., 2019).

Current clinical trials for combining therapies are centered on treatments for breast, bone, liver and pancreatic cancer, as well as tumor metastases. An overview of many clinical studies combining therapies and their therapeutic results is shown in Table 8.3 (Dahan et al., 2022; Sofuni et al., 2021; Wang, 2019; Wu et al., 2003).

It should be noted that the results of these combinatorial therapies are mostly effective for non-metastatic primary tumors which allows an improvement in the patient quality of life through control of tumor volume and pain reduction (Borasi et al., 2014). In the case of tumors that have metastasized, the main advantage of this combinatorial modality is given as a systemic treatment seeking to increase the life expectancy of patients. On the other hand, despite the positive results, clinical trials are liable to regulations and protocols to define aspects such as the type of primary therapy, drugs, types and stages of tumors, dose, as well as the modality and features of the HIFU procedures that will be used (ter Haar, 2016). Once these requirements can be optimized and validated, combination HIFU therapies will become an important alternative for the benefit of patients.

Table 8.3. Clinical studies of the combined application of HIFU.

Type of cancer	Number of patients	Treatment	Therapeutic outcomes
Osteosarcoma with metastasis	72	Chemotherapy (adriamycin) + HIFU	Higher survival rates after 2 or 3 years of treatment. High tumoral necrosis. Improvement in limb functions and decreasing Serum Alkaline phosphatase (ALP) rates.
Breast cancer	48	Modified radical mastectomy + HIFU	Coagulative necrosis in tumor cells with vascular disease could be observed. Spreading, invasion and tumor metastasis was absent.
Pancreatic cancer	176	Chemotherapy + HIFU	The results manifested an ablation rate > 90%, symptom relief > 50% and an increment in survival rate. All data compared with the control group.
Hepatocellular carcinomas	25	Transarterial Chemoembolization (TACE) + HIFU	There was a greater survival time for the HIFU group than the TACE group. Better results in the HIFU group than the TACE group for complete, partial and stable responses.
Breast cancer	48	Mastectomy + HIFU	Tumor-infiltrating DCs, macrophages and B lymphocytes increased significantly in the HIFU group.

Acronym/Abbreviation

2D	Two dimension
3D	Three dimension
ALP	Alkaline Phosphatase
°C	Celsius degrees
CE	European Committee
DCs	Dendritic Cells
FDA	Food and Drug Administration
FEM	Finite Element Method
FUS	Focused Ultrasound
HIFU	High Intensity Focused Ultrasound
Hz	Hertz
K	Kelvin
KZK	Khokhlov-Zabolotskaya-Kuznetsov equation
MRgHIFU	Magnetic Resonance guided HIFU
MRI	Magnetic Resonance Image
Pa	Pascal
PAA	Polyacrylamide Gel

PVA	Polyvinyl Alcohol Gel
PVC	Polyvinyl Chloride
PVCP	Polyvinyl Chloride Plastisol
RMS	Root Mean Square
s	seconds
SAR	Specific Absorption Rate
SDT	Sonodynamic Therapy
SFDA	State Food and Drug Administration
TACE	Transarterial Chemoembolisation
T-DM1	Trastuzumab Emtansine
US	Ultrasound
USgHIFU	Ultrasound (imaging) guided HIFU

Reference list

Almquist, S., Parker, D.L. and Christensen, D.A. (2016). Rapid full-wave phase aberration correction method for transcranial high-intensity focused ultrasound therapies. *Journal of Therapeutic Ultrasound*, 4(1): 1–11. https://doi.org/10.1186/S40349-016-0074-7/FIGURES/10.

Ambrogio, S., Baêsso, R.M., Bosio, F., Fedele, F., Ramnarine, K.V. et al. (2022). A standard test phantom for the performance assessment of magnetic resonance guided high intensity focused ultrasound (MRgHIFU) thermal therapy devices. *International Journal of Hyperthermia*, 39(1): 57–68. https://doi.org/10.1080/02656736.2021.2017023.

Arora, D., Cooley, D., Perry, T., Skliar, M., Roemer, R.B. et al. (2005). Direct thermal dose control of constrained focused ultrasound treatments: phantom and *in vivo* evaluation. *Physics in Medicine and Biology*, 50(8): 1919–1935.

Bamber, J.C. and Hill, C.R. (1979). Ultrasonic attenuation and propagation speed in mammalian tissues as a function of temperature. *Ultrasound in Medicine & Biology*, 5(2): 149–157. https://doi.org/10.1016/0301-5629(79)90083-8,

Barkin, J. (2011). High-intensity focused ultrasound. *The Canadian Journal of Urology*, 18(2): 5634–5643. https://doi.org/10.1016/B978-0-12-800077-9.00059-1,

Borasi, G., Melzer, A., Russo, G., Nahum, A., Zhang, Q. et al. (2014). Cancer therapy combining high-intensity focused ultrasound and megavoltage radiation. *International Journal of Radiation Oncology Biology Physics*, 89(4): 926–927. https://doi.org/10.1016/j.ijrobp.2014.03.025.

Borasi, G., Russo, G., Alongi, F., Nahum, A., Candiano, G.C. et al. (2013). High-intensity focused ultrasound plus concomitant radiotherapy: A new weapon in oncology. *Journal of Therapeutic Ultrasound*, 1(1): 2–5. https://doi.org/10.1186/2050-5736-1-6.

Brenin, D.R. (2019). Ablative treatment of breast cancer; Are we there yet? *Current Breast Cancer Reports*, 11(2): 43–50. https://doi.org/10.1007/s12609-019-0307-1.

Burchardt, E. and Roszak, A. (2018). Hyperthermia in cervical cancer – current status. *Reports of Practical Oncology & Radiotherapy*, 23(6): 595–603. https://doi.org/10.1016/J.RPOR.2018.05.006.

Chen, W., Zhu, H., Zhang, L., Li, K., Su, H. et al. (2010). Primary bone malignancy: Effective treatment with high-intensity focused ultrasound ablation. *Radiology*, 255(3): 967–978. https://doi.org/10.1148/radiol.10090374.

Choe, Y.S., Lee, W.M., Choi, J.S., Bae, J., Eom, J.M. et al. (2019). Clinical characteristics of patients with leiomyoma who undergo surgery after high intensity focused ultrasound

(HIFU). *Obstetrics and Gynecology Science*, 62(4): 258–263. https://doi.org/10.5468/ogs.2019.62.4.258.

Choi, M.J., Guntur, S.R., Lee, K.I.L., Paeng, D.G., Coleman, A. et al. (2013). A tissue mimicking polyacrylamide hydrogel phantom for visualizing thermal lesions generated by high intensity focused ultrasound. *Ultrasound in Medicine and Biology*, 39(3): 439–448. https://doi.org/10.1016/j.ultrasmedbio.2012.10.002.

Chopra, R., Burtnyk, M., N'Djin, W.A. and Bronskill, M. (2010). MRI-controlled transurethral ultrasound therapy for localised prostate cancer. *Http://Dx.Doi.Org/10.3109/02656736.2010.503670*, 26(8): 804–821. https://doi.org/10.3109/02656736.2010.503670.

Clarke, A.J., Evans, J., Truscott, J., Milner, R., Smith, M. et al. (1994). A phantom for quantitative ultrasound of trabecular. *Physics in Medicine and Biology*, 39(10): 1677–1687.

Cortela, G., Benech, N., Pereira, W.C.A. and Negreira, C. (2015). Characterization of acoustical properties of a phantom for soft tissues (PVCP and graphite powder) in the range 20–45°C. *Physics Procedia*, 70: 179–182. https://doi.org/10.1016/j.phpro.2015.08.107.

Dahan, M., Cortet, M., Lafon, C. and Padilla, F. (2022). Combination of focused Ultrasound, Immunotherapy, and Chemotherapy: new perspectives in breast cancer therapy. *Journal of Ultrasound in Medicine*, 1–15. https://doi.org/10.1002/jum.16053.

De Matheo, L.L., Maggi, L.E., Costa, J.F.S., Von Kruger, M.A., Pereira, W.C.A. et al. (2017). Comparison of two ultrasonic phantom manufacturing protocols made of polyvinyl chloride Plastisol. *Pan American Health Care Exchanges, PAHCE*, 1–4. https://doi.org/10.1109/GMEPE-PAHCE.2017.7972096.

Delannoy, B., Lasota, H., Bruneel, C., Torguet, R., Bridoux, E. et al. (1979). The infinite planar baffles problem in acoustic radiation and its experimental verification. *Journal of Applied Physics*, 50(8): 5189. https://doi.org/10.1063/1.326656.

Diederich, C.J. and Hynynen, K. (1999). Ultrasound technology for hyperthermia. *Ultrasound in Medicine and Biology*, 25(6): 871–887. http://www.sciencedirect.com/science/article/B6TD2-3WWV09B-1/2/6945a7eabcc7739e10741c45c33ed206.

Dodd, S.P., Cunningham, J.L., Miles, A.W., Gheduzzi, S., Humphrey, V.F. et al. (2006). Ultrasonic propagation in cortical bone mimics. *Physics in Medicine and Biology*, 51: 4635–4647. https://doi.org/10.1088/0031-9155/51/18/012.

Duraiswami, R. (2006). *The Wave Equation. The Helmoltz Equation. Boundary Conditions. Properties of Solutions*. Algorithms and Systems for Capture and Playback of Spatial Audio. University of Maryland, College Park. Lecture 3.

Ebbini, E.S. and Ter Haar, G. (2015). Ultrasound-guided therapeutic focused ultrasound: Current status and future directions. *International Journal of Hyperthermia*, 31(2): 77–89. https://doi.org/10.3109/02656736.2014.995238.

Elhelf, I.A.S., Albahar, H., Shah, U., Oto, A., Cressman, E. et al. (2018). High intensity focused ultrasound: The fundamentals, clinical applications and research trends. *Diagnostic and Interventional Imaging*, 99(6): 349–359. https://doi.org/10.1016/j.diii.2018.03.001.

Eranki, A., Mikhail, A.S., Negussie, A.H., Katti, P.S., Wood, B.J. et al. (2019). Tissue-mimicking thermochromic phantom for characterization of HIFU devices and applications. *International Journal of Hyperthermia*, 36(1): 518–529. https://doi.org/10.1080/02656736.2019.1605458.

Fan, T.B., Liu, Z.B., Zhang, Z., Zhang, D., Gong, X.F. et al. (2009). Modeling of nonlinear propagation in multi-layer biological tissues for strong focused ultrasound. *Chinese Physics Letters*, 26(8).

Gélat, P., Ter Haar, G. and Saffari, N. (2014). HIFU scattering by the ribs: Constrained optimisation with a complex surface impedance boundary condition. *Journal of Physics: Conference Series*, 498: 1–11. https://doi.org/10.1088/1742-6596/498/1/012004.

Ginter, S. (2000). Numerical simulation of ultrasound-thermotherapy combining nonlinear wave propagation with broadband soft-tissue absorption. *Ultrasonics*, 37(10): 693–696. https://doi.org/10.1016/S0041-624X(00)00012-3.
Goss, S.A. (1978). Comprehensive compilation of empirical ultrasonic properties of mammalian tissues. *The Journal of the Acoustical Society of America*, 64(2): 423. https://doi.org/10.1121/1.382016.
Guntur, S.R., Lee, K.I., Paeng, D.G., Coleman, A.J., Choi, M.J. et al. (2013). Temperature-dependent thermal properties of *ex vivo* liver undergoing thermal ablation. *Ultrasound in Med. & Biol.*, 39: 1771–1784.
Gutierrez, M.I., Martinez, R., Vera, A. and Leija, L. (2012). Technology in ultrasonic hyperthermia. pp. 41–83. *In*: Gao, X.-H. and Chen, H.-D. (eds.). *Hyperthermia: Recognition, Prevention and Treatment* (Issue 2). NOVA Publishers.
Hill, C.R., Bamber, J.C. and Haar, G.R. ter. (2004). *Physical Principles of Medical Ultrasonics* (2nd ed.). John Wiley & Sons, Ltd.
Hokland, S.L., Pedersen, M., Salomir, R., Quesson, B., Stodkilde-Jorgensen, H. et al. (2006). MRI-guided focused ultrasound: methodology and applications. *Medical Imaging, IEEE Transactions On*, 25(6): 723–731.
Hutchins, D.A. and Hayward, G. (1990). The radiated field of ultrasonic transducers. pp. 1–80. *In*: Thurston, R.N. and Pierce, A.D. (eds.). *Physical Acoustics* (Vol. 19). Academic Press.
Hwang, J.H. and Crum, L.A. (2009). Current status of clinical high-intensity focused ultrasound. *Conf. Proc. IEEE Eng. Med. Biol. Soc.*, 1: 130–133. http://www.ncbi.nlm.nih.gov/entrez/query.fcgi?cmd=Retrieve&db=PubMed&dopt=Citation&list_uids=19965122.
Illing, R. and Chapman, A. (2007). The clinical applications of high intensity focused ultrasound in the prostate. *International Journal of Hyperthermia*, 23(2): 183–191.
Illing, R.O., Kennedy, J.E., Wu, F., Ter Haar, G.R., Protheroe, A.S. et al. (2005). The safety and feasibility of extracorporeal high-intensity focused ultrasound (HIFU) for the treatment of liver and kidney tumours in a western population. *British Journal of Cancer*, 93(8): 890–895. https://doi.org/10.1038/sj.bjc.6602803.
IT'IS Foundation. (2022). *Tissue Properties*. https://itis.swiss/virtual-population/tissue-properties/database/acoustic-properties/.
Izadifar, Z., Izadifar, Z., Chapman, D. and Babyn, P. (2020). An introduction to high intensity focused ultrasound: Systematic review on principles, devices, and clinical applications. *Journal of Clinical Medicine*, 9(2): 460. https://doi.org/10.3390/jcm9020460.
Jenne, J.W., Preusser, T. and Günther, M. (2012). High-intensity focused ultrasound: Principles, therapy guidance, simulations and applications. *Zeitschrift Für Medizinische Physik*, 22(4): 311–322. https://doi.org/10.1016/j.zemedi.2012.07.001.
Ji, Y., Zhu, J., Zhu, L., Zhu, Y., Zhao, H. et al. (2020). High-intensity focused ultrasound ablation for unresectable primary and metastatic liver cancer: Real-world research in a chinese tertiary center with 275 cases. *Frontiers in Oncology*, 10: 1–14. https://doi.org/10.3389/fonc.2020.519164.
Kaczmarek, K., Mrówczyński, R., Hornowski, T., Bielas, R., Józefczak, A. et al. (2019). The effect of tissue-mimicking phantom compressibility on magnetic hyperthermia. *Nanomaterials*, 9(5): 803. https://doi.org/10.3390/nano9050803.
Lee, K.I., Sim, I., Kang, G.S. and Choi, M.J. (2008). Numerical simulation of temperature elevation in soft tissue by high intensity focused ultrasound. *Modern Physics Letters B*, 22(11): 803–807.
Maloney, E. and Hwang, J.H. (2015). Emerging HIFU applications in cancer therapy. *International Journal of Hyperthermia*, 31(3): 302–309. https://doi.org/10.3109/02656736.2014.969789.

Marinova, M., Wilhelm-Buchstab, T. and Strunk, H. (2019). Advanced pancreatic cancer: High-intensity focused ultrasound (HIFU) and other local ablative therapies. *RoFo*, 191(3): 216–227. https://doi.org/10.1055/a-0870-8974.

Martínez-Valdez, R., Ramos Fernández, A., Vera Hernandez, A. and Leija Salas, L. (2016). Design of a low power hybrid HIFU applicator for haemostasis based on acoustic propagation modelling. *International Journal of Hyperthermia*, 32(2). https://doi.org/10.3109/02656736.2015.1112437.

McGarry, C.K., Grattan, L.J., Ivory, A.M., Leek, F., Liney, G.P. et al. (2020). Tissue mimicking materials for imaging and therapy phantoms: A review. *Physics in Medicine and Biology*, 65(23). https://doi.org/10.1088/1361-6560/abbd17.

Menikou, G., Dadakova, T., Pavlina, M., Bock, M., Damianou, C. et al. (2015). MRI compatible head phantom for ultrasound surgery. *Ultrasonics*, 57: 144–152. https://doi.org/10.1016/j.ultras.2014.11.004.

Mikhak, Z. and Pedersen, P.C. (2002). Acoustic attenuation properties of the lung: An open question. *Ultrasound in Medicine and Biology*, 28(9): 1209–1216. https://doi.org/10.1016/S0301-5629(02)00561-6.

Moros, E.G., Dutton, A.W., Roemer, R.B., Burton, M., Hynynen, K. et al. (1993). Experimental evaluation of two simple thermal models using hyperthermia in muscle *in vivo*. *International Journal of Hyperthermia*, 9(4): 581–598.

Napoli, A., Anzidei, M., Ciolina, F., Marotta, E., Cavallo Marincola, B. et al. (2013). MR-guided high-intensity focused ultrasound: current status of an emerging technology. *CardioVascular and Interventional Radiology*, 36(5): 1190–1203. https://doi.org/10.1007/s00270-013-0592-4.

Napoli, A., Mastantuono, M., Marincola, B.C., Anzidei, M., Zaccagna, F. et al. (2013). Osteoid osteoma: MR-guided focused ultrasound for entirely noninvasive treatment. *Radiology*, 267(2): 514–521. https://doi.org/10.1148/radiol.13120873.

Peek, M.C.L. and Wu, F. (2018). High-intensity focused ultrasound in the treatment of breast tumours. *Ecancermedicalscience*, 12: 794. https://doi.org/10.3332/ecancer.2018.794.

Pennes, H.H. (1948). Analysis of tissue and arterial blood temperatures in the resting human forearm. *Journal of Applied Physiology*, 1(2): 93–122.

Ren, X.-L., Xiao-Dong Zhou, Jun, Z., Guang-Bin He, Zeng-Hui Han et al., (2007). Extracorporeal ablation of uterine fibroids with high-intensity focused ultrasound. *Journal of Ultrasound in Medicine*, 26(2): 201–212. https://doi.org/10.7863/jum.2007.26.5.702.

Ren, X.L., Zhou, X.D., Zhang, J., He, G.B., Han, Z.H. et al. (2007). Extracorporeal ablation of uterine fibroids with high-intensity focused ultrasound: imaging and histopathologic evaluation. *J. Ultrasound Med.*, 26(2): 201–212. https://doi.org/26/2/201 [pii].

Rodrigues, D.B., Stauffer, P.R., Vrba, D. and Hurwitz, M.D. (2015). Focused ultrasound for treatment of bone tumours. *International Journal of Hyperthermia*, 31(3): 260–271. https://doi.org/10.3109/02656736.2015.1006690.

Shaw, A. and Ter Haar, G.R. (2006). *Requirements for measurement standards in High Intensity Focused (HIFU) Ultrasound Fields.*

Sofuni, A., Asai, Y., Tsuchiya, T., Ishii, K., Tanaka, R. et al. (2021). Novel therapeutic method for unresectable pancreatic cancer—the impact of the long-term research in therapeutic effect of high-intensity focused ultrasound (HIFU) therapy. *Current Oncology*, 28(6): 4845–4861. https://doi.org/10.3390/curroncol28060409.

Stanislavova, N., Karamanliev, M., Ivanov, T., Yotsov, T., Zhou, K. et al. (2021). Is high-intensity focused ultrasound (HIFU) an option for neoadjuvant therapy for borderline resectable pancreatic cancer patients?—A systematic review. *International Journal of Hyperthermia*, 38(2): 75–80. https://doi.org/10.1080/02656736.2021.1909150.

ter Haar, G. and Coussios, C. (2007a). High intensity focused ultrasound: Past, present and future. *International Journal of Hyperthermia*, 23(2): 85–87. http://www.ncbi.nlm.nih.gov/entrez/query.fcgi?cmd=Retrieve&db=PubMed&dopt=Citation&list_uids=17578334.

ter Haar, G. and Coussios, C. (2007b). High intensity focused ultrasound: Physical principles and devices. *International Journal of Hyperthermia*, 23(2): 89–104. http://www.ncbi.nlm.nih.gov/entrez/query.fcgi?cmd=Retrieve&db=PubMed&dopt=Citation&list_uids=17578335.

ter Haar, Gail. (1999). Therapeutic ultrasound. *European Journal of Ultrasound*, 9(1): 3–9. http://www.sciencedirect.com/science/article/B6T6G-3W6F38Y-2/1/abf509016534c6e94c78826572a52bf2.

ter Haar, Gail. (2016). HIFU tissue ablation: Concept and devices. *Advances in Experimental Medicine and Biology*, 880: 3–20. https://doi.org/10.1007/978-3-319-22536-4_1.

Vera, A., Leija, L. and Munoz, R. (2008). Ultrasonic hyperthermia. pp. 467–495. *In*: Vives, A.A. (ed.). *Piezoelectric Transducers and Applications*. Springer-Verlag Berlin Heidelberg. https://doi.org/978-3-540-77508-9.

Vera, Arturo, Leija, L. and Muñoz, R. (2008). Ultrasonic hyperthermia. In *Piezoelectric Transducers and Applications* (pp. 467–495).

Vives, A.A. (2008). Piezoelectric transducers and applications. In *Piezoelectric Transducers and Applications*. Springer Berlin Heidelberg. https://doi.org/10.1007/978-3-540-77508-9.

von Minckwitz, G., Huang, C.-S., Mano, M.S., Loibl, S., Mamounas, E.P. et al. (2019). Trastuzumab emtansine for residual invasive HER2-positive breast cancer. *New England Journal of Medicine*, 380(7): 617–628. https://doi.org/10.1056/nejmoa1814017.

Wang, C. (2019). Therapeutic effects of adriamycin combined with high-intensity focused ultrasound on osteosarcoma. *Journal of BUON*, 24(2): 826–831.

Wootton, J.H., Hsu, I.C.J. and Diederich, C.J. (2011). Endocervical ultrasound applicator for integrated hyperthermia and HDR brachytherapy in the treatment of locally advanced cervical carcinoma. *Medical Physics*, 38(2): 598. https://doi.org/10.1118/1.3512803.

Wootton, J.H., Prakash, P., Hsu, I.C.J. and Diederich, C.J. (2011). Implant strategies for endocervical and interstitial ultrasound hyperthermia adjunct to HDR brachytherapy for the treatment of cervical cancer. *Physics in Medicine & Biology*, 56(13): 3967. https://doi.org/10.1088/0031-9155/56/13/014.

Wu, F., Wang, Z.B., Cao, Y. De, Chen, W.Z., Bai, J. et al. (2003). A randomised clinical trial of high-intensity focused ultrasound ablation for the treatment of patients with localised breast cancer. *British Journal of Cancer*, 89(12): 2227–2233. https://doi.org/10.1038/sj.bjc.6601411.

Wydra, A. and Maev, R.G. (2013). A novel composite material specifically developed for ultrasound bone phantoms: Cortical, trabecular and skull. *Physics in Medicine and Biology*, 58(22): 303–319. https://doi.org/10.1088/0031-9155/58/22/N303.

Zell, K., Sperl, J.I., Vogel, M.W., Niessner, R., Haisch, C. et al. (2007). Acoustical properties of selected tissue phantom materials for ultrasound imaging. *Physics in Medicine and Biology*, 52(20): 475–484. https://doi.org/10.1088/0031-9155/52/20/N02.

Zhang, L. and Wong, F. (2020). A high-intensity focused ultrasound surgery theater design in a private clinic. *Gynecology and Minimally Invasive Therapy*, 9(1): 1–5. https://doi.org/10.4103/GMIT.GMIT_108_19.

Zhou, Y., Zhai, L., Simmons, R. and Zhong, P. (2006). Measurement of high intensity focused ultrasound fields by a fiber optic probe hydrophone. *The Journal of the Acoustical Society of America*, 120(2): 676–685. https://doi.org/10.1121/1.2214131.

Zubair, M., Adams, M.S. and Diederich, C.J. (2021). Deployable ultrasound applicators for endoluminal delivery of volumetric hyperthermia. *Https://Doi.Org/10.1080/02656736.2021.1936216*, 38(1): 1188–1204. https://doi.org/10.1080/02656736.2021.1936216.

CHAPTER 9

Biological Effects of Thermal Therapies (EM Waves and Mechanical Waves)

Christian Chapa González,[1,*] *Citlalli J. Trujillo-Romero*[2] and *Raquel Martínez-Valdez*[3]

Thermal therapies based on the exposition of the tumor to electromagnetic (EM) fields or ultrasonic energy produce several thermal and biological effects. Thermal therapies are based on the application of non-ionizing radiations and their expected outcome is tissue heating. However, increase in the tumor temperature is required up to therapeutic levels (hyperthermia or thermal ablation) without affecting the surrounding healthy tissue. Thermal tissue injuries will always depend on the energy/powers applied as well as the temperature attained and treatment time. The temperature increase produces several biological effects not only on the healthy tissue but also on the tumor. Physiological changes such as blood perfusion, vascular permeability, and metabolism are modified by the temperature increase. These conditions, together with the rates of cell survival due to heat, tumor conditions, and tissue thermotolerance are the main reasons for thermotherapy success.

[1] Autonomous University of Ciudad Juarez. Av. del Charro 450 nte. Ciudad Juárez, Chihuahua, México. C.P. 32310.
[2] Division of Medical Engineering Research, National Institute of Rehabilitation LGII. Calz. Mexico Xochimilco No. 289, Col. Arenal de Guadalupe, Mexico City, 14389, Mexico.
[3] Biomedical Engineering Program, Polytechnic University of Chiapas. Carretera Tuxtla Gutiérrez – Portillo Zaragoza km 21+500, Col. Las Brisas, Suchiapa, Chiapas, 29150, Mexico.
Emails: cjtrujillo@inr.gob.mx; rmartinez@ib.upchiapas.edu.mx
* Corresponding author: christian.chapa@uacj.mx

9.1 Introduction

The use of thermal therapies based on the exposure of biological tissues to electromagnetic (EM) waves or mechanical waves has shown great potential in the treatment of various diseases, including cancer. These therapies rely on the ability of thermal energy to induce biological changes in the target tissue, leading to tumor destruction or growth inhibition. However, the effectiveness of these therapies depends on achieving the appropriate temperature increase in the target tissue while minimizing damage to the surrounding healthy tissue.

This section of the book focuses on the thermal and non-thermal effects induced by EM and mechanical waves in biological tissues. The section aims to provide a comprehensive overview of the underlying physical and biological mechanisms of thermal therapies and their potential applications in clinical settings. The section will cover topics such as tissue injury, exposure guidelines for electromagnetic radiation, and non-thermal effects. It will also highlight the challenges and limitations of thermal therapies and the need for standardized protocols and safety guidelines to ensure their safe and effective clinical application.

9.2 Thermal effects

Thermal effects are characteristic of the dielectric heating brought on by common non-ionizing radiation (NIR), such as microwaves. The only effect of NIR that makes sense is temperature increase since it does not have enough quantum energy to interact with atoms' outer orbitals or to disrupt intra-or extra molecular interactions. It has been established that this occurs, even for very small temperature changes, because cells have a sophisticated system to respond to temperature changes, including chemical cascades and heat shock proteins. Some compounds, like proteins, may get denaturized when exposed to significant temperature variations for sufficiently long periods (The Edumed Institute for Education in Medicine and Health, 2010).

Thermal therapies refer to all those treatments in which thermal energy is transferred by using an external source (Radiofrequency, Microwaves, Ultrasound and more), to the human body. From the clinical and oncological point of view, the main objective of thermal therapy is achieving an effective temperature increase over the region under treatment (tumor), without affecting the surrounding healthy tissue. Tissue damage will depend mainly on the energy/powers applied. The temperature increases reached over the tissue can generate biological effects, which in some cases can become adverse health effects. The strength of the Electric (*E*) field and the quantity of energy in the photons play an important role in how electromagnetic (EM) waves affect biological systems. Even at high intensities, non-ionizing radiation cannot ionize any living thing; nevertheless, they do have other biological effects, like tissue temperature increase. Biological effects that EM waves can have occasionally, but not always, are negative impacts on human health. Therefore, it is crucial to comprehend how these two differ (biological and adverse effects) (WHO, 1998):

- **Biological effect:** It happens when exposition to EM waves results in notable physiological changes in the biological system, like tissue heating in hyperthermia treatment.
- **Adverse health effect:** It happens when a biological effect exceeds the usual range; consequently, the biological system cannot compensate for it. Therefore, a negative health state is presented.

9.2.1 Biological aspects

Several modifications in tissue physiology are brought on by heat, including a rise in blood flow (perfusion), vascular permeability, and metabolic activity. However, perfusion is the most crucial physiological factor in thermotherapies. Mainly, because any tissue that is heated experiences a variety of physiological changes, the bulk of which are brought on by modifications in blood flow. The ability to heat tissues is significantly influenced by the tissue blood supply because blood flow is the main mechanism used to remove heat from tissues. Blood flow measurements in some solid tumors may be higher than in healthy tissues, the tumor blood supply is typically more primitive and chaotic than that of normal tissues, which can lead to regions that are deficient in nutrients, oxygenated poorly, and extremely acidic. Cells under these circumstances are typically more susceptible to the cytotoxic heat effects. The main side-effect of thermotherapies is either an elevated body core temperature or elevated organ temperatures. Moreover, the human body needs to regulate its internal temperature because it can impact many cellular structures and metabolic pathways. When we think about the biological aspects of thermotherapies, it is important to think about how electromagnetic (EM) fields interact with living systems at molecular, subcellular, cellular, and organ levels as well as throughout the whole body. Therefore, it must be clear that the outcome of the interaction of the human body with the thermotherapies will depend strictly on the response of the living system and the heat source. The biological effect produced by the exposition of EM fields can be divided into 3 main categories. (1) thermal, (2) a-thermal and (3) non-thermal effects. However, in thermotherapies, the thermal effects play the most important role (Habash et al., 2006).

Since tissues are electrically conductive, RF electric fields have a thermal influence on them. The temperature rises because the E field in the tissue produces currents, which results in the absorption of dissipated energy. The interaction of EM waves with the human body is fundamentally related to three factors: (1) cellular survival, which is influenced by treatment temperature and duration; (2) tumor conditions, like hypoxia, low pH, poor nutrition and sensitivity to heat of tumor/healthy tissue which may be advantageous for thermotherapies; and (3) thermotolerance, which is a property that cells develop over time (James R Lepock, 2003) (Gianfranco Baronzio et al., 2014). These factors are strongly related to the temperature increase in the biological tissues due to the thermotherapies.

On the other hand, microwaves (MW) with frequencies higher than 10,000 MHz (short wavelengths) merely heat the skin's surface and do not penetrate underneath it. Long-wavelength microwave radiation with a frequency under 150 MHz enters the body with extremely minimal energy losses. The testicles and the eyes are the bodily parts that are most vulnerable to thermal impacts for frequencies between 150 and 10,000 MHz. Moreover, the testicles are extremely vulnerable to temperature increases (McRee, 1974).

9.2.2 Biological tissues and temperature increase

The interaction of external heat sources with tissues is mainly due to three fundamental aspects, described below (Gianfranco Baronzio et al., 2014) (James R Lepock, 2003) (Hildebrandt and Wust, 2007):

1) Cell survival.
2) Tumor conditions.
3) Thermotolerance developed by cells.

Cell survival is closely related to the temperature attained and treatment time or exposure to the heat source. Temperatures above 40°C are considered cytotoxic; therefore, they can cause cell death. However, this will depend mainly on the exposure time (James R Lepock, 2003). Research has shown that the sensitivity of tumor cells to high temperatures is higher than the one of healthy cells (Chicheł et al., 2007). Lethal effects may occur at specific exposure levels and timeframes, therefore there is always a temperature that may be applied for a determined time duration to produce cell death. The higher the final temperature, the lower the exposure time and vice versa. For several biological systems, the correlation between temperature and treatment time during hyperthermia has been documented (Roizin-Towle and Pirro, 1991). The temperature has an impact on a cell's growth rate, which typically rises as temperature rises. However, there is a maximum temperature at which growth abruptly halts (40°C–41°C) (James R Lepock, 2003) (James R Lepock, 2005) (Sawaji et al., 2002), which benefits cancer treatments. Although how high temperatures kills cancer cells has been studied for several years, the precise mechanism that causes damage when applying hyperthermia remains

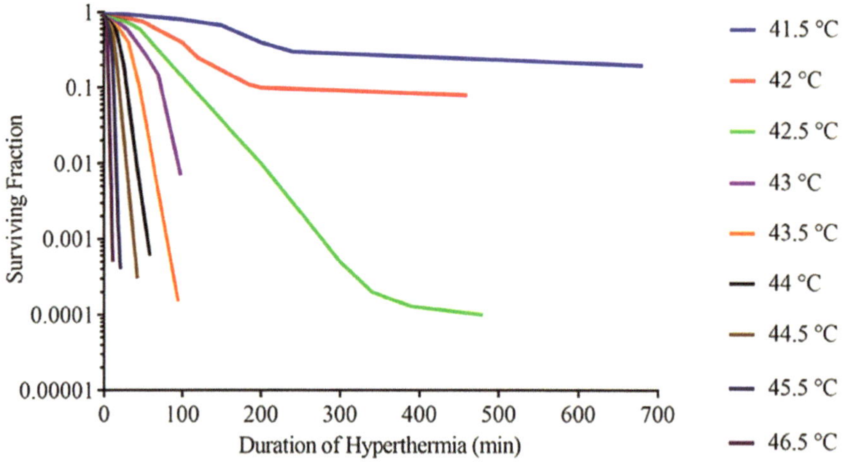

Figure 9.1. Survival of Chinese hamster cells as a function of temperature and time of exposure or treatment. Taken from ref. (Chang et al., 2018) under CC BY.

unknown (Miles, 2006). However, it is widely believed that tumor cells perish due to the thermal denaturalization of vital cellular targets. To cause irreversible damage to cancer cells, it is necessary that during hyperthermia therapy temperatures between 42°C and 45°C are achieved. In addition, these temperatures must be maintained for periods between 30 min–60 min. Several authors have reported the relationship between treatment time, and cellular response in hyperthermia for various biological systems (Gerweck, 1985) (James R Lepock, 2003). Figure 9.1 depicts the survival of hamster cells depending on temperature and treatment time.

1) When temperatures (hyperthermia) are reached just above the maximum temperature that allow cell growth, three important cellular responses are generated, which are: cytotoxicity, radio sensitization, and thermotolerance. These responses are due to alterations in the molecular pathways, caused by the increase in temperature (James R Lepock, 2005). It is known that a temperature between 42°C–45°C applied between 30 min–60 min causes irreversible damage. At temperatures above 50°C, the time needed to generate irreparable cell damage decreases exponentially. Temperatures between 100°C–300°C produce the vaporization of water contained in tissues; while temperatures above 200°C produce the carbonization of tissues (Habash et al., 2006).

   Protein damage is one of the most relevant molecular effects among the biological effects produced by hyperthermia, reaching temperatures between 39°C–45°C. In this temperature range, activation of protein denaturation and dead cells due to heat is observed (Dewey, 2009).

Although protein damage is strongly linked to the biological effects of hyperthermia, little is known about what ultimately causes cell death. In cell survival curves a different slope is observed for each of the temperatures reached depending on the temperature and exposure time; this can be caused by different mechanisms of cell destruction or due to the thermotolerance of the cells; that is, due to the resistance to heat that the cells develop when exposed to these temperatures for a long time.

The unfolding and aggregation of proteins are caused by the rise in temperature. Because there are so many proteins and so much DNA packed inside the nucleus, protein aggregation and misfolding should have a big effect there. Additionally, proteins in the nucleus are essential for controlling DNA replication, transcription, and repair. Changes in these proteins' structure and function should, therefore, have a significant impact on DNA replication and result in chromosomal abnormalities, genomic instability, incorrect chromosome segregation, and cell death (Lepock, 2004). DNA damage is not directly caused by hyperthermia (Roti, 2008) however cells in any stage of the cell cycle may be stopped by heat shock. It should be noted that temperature duration and circumstances have an important effect on cell cycle transitions, arrests, and cell death. Temperature sensitivity is reliant on the cell type and, in part, on the cell phase of the cell cycle (Kühl and Rensing, 2000). Consequently, the same heat stress may result in different effects when several cell lines are studied or when cells are at different stages of the cell cycle. The type of temperature-induced cell death is dependent on the type of cell and the final temperature (Roti, 2008).

2) The tumor conditions are mainly hypoxia, low pH, poor nutrition, and sensitivity to heat of healthy tissues and tumors, among others. The pH of tumors is typically lower than that of healthy tissues. Tumors are more acidic than healthy tissue; moreover, the pH drops much more in the hyperthermia treatment. This increased acidity enhances the thermal injury caused to the cancer cells during hyperthermia, together with inadequate nutrition and hypoxia brought on by restricted blood flow (Vaupel et al., 1989) (Østergaard et al., 2013).

The hypoxic state of cancer cells makes them more resistant to hyperthermia treatment. Increased blood perfusion, metabolism, oxygenation, among others. are the effects of tissue heating. Nevertheless, blood perfusion is one of the most important parameters because different biological changes occur when it is modified due to the temperature increase (Elming et al., 2019). Generally, blood perfusion in a solid tumor is lower compared to that present in healthy tissue (GianFranco Baronzio et al., 2013) (Jain, 1988). The vascular beds of a tumor tend to be chaotic and poorly

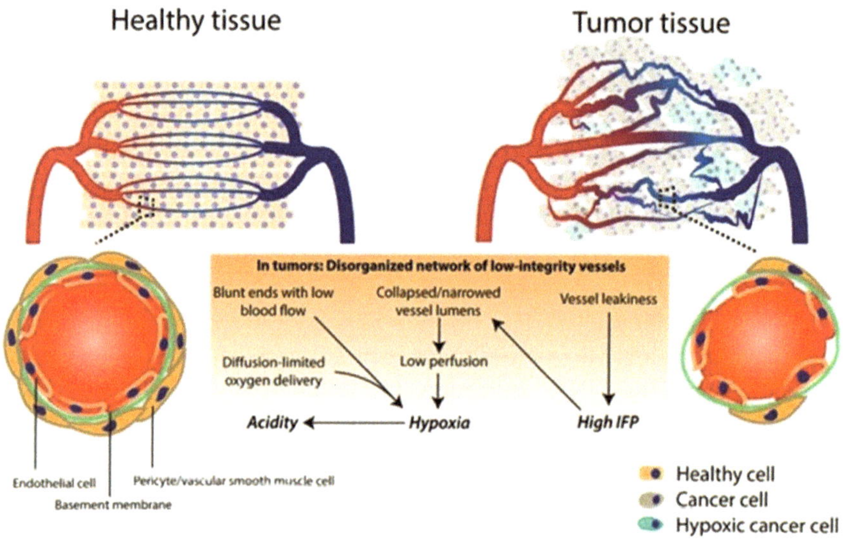

Figure 9.2. Comparison of blood vasculature in healthy tissue and tumor. Taken from (Schaaf et al., 2018) under Creative Commons Attribution 4.0 International License.

organized, causing their blood supply to be temporally and spatially unbalanced. This causes different regions of a tumor to be hypoxic/acidic and resistant to chemo- and radiotherapy (GianFranco Baronzio et al., 2013). However, this hostile tumor environment does not interfere with the action of hyperthermia. In fact, some studies have shown that hyperthermia could specifically benefit from the hypoxic state of cells (Gianfranco Baronzio et al., 2014). Several studies have shown that if a tumor is heated to temperatures between 41°C and 43°C, blood flow changes by a factor of approximately two; however, blood flow in healthy tissue increases by a factor of up to 10 when exposed to temperatures between 42°C–45°C (Gerweck, 1985) (Ward-hartley, 1984). These indicate that in hyperthermia the dissipation of heat produced by blood flow is lower in tumors than in healthy tissues (Østergaard et al., 2013. Figure 9.2 shows the comparison of the blood vasculature where the tumor microenvironment presents a chaotic blood vessel network, with scarce pericyte coverage and loose interendothelial junctions. In the same way, the blunted or collapsed vessels give rise to localized regions in the tumor where nutrients (including oxygen) are scarce. Consequently, tumors reach higher temperatures and greater damage is produced.

On the other hand, some studies have shown that metabolic state and energy deprivation increase sensitivity to heat (Gianfranco Baronzio et al., 2013) (Hahn, 1974). Table 9.1 describes some of the effects produced

Table 9.1. Comparison of the effects produced by chemotherapy, radiotherapy, and hyperthermia (Gianfranco Baronzio et al., 2014).

Cells of the tumor structure affected	Chemotherapy	Radiotherapy	Hyperthermia
Oxygenation of cells	++	+++	+++
Hypoxic cells	+-	---	+++
Vascular structure	+	++	+++
Estroma	+	+	++
Microcirculation	-	+	++

by two of the most common cancer treatments (radio and chemotherapy) and hyperthermia on the tumor cells.

Hyperthermia can kill hypoxic cells directly, depending on the temperature reached and the treatment time, i.e., the generated effect is larger when applied at a higher temperature and for a longer period of time (Elming et al., 2019) (Roizin-Towle and Pirro, 1991). Temperatures below 42°C generate low levels of cell death. If a greater effect is to be achieved it is necessary to prolong the exposure time (> 1 hr) (Elming et al., 2019). Cell death increases significantly when temperatures above 42°C are reached and if cells are deprived of oxygen and/or pH (Gerweck, 1985). These adverse microenvironmental conditions are typical of hypoxic regions of a tumor.

Necrosis and apoptosis are two of the many mechanisms by which heat kills cells. Moreover, chromosome abnormalities, cytoskeletal damage, modifications to membrane fluidity and transport, and metabolic adjustments are some biological impacts of cell heating (James R Lepock, 2003). However, the most important effect on cell death caused by hyperthermia is protein denaturation. It is closely related to cell death caused by the time-temperature ratio that is reached at temperatures above 42.5°C. At temperatures above 43°C and below cell apoptosis, necrosis predominates (Elming et al., 2019). Several studies have shown that cell destruction caused by temperature increases substantially if cells have a low pH (Overgaard and Bichel, 1977). Hyperthermia can also induce vascular damage, in this way tumor cells die indirectly.

3) The term "thermotolerance" refers to the improved thermal resistance that cells develop as a result of prior heat treatment. Thermotolerance is crucial since it is one of the main factors affecting how sensitive cells respond to thermotherapies. That is, thermotolerance is a general phenomenon that indicates a transient resistance to thermal stress (Chicheł et al., 2007). Heat shock proteins are the cause of the thermotolerance effect. After experiencing a heat shock, both healthy and cancerous cells produce heat shock proteins (Kampinga, 2006). Thermotolerance causes thermally treated cells to become more heat resistant, which is problematic because thermotolerance guards against hyperthermia in cells. The impact of this phenomenon, however, only lasts a few days. The onset of thermotolerance in cells might vary. Depending on the intensity of the initial treatment, cells can develop thermotolerance at various times. Therefore, it is important to schedule sufficient time spans between hyperthermia treatments to prevent the thermotolerance effect. This phenomenon makes it impossible to apply two different sessions of hyperthermia in a period of less than 48–72 hours; that is, you must wait until the thermal resistance decays to a negligible level (Chicheł et al., 2007). Higher temperatures and exposure times of tumor tissue produce a more lethal effect while reducing induced thermotolerance.

9.2.3 Tissue injury

The growth rate of cells is the most obvious characteristic that temperature can affect. There are three important cellular reactions to thermotherapy in the hyperthermic area above the maximum growth temperature, namely cytotoxicity, radio sensitization, and thermotolerance. These alterations at the cellular level are the result of changes in molecular pathways brought on by changes in temperature. According to traditional hyperthermia, cellular damage must occur at temperatures between 42°C–45°C applied for intervals of 30 to 60 minutes. The required time to produce irreparable cellular damage decreases exponentially as tissue temperature increases to 50°C. Cell death occurs as a result of protein denaturation. Between 100°C and 300°C, tissue water vaporization is overlaid in this process; moreover, between 300°C and 1000°C carbonization, charring, and smoke production also take place. Table 9.2 describes the physical and biological effects that can occur in

Table 9.2. Physiological effects produced in biological tissues due to an increase in temperature.

Temperature (°C)	**Exposition time**	**Biological and physical effect**
< –50	> 10 min	Whole cellular destruction and freezing.
0–25		Reduction in blood flow, metabolic rate, death by hypothermia, cell permeability decrease.
40–46	30–60 min	Enhanced blood flow, permeability, pH, development of thermotolerance, hyperthermic death, optical tissue properties are modified by the temperature.
47–50	> 10 min	Denaturation of proteins, coagulation, and necrosis.
> 50	2 min	Cells die, necrosis and coagulation.
60–100	< 1 min	Denaturation of proteins, coagulation, and ablation.
>100	< 1 min	Shrinking of the cell and the extracellular steam vacuole.

biological tissues due to an increase in their temperature (Habash et al., 2006) (James R Lepock, 2005) (Stauffer, 2005).

On the other hand, heat-induced tissue damage happens in two separate phases (Habash et al., 2006).

Phase 1: Refers to the direct thermal injury that is determined by the amount of energy applied to the tumor, its biology, and microenvironment. The cell membrane in this instance is the cell component most susceptible to heat damage.

Phase 2: Refers to the secondary injury produced after the application of focused hyperthermia, which causes progressive tissue damage. Numerous variables contribute to this injury, including microvascular damage, ischemia, the activation of apoptosis, altered cytokine production, and immune response regulation. The effects of heat depend on the temperature attained, which is associated to the total amount of applied energy, the rate of heat removal or dissipation, and the tissue's thermal sensitivity.

9.3 Non-thermal effects

During ultrasound-based thermal therapies, the biological effects produced by the interaction between acoustic waves and tissue are known as thermal and non-thermal effects. Thermal effects are caused by the temperature increase in tissue due to ultrasound as the pressure fluctuations act on microscopic tissue structures leading to a shearing motion (ter Haar and Coussios, 2007). Nonetheless, mechanical effects are present during ultrasound exposure which are accompanied by heating as a result of wave-tissue interaction (Baker et al., 2001). These effects are taken advantage of in applications such as histotripsy, sonophoresis, sonoporation, lithotripsy, ultrasonic-enhanced drug delivery (Braunstein et al., 2022; Hill et al., 2004), and sonoluminescence (Hill et al., 2004).

When a soundwave travels through a viscous-elastic medium, such as soft tissues, the oscillatory action of acoustic pressure makes gas bubbles of variable submicroscopic dimensions due to rarefaction and compression phases in the medium (Hill et al., 2004; ter Haar and Coussios, 2007). During the rarefaction or negative pressure phase, the bubble or cavity allows more gas to diffuse into it (it expands); when the half-cycle passes, i.e., in the compression or positive pressure phase, a lower amount of gas diffuses out of the bubble (it collapses) (Hill et al., 2004; Nyborg, 2001). After several cycles, the bubble reaches a resonant size in accordance with the ultrasonic wavelength, then it continues expanding and collapsing, which may result in period-doubling oscillations of the cavity wall (ter Haar and Coussios, 2007); this effect is known as stable or non-inertial cavitation (Hill et al., 2004). On the other hand, at higher pressure amplitudes, bubbles grow abruptly from their initial size and may collapse to disintegrate into small fragments (Hill et al., 2004; ter Haar and Coussios, 2007); this effect is known as inertial cavitation.

Bubble formation is not always due to pressure oscillations, when the tissue temperature increases abruptly near boiling levels, vapor bubbles can grow rapidly to a large size and explode (Khokhlova et al., 2006). Both boiling bubbles and cavitation modify the ultrasonic beam pattern due to scattering and reflections (Chavrier et al., 2000; Khokhlova et al., 2006) which may result in errors in the measurement of acoustic properties of the medium (Hill et al., 2004). Either cavitation is inertial or non-inertial and enhanced heating deposition near the cavities will occur because the acoustic energy is absorbed easily (ter Haar and Coussios, 2007). The formation of bubbles or cavities in a medium due to its exposure to an ultrasonic field is usually defined as acoustic cavitation; and it depends on acoustic pressure amplitude, frequency, temperature and the transducer drive mode (continuous wave or pulsed), among others (Hill et al., 2004).

During ultrasound thermal ablation, high acoustic pressure is efficiently localized into the body to attain a specific target. The well-localized energy results in a rapid temperature increase; therefore, cavitation may occur causing damage to surrounding healthy tissues. In ultrasonic imaging-guided therapy, formation of cavities can be detected because the scattered energy produces a hyperechoic (brighter) region in the B-mode image (Chavrier et al., 2000; Khokhlova et al., 2006). Conversely, although magnetic resonance imaging (MRI) allows tracking thermal maps during sonication, thermometry is affected by cavitation and, in some cases, leads to stopping an in situ ablation therapy treatment (Hu et al., 2020). Nevertheless, by injecting an ultrasound contrast agent in the sonicated region approximately 30 min after the treatment has finished, tissue damage can be evaluated, such is the case of the blood-brain barrier (BBB) opening enhancing drug delivery into the brain (Hu et al., 2020; Ji et al., 2021).

The presence of cavities, vapor gas bubbles, or microbubbles during sonication acts as a secondary source of sound (Hill et al., 2004; Hodnett and Zeqiri, 2004) because ultrasonic energy is trapped inside them (ter Haar and Coussios, 2007) and then it is redistributed as spherical waves with a different acoustic spectrum (Chavrier et al., 2000; Hodnett and Zeqiri, 2004; Ji et al., 2021). The frequency components emitted by cavitation are comprised of subharmonics, the second harmonic of the fundamental frequency, ultra-harmonics, and broadband signals (Braunstein et al., 2022; Chavrier et al., 2000; Hill et al., 2004; Hodnett and Zeqiri, 2004) which can be monitored by a hydrophone or a passive cavitation detector (Hodnett and Zeqiri, 2004; Ji et al., 2021). These acoustic emissions could be considered indicators to determine the cavitation dose (ter Haar and Coussios, 2007) or detect the threshold of inertial cavitation activity (Hodnett and Zeqiri, 2004). Increase in membrane permeability for BBB delivery of therapeutic drugs to the central nervous system (Hu et al., 2020), enhance thermal effects, drug or gene delivery (Braunstein et al., 2022; Chavrier et al., 2000; Hill et al., 2004; Hu et al., 2020; ter Haar and Coussios, 2007) are some effects caused by cavitation. Cavitation induced by focused ultrasound has been shown to sensitize cell cancer to radiotherapy application, reduced cell clonogenic survival (head, neck, and prostate cancer cells), and a reduction of metabolic activity and loss of the cell invasion capability (Hu et al., 2020). On the other hand, escaping vapor bubbles might induce undesirable vessel rupture (Chen et al., 2003), lesion formation outside the target area, and unwanted red blood cell extravasation due to cavitation (Chen et al., 2003; Ji et al., 2021).

Radiation forces and acoustic microstreaming are other non-thermal mechanisms that can occur during ultrasonic exposure (Baker et al., 2001; Hill et al., 2004; ter Haar and Coussios, 2007). Radiation forces are present in the boundaries of the inhomogeneous structures of the propagation medium, and have a steady state component useful for measuring average spatial intensities; also, they are responsible for blood cell stasis (Hill et al., 2004). Microstreaming is the result of fluid motion during acoustic energy absorption by a liquid. The gradient velocities formed by the acoustic streaming may increase heat transfer, damage molecules and cells located in the vicinity of the streaming fields (Hill et al., 2004), alter membrane permeability, and stimulate cell activity (Baker et al., 2001).

Ultrasound ablation therapy effects in prostate cancer, after histopathologic analysis, showed that epithelium nuclei and stroma cells were pyknotic or absent, indicating cell necrosis (Van Leenders et al., 2000). It was also observed that carcinoma did not show nucleic or cell membranes, and organelles were not identified, but chromatin with a good shape pattern had appeared (Van Leenders et al., 2000). On the other hand, histologic damage due to high-intensity focused ultrasound (HIFU) was compared with heat therapy in human prostate cancer cell lines; this study showed that some enzymes and antibody activation are related to instantaneous temperature (Ide et al., 2008). Heat treatment has shown that temperature increase induces early apoptosis in the cell lines LNCaP y DU-145 (Ide et al., 2008). (Biermann et al., 2010) reported the effects of HIFU in 25 benign and malignant prostate biopsies using different activation enzymes.

In search of a relation between acoustic intensity and HIFU exposure time to be considered for treatment, liver line cells H22 were used to show the mortality rate (Wang et al., 2005). Another research group (Fan et al., 2012) realized experimentation in 50 rabbits to observe the combined effects of HIFU and ultrasonic contrast agents. The results showed that cancerous cells' nuclei were abnormally bigger, the chromatin was thicker, and the nuclei were visible; besides, necrotic tissue was homogeneous and did not present cell structures. Coagulative necrosis without living cell debris was found in solid malignant tumors in samples obtained from 164 patients undergoing HIFU treatment (Feng Wu et al., 2001). It was also observed that the lesion contour area was replaced by mature fibrous tissue, while the lesion was partially absorbed and replaced by new tissue (Feng Wu et al., 2001).

Wu et al. (Feng Wu et al., 2004) report an analysis made in blood samples; results suggest that HIFU may have activated cell immunity of T lymphocytes. This, in the long term, could develop a defense against cell cancer from the same origin, while primary cancer is treated. Other studies (Ren et al., 2007; Feng Wu et al., 2004) have determined coagulative necrosis in malignant tumor biopsies after observing distortion of tumor cells, pyknotic nuclei, nuclei reduction, cell debris, and hyalinization. Endothelial death in 187 uterine fibrosis cases was determined by cytomembrane disruption, karyorrhexis, and cyst implosion (Ren et al., 2007). On the other hand, necrotic changes like edema in the stroma and vascular congestion, coagulative necrosis with loss of surface epithelium, and hemorrhage were observed in the treatment of cataracts in 18 rabbit eyes (Aptel et al., 2010).

Results from studies realized in breast cancer after HIFU show that cancerous cells lost their capability to proliferate, invade and metastasize (Wu et al., 2003), and coagulate necrosis (Niu et al., 2010; Feng Wu et al., 2005). Besides, cell death was detected (Wu et al., 2006, 2007; Feng Wu et al., 2007), organelles disappeared (Niu et al., 2010; Wu et al., 2006), and plasmatic membranes and intracytoplasmic organelles got destroyed (Wu et al., 2006).

In acoustic hemostasis, thermal coagulation in the vessel walls and mechanical effects (microstreaming and radiation forces) may influence vessel occlusion to stop the bleeding (Vaezy et al., 2001; Vaezy and Zderic, 2007). Acoustic microstreaming in the focal zone prevents bleeding by creating stream flows opposing blood flow causing it to stay within the vessel (Goertz, 2015). Higher acoustic intensities are needed in big vessels in order to cause thermal injury during blood flow; as a result, the intensity level, time of exposure, and frequency must be carefully selected in order to generate hemostasis (Goertz, 2015).

9.4 Exposure guidelines for electromagnetic radiation

There is currently a great deal of worry about electromagnetic radiation (EMR) exposure from mobile phones and other EM equipment, including that used for medical diagnosis and treatment. Medical exposures can result from tools used in thermotherapies to treat cancerous tumors, electrosurgical tools used to cut and weld tissues, and diagnostic tools like magnetic resonance imaging (MRI). Guidelines for EM exposure are based on the evaluation of the biological impact and the potentially harmful implications on human health. Any observable changes in a biological system as a result of exposure to EM fields are referred to as biological effects. However, not all biological effects will necessarily be bad. Based on a rigorous analysis of the scientific data that is currently available, the exposure levels that are likely to be deemed hazardous to human health are established. Maximum permitted exposure (MPE) values, often known as threshold levels, are standards for human exposure to EM fields. Over the past decade , recommendations for RF exposure limits have undergone constant change.

Basic restrictions on electromagnetic field (EMF) exposure are defined according to the negative impact on health. The EMF frequency affects these fundamental restrictions, which can be described by the current density, the specific absorption rate (SAR), or the power density. According to the International Commission on Non–Ionizing Radiation Protection (ICNIRP), "protection against adverse health effects requires that these restrictions are not exceeded" (The Edumed Institute for Education in Medicine and Health, 2010). The exposure levels known as threshold values are those where the known adverse health effects start to appear, are multiplied by a safety factor of 10 to determine the value for the fundamental limits for employees. The basic limits for exposure to the general public are determined by dividing the threshold values by 50, implying that the exposure guidelines for the general population are five times stricter than the exposure standards for the workplace.

The basic restrictions are physical quantities that are established by the interaction mechanisms that result in negative health impacts. They are, however, a challenge for field measurement; therefore, these reference levels are derived from the fundamental constraints utilizing computational models and measurement techniques. The fundamental limitations up to 10 GHz relate to whole-body and localized SAR as well as current density. In terms of electric and magnetic fields, magnetic flux density, and power density, their respective reference levels are provided. The reference levels for frequencies between 10 GHz and 300 GHz are represented in terms of power density. When the EM field is measured at a distance greater than one wavelength from the antenna at frequencies below 10 MHz, the electric and magnetic fields are uncoupled from one another; therefore, both fields should be measured to determine compliance. If the electromagnetic field (EM field) is measured at a distance larger than one wavelength from the antenna the continuous relationship of the electric and magnetic fields must be accounted for and for this reason, only one field must be measured. Moreover, for frequencies above 10 MHZ, the electric and magnetic fields are coupled; within the far field, both are related

Table 9.3. Basic restrictions and threshold levels reported by the ICNIRP for general public exposure to electromagnetic fields at different ranges of frequencies.

Basic restrictions				Thresholds levels		
Frequency (MHz)	Whole-body average SAR (W/kg)	Localized SAR in head and trunk (W/kg)	Localized SAR in limbs (W/kg)	*E* field (V/m)	*H* field (A/m)	B field (μT)
88–108	0.08	–	–	28	0.073	0.092
54–88 174–216	0.08	–	–	28	0.073	0.092
407–806	0.08	–	–	29.8	0.08	0.099
806–869	0.08	2	4	40	0.1	0.13
890–960	0.08	2	4	41	0.11	0.14
1710–1880	0.08	2	4	56.9	0.15	0.19
1850–1900	0.08	2	4	60.5	0.16	0.20

by the impedance of the medium. Nevertheless, in the near field, both fields are decoupled, therefore, as a conservative approach, the levels of both fields for far fields could be used because their values could not exceed the SAR limits. Table 9.3 shows the basic restrictions and the threshold levels reported by the ICNIRP for general public exposure to electromagnetic fields at the most common frequencies used in telecommunication services (The Edumed Institute for Education in Medicine and Health, 2010). Different medical applications use a variety of EM fields (thermotherapies, magnetic resonance imaging, among others) that most of the time exceed the exposure levels, even with values much higher than the maximum permitted exposure. In cases such as thermotherapies, where the patients are exposed to EM fields, the limits are not considered due to the risk/benefit ratios (Habash et al., 2006).

9.5 Conclusion

The interaction of either electromagnetic or ultrasonic fields with biological tissues can produce heat. The thermal effects in tissue depend on exposure time in a specific temperature range; which can produce destruction (freezing), growth, improve blood perfusion, denaturation, cell structural damage and carbonization. In the case of EMR, the tissue damage strongly depends on the frequency, radiated energy intensity and SAR. On the other hand, tissue damage during ultrasonic ablation may be provoked by thermal and non-thermal effects, and depends on frequency and acoustic intensity deposition in the target. The presence of vapor bubbles or cavities (non-thermal effects) may enhance the temperature deposition, but it also can produce unwanted tissue damage in the surrounding areas of the thermal lesion. However, desirable non-thermal effects are pursuits in applications like lithotripsy, histotripsy, and gene and drug delivery. The resulting thermal bioeffects due to either type of radiated energy has demonstrated damage production at different cell structure levels and sensitized cells for further radio- or chemotherapy application. Nonetheless, energy application protocols and regulations must be standardized to guarantee the safe clinical application of thermotherapies.

Acronym/Abbreviation

BBB	Blood-Brain Barrier
DNA	Deoxyribonucleic Acid
E	Electric Field
EM	Electromagnetic
EMF	Electromagnetic Field
EMR	Electromagnetic Radiation
H	Magnetic Field
HIFU	High-Intensity Focused Ultrasound
ICNIRP	International Commission on Non−Ionizing Radiation Protection
MPE	Maximum Permitted Exposure
MRI	Magnetic Resonance Imaging
MW	Microwaves
NIR	Non-Ionizing Radiation
RF	Radiofrequency
SAR	Specific Absorption Rate

Reference list

Aptel, F., Charrel, T., Palazzi, X., Chapelon, J.-Y., Denis, P. and Lafon, C. (2010). Histologic effects of a new device for high-intensity focused ultrasound cyclocoagulation. *Investigative Ophthalmology & Visual Science*, 51(10): 5092–5098. https://doi.org/10.1167/iovs.09-5135.

Baker, K.G., Robertson, V.J. and Duck, F.A. (2001). A review of therapeutic ultrasound: Biophysical effects. *Physical Therapy*, 81(7): 1351–1358.

Baronzio, GianFranco, Cerreta, V., Baronzio, A., Freitas, I., Mapelli, M. and Gramaglia, A. (2013). *Thermo-Chemo-Radiotherapy Association: Biological Rationale, Preliminary Observations on Its Use on Malignant Brain Tumors*.

Baronzio, Gianfranco, Kiselevsky, M., Ballerini, M., Cassuti, V., Schwartz, L., Freitas, I., Fiorentini, G. and Parmar, G. (2013). Hypoxia immunity, metabolism, and hyperthermia. *Conference Papers in Medicine*, 2013: 1–5. https://doi.org/10.1155/2013/528909.

Baronzio, Gianfranco, Parmar, G., Ballerini, M. and Szasz, A. (2014). A brief overview of hyperthermia in cancer treatment. *Journal of Integrative Oncology*, 03(01). https://doi.org/10.4172/2329-6771.1000115.

Biermann, K., Montironi, R., Lopez-Beltran, A., Zhang, S.B. and Cheng, L.A. (2010). Histopathological findings after treatment of prostate cancer using High-Intensity Focused Ultrasound (HIFU). *Prostate*, 70(11): 1196–1200.

Braunstein, L., Brüningk, S.C., Rivens, I., Civale, J. and Haar, G. ter. (2022). Characterization of acoustic, cavitation, and thermal properties of poly(vinyl alcohol) hydrogels for use as therapeutic ultrasound tissue mimics. *Ultrasound in Medicine & Biology*, 48(6): 1095–1109. https://doi.org/10.1016/J.ULTRASMEDBIO.2022.02.007.

Chang, D., Lim, M., Goos, J.A.C.M., Qiao, R., Ng, Y.Y., Mansfeld, F.M., Jackson, M., Davis, T.P. and Kavallaris, M. (2018). Biologically targeted magnetic hyperthermia: Potential and limitations. *Frontiers in Pharmacology*, 9(AUG). https://doi.org/10.3389/fphar.2018.00831.

Chavrier, F., Chapelon, J.Y., Gelet, A. and Cathignol, D. (2000). Modeling of high-intensity focused ultrasound-induced lesions in the presence of cavitation bubbles. *Journal of the Acoustical Society of America*, 108(1): 432–440.

Chen, W.S., Lafon, C., Matula, T.J., Vaezy, S. and Crum, L.A. (2003). Mechanisms of lesion formation in high intensity focused ultrasound therapy. *Acoustics Research Letters Online-Arlo*, 4(2): 41–46.

Chicheł, A., Skowronek, J., Kubaszewska, M. and Kanikowski, M. (2007). Hyperthermia - Description of a method and a review of clinical applications. *Reports of Practical Oncology and Radiotherapy*, 12(5): 267–275. https://doi.org/10.1016/S1507-1367(10)60065-X.

Dewey, W.C. (2009). Arrhenius relationships from the molecule and cell to the clinic. *International Journal of Hyperthermia*, 25(1): 3–20. https://doi.org/10.1080/02656730902747919.

Elming, P.B., Sørensen, B.S., Oei, A.L., Franken, N.A.P., Crezee, J., Overgaard, J. and Horsman, M.R. (2019). Hyperthermia: The optimal treatment to overcome radiation resistant hypoxia. In *Cancers* (Vol. 11, Issue 1). MDPI AG. https://doi.org/10.3390/cancers11010060.

Fan, Z., Luo, W., Song, Z., Zheng, W., Hu, H., Du, L. and Zhou, X. (2012). Effect of healthy tissue ablation surrounding VX2 rabbit liver tumors by high-intensity focused ultrasound combined with an ultrasound contrast agent. *Journal of Ultrasound in Medicine*, 31(6): 863–871. http://www.ncbi.nlm.nih.gov/pubmed/22644682.

Gerweck, L.E. (1985). *Hyperthermia in Cancer Therapy: The Biological Basis and Unresolved Questions*.

Goertz, D.E. (2015). An overview of the influence of therapeutic ultrasound exposures on the vasculature: High intensity ultrasound and microbubble-mediated bioeffects. In *International Journal of Hyperthermia*, 31(2): 134–144. Informa Healthcare. https://doi.org/10.3109/02656736.2015.1009179.

Habash, R.W.Y., Bansal, R., Krewski, D. and Alhafid, H.T. (2006). Thermal therapy, Part 1: An introduction to thermal therapy. In *Critical Reviews in Biomedical Engineering*, 34(6). https://doi.org/10.1615/CritRevBiomedEng.v34.i6.20.

Hahn, G.M. (1974). Metabolic aspects of the role of hyperthermia in mammalian cell inactivation and their possible relevance to cancer treatment. *Cancer Research*, 34(11).

Hildebrandt, B. and Wust, P. (2007). The biologic rationale of hyperthermia. *Cancer Treatment and Research*, 134(Table 1): 171–184.

Hill, C.R., Bamber, J.C. and Haar, G. R. ter. (2004). *Physical Principles of Medical Ultrasonics* (2nd ed.). John Wiley & Sons, Ltd.

Hodnett, M. and Zeqiri, B. (2004). A detector for monitoring the onset of cavitation during therapy-level measurements of ultrasonic power. *Journal of Physics: Conference Series*, 2: 103–108.

Hu, S., Zhang, X., Unger, M., Patties, I., Melzer, A. and Landgraf, L. (2020). Focused ultrasound-induced cavitation sensitizes cancer cells to radiation therapy and hyperthermia. *Cells*, 9(12). https://doi.org/10.3390/cells9122595.

Ide, H., Nakagawa, T., Terado, Y., Yasuda, M., Kamiyama, Y., Muto, S. and Horie, S. (2008). DNA damage response in prostate cancer cells after high-intensity focused ultrasound

(HIFU) treatment. *Anticancer Res.*, 28(2A): 639–643. http://www.ncbi.nlm.nih.gov/pubmed/18507002.

Jain, R.K. (1988). Determinants of tumor blood flow: A review. *Cancer Research*, 48(10).

Ji, R., Karakatsani, M.E., Burgess, M., Smith, M., Murillo, M.F. and Konofagou, E.E. (2021). Cavitation-modulated inflammatory response following focused ultrasound blood-brain barrier opening. *Journal of Controlled Release*, 337. https://doi.org/10.1016/j.jconrel.2021.07.042.

Kampinga, H.H. (2006). Cell biological effects of hyperthermia alone or combined with radiation or drugs: A short introduction to newcomers in the field. *International Journal of Hyperthermia*, 22(3): 191–196. https://doi.org/10.1080/02656730500532028.

Khokhlova, V.A., Bailey, M.R., Reed, J.A., Cunitz, B.W., Kaczkowski, P.J. and Crum, L.A. (2006). Effects of nonlinear propagation, cavitation, and boiling in lesion formation by high intensity focused ultrasound in a gel phantom. *Journal of the Acoustical Society of America*, 119(3): 1834–1848.

Kühl, N.M. and Rensing, L. (2000). Heat shock effects on cell cycle progression. *Cellular and Molecular Life Sciences*, 57(3): 450–463. https://doi.org/10.1007/PL00000707.

Lepock, J.R. (2004). Role of nuclear protein denaturation and aggregation in thermal radiosensitization. *International Journal of Hyperthermia*, 20(2): 115–130. https://doi.org/10.1080/02656730310001637334.

Lepock, James R. (2003). Cellular effects of hyperthermia: Relevance to the minimum dose for thermal damage. *Int. J. Hyperthermia*, 19(3): 252–266. https://doi.org/10.1080/0265673031000065042.

Lepock, James R. (2005). How do cells respond to their thermal environment? *International Journal of Hyperthermia*, 21(8): 681–687. https://doi.org/10.1080/02656730500307298.

McRee, D.I. (1974). Biological effects of microwave radiation. *Journal of the Air Pollution Control Association*, 24(2): 122–127. https://doi.org/10.1080/00022470.1974.10469899.

Miles, C.A. (2006). Relating cell killing to inactivation of critical components. *Applied and Environmental Microbiology*, 72(1): 914–917. https://doi.org/10.1128/AEM.72.1.914-917.2006.

Niu, L., Wang, Z., Zou, W., Zhang, L., Xiang, L., Zhu, H., Chen, W., Bai, J. and Wu, J. (2010). Pathological changes on human breast cancer specimens ablated *in vitro* with high-intensity focused ultrasound. *Ultrasound Med. Biol.*, 36(9): 1437–1444. https://doi.org/10.1016/j.ultrasmedbio.2010.05.016.

Nyborg, W.L. (2001). Biological effects of ultrasound: Development of safety guidelines. Part II: General review. *Ultrasound in Medicine and Biology*, 27(3): 301–333.

Østergaard, L., Tietze, A., Nielsen, T., Drasbek, K.R., Mouridsen, K., Jespersen, S.N. and Horsman, M.R. (2013). The relationship between tumor blood flow, angiogenesis, tumor hypoxia, and aerobic glycolysis. *Cancer Research*, 73(18): 5618–5624. https://doi.org/10.1158/0008-5472.CAN-13-0964.

Overgaard, J. and Bichel, P. (1977). The influence of hypoxia and acidity on the hyperthermic response of malignant cells *in vitro*. *Radiology*, 123(2): 511–514. https://doi.org/10.1148/123.2.511.

Ren, X.L., Zhou, X.D., Zhang, J., He, G.B., Han, Z.H., Zheng, M.J., Li, L., Yu, M. and Wang, L. (2007). Extracorporeal ablation of uterine fibroids with high-intensity focused ultrasound: Imaging and histopathologic evaluation. *J. Ultrasound Med.*, 26(2): 201–212. https://doi.org/26/2/201 [pii].

Roizin-Towle, L. and Pirro, J.P. (1991). The response of human and rodent cells to hyperthermia. *International Journal of Radiation Oncology, Biology, Physics*, 20(4): 751–756. https://doi.org/10.1016/0360-3016(91)90018-Y.

Roti, J.L. (2008). Cellular responses to hyperthermia (40–46°C): Cell killing and molecular events. *International Journal of Hyperthermia*, 24(1): 3–15. https://doi.org/10.1080/02656730701769841.

Sawaji, Y., Sato, T., Takeuchi, A., Hirata, M. and Ito, A. (2002). Anti-angiogenic action of hyperthermia by suppressing gene expression and production of tumour-derived vascular endothelial growth factor *in vivo* and *in vitro*. *British Journal of Cancer*, 86: 1597–1603. https://doi.org/10.1038/sj/bjc/6600268.

Schaaf, M.B., Garg, A.D. and Agostinis, P. (2018). Defining the role of the tumor vasculature in antitumor immunity and immunotherapy article. *Cell Death and Disease*, 9(2). https://doi.org/10.1038/s41419-017-0061-0.

Stauffer, P.R. (2005). Evolving technology for thermal therapy of cancer. *International Journal of Hyperthermia*, 21(8): 731–744. https://doi.org/10.1080/02656730500331868.

ter Haar, G. and Coussios, C. (2007). High intensity focused ultrasound: Physical principles and devices. *International Journal of Hyperthermia*, 23(2): 89–104. http://www.ncbi.nlm.nih.gov/entrez/query.fcgi?cmd=Retrieve&db=PubMed&dopt=Citation&list_uids=17578335.

The Edumed Institute for Education in Medicine and Health. (2010). *Non-Ionizing Electromagnetic Radiation in the Radiofrequency Spectrum and its Effects on Human Health* (Issue June).

Vaezy, S., Martin, R. and Crum, L. (2001). High intensity focused ultrasound: A method of hemostasis. *Echocardiography*, 18: 309–315.

Vaezy, S. and Zderic, V. (2007). Hemorrhage control using high intensity focused ultrasound. *International Journal of Hyperthermia*, 23(2): 203–211.

Van Leenders, G.J., Beerlage, H.P., Ruijter, E.T., de la Rosette, J.J. and van de Kaa, C.a. (2000). Histopathological changes associated with high intensity focused ultrasound (HIFU) treatment for localised adenocarcinoma of the prostate. *Journal of Clinical Pathology*, 53(5): 391–394. http://www.pubmedcentral.nih.gov/articlerender.fcgi?artid=1731195&tool=pmcentrez&rendertype=abstract.

Vaupel, P., Kallinowski, F. and Okunieff, P. (1989). Blood flow, oxygen and nutrient supply, and metabolic microenvironment of human tumors: A review. *Cancer Research*, 49(23).

Wang, X., Yuan, S., Lu, Y., Zhang, J., Liu, B., Zeng, W., He, Y. and Fu, Y. (2005). Growth inhibition of high-intensity focused ultrasound on hepatic cancer *in vivo*. *Journal of Gastroenterology*, 11(28): 4317–4320.

Ward-hartley, K. (1984). Tumor blood flow-characterization, modifications, and role in hyperthermia. *IEEE Transactions on Sonics and Ultrasonics*, SU-31(5).

WHO. (1998). Electromagnetic fields and public health. Physical properties and effects on biological systems. *Fact Sheet*, May, 1–4 TS-Reference Manager DB p4 (via RIS-Export).

Wu, F., Wang, Z.B., Cao, Y.D., Chen, W.Z., Zou, J.Z., Bai, J., Zhu, H., Li, K.Q., Jin, C.B., Xie, F.L., Su, H.B. and Gao, G.W. (2003). Changes in biologic characteristics of breast cancer treated with high-intensity focused ultrasound. *Ultrasound Med. Biol.*, 29(10): 1487–1492. https://doi.org/10.1016/S0301-5629(03)01034-2.

Wu, F., Wang, Z.B., Cao, Y.D., Xu, Z.L., Zhou, Q., Zhu, H. and Chen, W.Z. (2006). Heat fixation of cancer cells ablated with high-intensity-focused ultrasound in patients with breast cancer. *Am. J. Surg.*, 192(2): 179–184. https://doi.org/10.1016/j.amjsurg.2006.03.014.

Wu, F., Wang, Z.B., Cao, Y.D., Zhu, X.Q., Zhu, H., Chen, W.Z. and Zou, J.Z. (2007). "Wide local ablation" of localized breast cancer using high intensity focused ultrasound. *J. Surg. Oncol.*, 96(2): 130–136. https://doi.org/10.1002/jso.20769.

Wu, Feng, Chen, W.-Z., Bai, J., Zou, J.-Z., Wang, Z.-L., Zhu, H. and Wang, Z.-B. (2001). Pathological changes in human malignant carcinoma treated with high-intensity focused ultrasound. *Ultrasound in Medicine & Biology*, 27(8): 1099–1106. https://doi.org/10.1016/S0301-5629(01)00389-1.

Wu, Feng, Wang, Z.-B., Cao, Y.-D., Zhou, Q., Zhang, Y., Xu, Z.-L. and Zhu, X.-Q. (2007). Expression of tumor antigens and heat-shock protein 70 in breast cancer cells after high-intensity focused ultrasound ablation. *Annals of Surgical Oncology*, 14(3): 1237–1242. https://doi.org/10.1245/s10434-006-9275-6.

Wu, Feng, Wang, Z.-B., Zhu, H., Chen, W.-Z., Zou, J.-Z., Bai, J., Li, K.-Q., Jin, C.-B., Xie, F.-L. and Su, H.-B. (2005). Extracorporeal high intensity focused ultrasound treatment for patients with breast cancer. *Breast Cancer Research and Treatment*, 92(1): 51–60. https://doi.org/10.1007/s10549-004-5778-7.

Wu, Feng, Wang, Z.B., Chen, W.Z., Wang, W., Gui, Y., Zhang, M., Zheng, G., Zhou, Y., Xu, G., Li, M., Zhang, C., Ye, H. and Feng, R. (2004). Extracorporeal high intensity focused ultrasound ablation in the treatment of 1038 patients with solid carcinomas in China: An overview. *Ultrasonics Sonochemistry*, 11(3-4): 149–154. https://doi.org/10.1016/j.ultsonch.2004.01.011.

CHAPTER 10

Photothermal Techniques in Cancer Detection-Photoacoustic Imaging

Lara Hernández Gemima[1] and
Flores Cuautle Jose de Jesus Agustin[2,*]

Photothermal techniques have been an outstanding tool in the thermal characterization of materials based on the principle of energy absorption in the sample and the variations that it presents. In photoacoustic techniques, light excitation is absorbed by the sample, and temperature variations occur in the sample that causes a photothermal effect. The thermoelastic expansion due to excitation is then detected by using ultrasound detectors.
The combination of light excitation and ultrasound detection gives information about the sample's optical absorption properties without optical scattering. The thermoelastic expansion information is obtained using an ultrasound detector in different configurations. The information collected by the ultrasound detectors makes it possible to reconstruct images from the thermal expansion produced by the light excitation.

10.1 Introduction

The photothermal techniques give information about the thermal and optical sample properties in a non-invasive way. Those techniques are based on electromagnetic radiation (EM) absorption, provoking a temperature change

[1] Tecnológico Nacional de México/I.T. Orizaba, Orizaba, Ver. México.
[2] CONACYT-Tecnológico Nacional de México/I.T. Orizaba, Orizaba, Ver. México.
Email: larag_139@hotmail.com
* Corresponding author: jflores_cuautle@hotmail.com

in the sample. When a frequency-modulated light excites the sample, the sample temperature changes simultaneously. The sample is subject to frequency-dependent expansion-contraction because thermoelastic effects lead to acoustic waves generation, namely photoacoustic effect.

The photoacoustic effect is mathematically modeled using the pressure wave propagation equation with a term representing the thermal energy deposited by EM radiation per time per volume, analyzed in terms of the velocity potential to determine the acoustic pressure. Photoacoustic signals are detected using a wide-band ultrasound transducer; because of the employed transducer, the already developed ultrasound imaging techniques are used in photoacoustic imaging. Some photoacoustic variants are photoacoustic thermography, photoacoustic microscopy, and photoacoustic endoscopy.

10.2 The photoacoustic techniques

Photoacoustic techniques have become very popular in recent years; particularly, the number of articles on photoacoustics in cancer treatment has quadrupled in the last ten years, as can be seen from a search in databases using the keywords "photoacoustics" and "cancer treatment." An analysis from this search reveals three clusters, each with a central topic: photothermal therapy, tissue, and photodynamic therapy. Among the applications of photoacoustic detection found in the literature angiogenesis (García-Figueiras et al., 2015; Horiguchi et al., 2017; Quiros-Gonzalez et al., 2018; Wu et al., 2016; Yao et al., 2012a, 2012b), melanoma (Neuschmelting et al., 2016; Park et al., 2020; Stoffels et al., 2019; Sunar et al., 2013), breast (Lin et al., 2021; Piras et al., 2011; Thomas et al., 2008), prostate (Borkar et al., 2020; Horiguchi et al., 2017; Peng et al., 2016; Xu et al., 2017), and ovarian cancer (Aguirre et al., 2011; Alqasemi et al., 2013; Salehi et al., 2014) stand out as the most frequent applications of the photoacoustic techniques in cancer detection.

The photoacoustic (PA) technique can be defined as a combined technique comprised of ultrasound and optical imaging techniques; PA combines the advantages of each technique mentioned above. When the pure optical imaging method is used, the optical scattering imposes a limit on the image resolution; therefore, when dealing with high-scattering tissues, the spatial resolution decreases exponentially with depth. On the other hand, ultrasound provides spatial resolution limited by acoustic properties; this means the image diminishes in contrast, as the acoustic impedance between the interfaces of two adjacent layers reduces. The low contrast offered by pure ultrasound does not allow their use to detect early-stage tumors.

The photoacoustic effect combines photons and acoustic vibrations, that is, electromagnetic (EM) radiation and, on the other hand, acoustic waves, which consist of a radiant source inside the sample (source of electromagnetic radiation). In the area of interest, the material absorbs the radiation at a specific wavelength, so that the said material is heated. As a result of this heating, there is a thermal expansion that results in pressure changes that can propagate in and around the area. Suppose a detector is placed, which may be the case of an ultrasonic transducer. In that case, we can detect this acoustic signal and identify the size of the material it absorbs and its position. When the length of the pulse is sufficiently shorter compared to the thermal diffusion time, also named thermal confinement and can be expressed using Equation 10.1, it causes dilation in the analysis zone due to the augmentation in temperature that occurs when the modulated light beam radiates the absorbent medium; after this, a relaxation follows as Equation 10.2 shows. This temperature increase and relaxation is related to vibration phenomena, as Figure 10.1 inset shows. The local increase in temperature, of a few tenths of a degree, is associated with a thermal expansion, resulting in an acoustic wave described by Equation 10.3.

Depending on the experimental conditions under which the measurement is carried out, the acoustic wave can be detected by commonly used ultrasound detectors in imaging. A standard configuration for the PA transducer consists of a light source, a piezo-detector mounted in the same encloser, and acoustic lenses to direct the echo from the sample to the piezo transducer; a coupling layer enhances the echo transduction into the ultrasound detectors. When the image is formed using only the depth-dependent information, the configuration is named PA depth profiling; when either the sample or the transducer is moving, the configuration is termed PA tomography. A particular tomography configuration is built using an ultrasound detector array; then, the information is obtained from each detector placed at different places. The data collected by the ultrasound detectors makes it possible to reconstruct images from the thermal expansion produced by light excitation.

$$\rho C_p \frac{\partial T(\vec{r},t)}{\partial T} = H(\vec{r},t) \tag{10.1}$$

$$\nabla \cdot u(\vec{r},t) = -\frac{p(\vec{r},t)}{\rho v_s} + \beta T(\vec{r},t) \tag{10.2}$$

$$\left(\nabla^2 - \frac{1}{v_s^2}\frac{\partial^2}{\partial t^2}\right) p(\vec{r},t) = -\frac{\beta}{C_p}\frac{\partial H(\vec{r},t)}{\partial t} \tag{10.3}$$

C_p is the specific heat capacity at a constant pressure, ρ is the density, $T(\vec{r}, t)$ the temperature rise due to the irradiated modulated light beam and $H(\vec{r}, t)$ is the thermal energy deposited by EM radiation per time per volume (commonly referred as the heating function), $u(\vec{r}, t)$ the acoustic displacement as a function of position and time for PA can be $\nabla \cdot u(\vec{r},t) = -\frac{p(\vec{r},t)}{\rho}$, β is the isobaric volume thermal expansion coefficient, v_s is the speed of sound at position, $p(\vec{r}, t)$ the sound pressure at a given $\vec{r}$, t, i.e., position and time in an acoustically homogeneous medium determined by focal EM absorption $\left(p(\vec{r},t) = -\rho\frac{\partial\varphi(\vec{r},t)}{\partial t}\right.$, displacement potential).

When the acoustic inhomogeneities are considered, it is necessary to consider the individual scattered waves resulting from each inhomogeneity. A generator can be regarded as a "point source" if its size is substantially smaller than the thermal sound wavelength. Therefore, the medium can be considered a continuous distribution of inhomogeneities non and otherwise acoustically uniform material. It may be seen as a bundle of point-scattered generators if the incidence pressure field and the sources are known. In that case, it is possible to estimate the total area at the detector by summing the individual dispersed waves caused by interactions between a pressurized field and the constituents of the entire medium. "The Green's function is a method for solving the inhomogeneous Helmholtz equation for p that requires rewriting the differential equation as an integral equation for pressure" (Shung and Thieme, 1992). Green's function G provides the observed pressure field at the detector's position r_0 due to a scattering point generator at r.

The wave equation with inhomogeneities in three dimensions can be Fourier and Laplace transformed in space and time respectively and the Green's function can be expressed by Equation 10.4 (Kah and Slaney, 1999; Minghua and Wang, 2002; Norton and Linzer, 1981; Weber and Arfken, 2004; Xu and Wang, 2006)

$$p(\vec{r},t) = \frac{\eta}{4\pi}\iiint \frac{d^3r}{R}\frac{\partial H(\vec{r'},t')}{\partial t'}\Big|_{t'=t-\frac{R}{v_s}} \tag{10.4}$$

where, $R = |r - r'|$, t' is the time delay with $\eta = \beta/C_p$. In time, the excitation wave leads the response by a factor of $\frac{R}{v_s}$; this means the response concerning the excitation is causally delayed by a value of time it takes for the wave to travel from the point where it is excited, which is where the wave is produced up to the point where the wave is measured, divided by the wave speed.

The characteristics of the thermal wave depend on the degree of light absorption by the area under analysis; this absorption is mainly due to the optical absorbers. Figure 10.1 shows a scheme of the thermal wave generation as a result of light absorption. For this reason, the acoustic wave relies on the optical properties of the area of interest and the characteristics

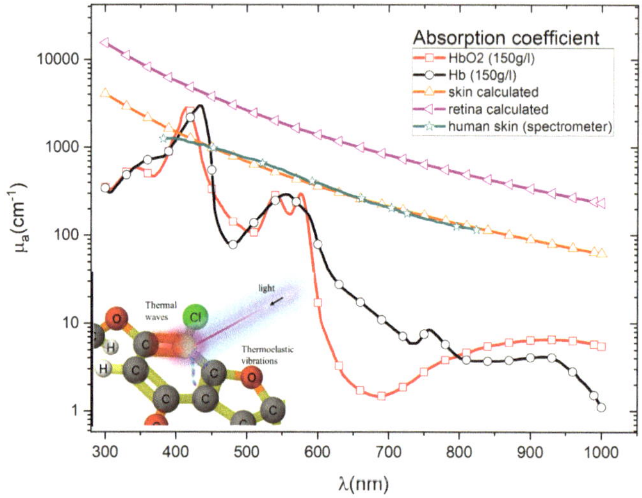

Figure 10.1. The absorption coefficient for different body components (Blakney and Dinwoodie, 1975; Jacques, 1996; Meglinski and Matcher, 2002; Sardar et al., 2009) inset: the thermoelastic wave generation mechanism.

of the excitation light (Oraevsky and Karabutov, 2000). Photoacoustic and ultrasound information can be collected simultaneously and related to each other using an ultrasound sensor. This form of processing helps overcome the limitations of ultrasound or purely optical techniques when used separately.

Since most human structures have distinct optical absorption spectra, a valuable endogenous contrast for photoacoustic imaging is already available. Figure 10.1 shows the absorption coefficient for different body components. Using a dual-wavelength scan and photoacoustic imaging, the oxygen saturation of a mouse ear vasculature at the level of a single red blood cell has been demonstrated. The aptness of photoacoustic imaging to measure the metabolic oxygen rate, which properly represents the oxygen consumption dynamics, has also proven to be very promising. This measure gives more information than static oxygen concentration because it more precisely reflects the functioning of the tissue or tumor.

The photoacoustic technique integrates the ultrasound's excellent resolution of the ultrasonic transducers with the large optical contrast sought in the regions of interest. Therefore, the photoacoustic technique is the option of choice for cancer imaging.

10.3 Ultrasound resolution

Ultrasound has been widely used in biomedical imaging applications; it relies on generating a pulse or train of ultrasound (US) pulses and analyzing the signal produced by the echoes caused when the US hits the object or an acoustic boundary (Hughes, 2001). The US resolution is intimately related to the wavelength and is defined by its frequency and sound velocity in the propagation medium; the expression for the wavelength is expressed by Equation 10.5

$$\lambda = \frac{v_s}{f} \tag{10.5}$$

v_s being the sound velocity in the medium and f the excitation frequency. There are two practical ways to excite the US transducer applying a sinusoidal signal or using an electrical spike. When the US transducer is excited with a sinusoidal signal, it is common to use a pulse train, which means applying a certain number of pulses to the transducer (Aldrich, 2007; Chan and Perlas, 2011). The number of applied pulses is named pulse repetition frequency (PRF). PRF also influences the US resolution (Chan and Perlas, 2011); assuming n is the number of excitation pulses per unit time, the US spatial resolution can be described by Equation 10.6

$$\Delta x = \frac{n\lambda}{2} = \frac{nv_s}{2f} \tag{10.6}$$

As Equation 10.6 demonstrates, the spatial resolution can be improved by either reducing the number of pulses or increasing the frequency of the US. The acoustic impedance of any medium is expressed by Equation 10.7 in terms of medium density

$$Z = \rho v_s \tag{10.7}$$

ρ being the density of the medium, when the ultrasound wave passess though the boundary formed by two layers with different acoustic impedance, the reflection, transmission, and scatter phenomena appear.

The resonant frequency of a piezoelectric transducer relies on the piezoelectric thickness, the material, polarization, the coupling among transducer components, and the type of electric contact that plays an important role (Lara-Hernandez et al., 2017; Wissmeyer et al., 2018). Due to the vast range of cancer cell sizes up to the micron scale, ultrasound detectors must have core bandwidths in the sub-MHz to the hundreds-MHz range.

Although the higher ultrasound transducer frequencies achieve a higher image resolution it lowers the depth resolution. It is also necessary to take care of the maximum laser illumination intensity in the organs/tissues allowed by the ANSI safety standards, for which there are standards in the eye and skin (Ratto et al., 2009). It is worth mentioning that radiation with high photon energy may be hazardous to humans at wavelengths below the visible region. However, for some organs/tissues there is no maximum limit legally allowed until this time, such as the brain, prostate, liver, or heart tissue (Bell, 2020). Optical ultrasonic detection is essential for photoacoustic imaging because it permits non-contact detection avoiding a coupling agent.

10.4 Photoacoustic time-resolved sensitivity

High resolution is one of the benefits of optoacoustic imaging when analyzing heterogeneous structures such as the tissues of living beings. The resolution presented by this technique depends on various factors associated with the detection, source of excitation, and coupling between the tissue and the detector. The type of detector used to convert acoustic waves into electronic signals can be grouped into piezoelectric transducers (Oraevsky et al., 1995; Oraevsky, 1994) and optical interferometers (Diebold et al., 1991; Guenther Paltauf et al., 1996; Paltauf et al., 1996).

Remembering that an essential part of the photoacoustic output is the result of the optical absorption of the different structures present in the tissues, it is reasonable to assume that cells with a high coefficient of optical absorption in the excitation wavelength behave as sources of origin for thermomechanical waves. When it is considered that the source of the thermomechanical wave is a sphere of radius R_0 and that the heat generated is uniformly distributed in the said volume and representing the absorbed optical energy by the sphere as E_{abs}, it is possible to determine the acoustic pulse resulting from rapid heating using the Equation 10.8.

$$p'(t'=t-\frac{R}{v_s})=\frac{1}{4\pi R}\frac{\beta v_s^2}{c_p}\frac{3E_{abs}}{2R_0^3}\{-v_s t',|\,v_s t'\,|\le R_0;\ 0,|\,v_s t'\,|\ge R_0 \tag{10.8}$$

10.5 Photoacoustic imaging

The imaging procedure begins with the transmission of laser pulses illuminating the target region. By this process the sample absorbs light causing thermal expansion, which causes a certain pressure generating sound waves that are captured by a transducer or by an array of ultrasonic transducers. The information of the recorded signals and images of the area is created, and an optical absorption map is generated.

In the optical generation of PA signals, the laser must be modulated to obtain the necessary intensity, depending on the type, dimensions, and area of the tissue to be studied. This technique is used for cancer diagnosis and tumor delimitation during pre and post-surgical treatment (Hult et al., 2020; Xu and Wang, 2006). It is worth mentioning that cancer cells behave in different ways depending on the area where they are located.

10.6 Photoacoustics in bone analysis

Photoacoustics are sensitive to optical properties; therefore, performing PA backpropagating can give images sensitive to the organic parts of the bone and can detect variations in its composition making it an option to study bones in inflammatory arthritis (Jo et al., 2018) and differentiate cancellous and cortical bones (Shubert and Lediju Bell, 2018).

In photoacoustic imaging, obtaining the most significant amount of data from the transducers is necessary to generate images. The transducer data depends on the various transducer arrangements, the incidence angle, and the medium where the sample is located. In the cases where there is only one transducer or a linear array, it is possible to speak of a plane wave, which is said to be a 1D study, shown in Figure 10.2(a) and (b). When there is a linear array, it is possible to acquire the signal from one transducer at a time or

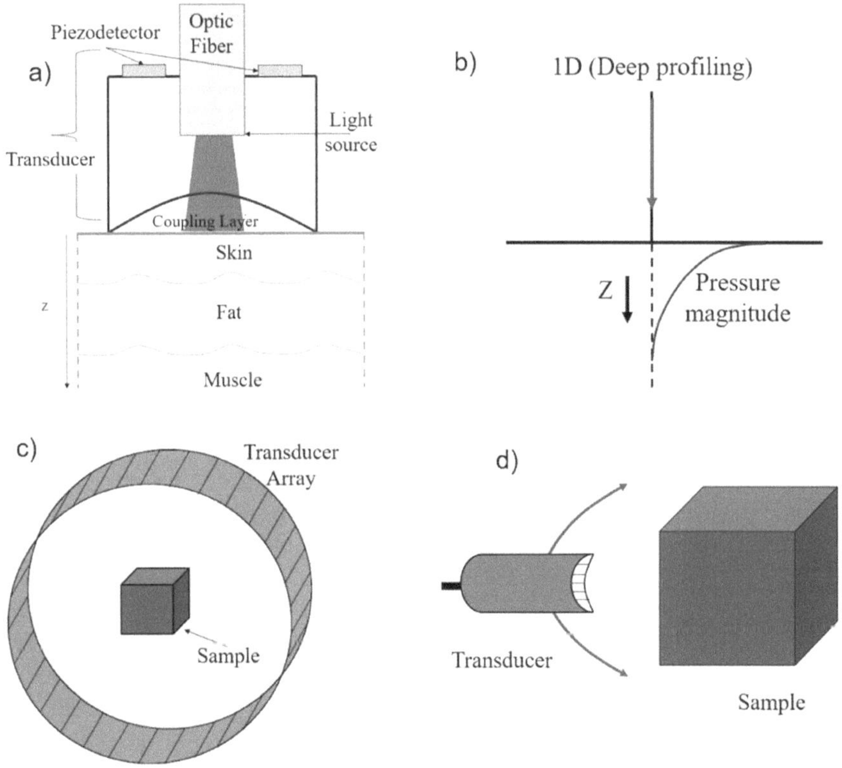

Figure 10.2. (a) diagram of a typical photoacoustic imaging system, (b) acoustic pressure ratio as a function of depth, (c) circular array for photoacoustic imaging, (d) mobile ultrasound array on the sample.

from all the transducers simultaneously; the distinction in this scenario is the intricacy of the accompanying electronics.

When the signal from all the transducers is acquired simultaneously, it is necessary to have enough amplification stages and to process the signals from all the transducers simultaneously. When the signal is acquired from one transducer at a time, it is only necessary to amplify and process one signal. However, it is necessary to include a multiplexer to obtain the signal of the corresponding transducer; the signal acquisition time increases when the signal is acquired from one transducer at a time, compared to acquiring the signal simultaneously. When the region is enclosed in a circumferential arrangement (Figure 10.2c) or the row arrangement has a circular motion (Figure 10.2d), assuming that the sources generate omnidirectional waves, photoacoustic images can be obtained either in 2D or 3D.

The image reconstruction models used in 1D and 2D photoacoustics are those used in traditional ultrasound and have already been reported in the literature (Hsu et al., 2021; John and Barhumi, 2022; Xu and Wang, 2006).

10.7 Cancer detection zones

A laser can be used at a specific wavelength to detect cancer cells, where the skin practically does not absorb and where the blood is highly absorbent; then, the vascularizations generated around a tumor make it highly absorbent and cause a high contrast of the tumor with its surroundings. This option will depend on where the cancer cells are located.

Before, during, and after treatment, photoacoustic imaging has been effectively used to assess tumor vasculature growth and neovascularization longitudinally. Photoacoustic imaging can track the temporal advancement of two kinds of human colorectal malignancies, and the metabolic rate of oxygen during hyperthermia and cryotherapy, and the initial phase of melanoma and human glioblastoma in vivo, according to data from preclinical studies.

10.7.1 Melanoma

Melanoma can be defined as a dermis invasion by atypical melanocytes with altered morphology like ovoid, naevoid, or spindled. Malignant melanoma is a skin cancer form and is considered the deadliest type. Since sun exposure is considered a risk factor, melanoma prevalence has risen, but at the same time, awareness and screening have led to reduced time diagnosis (Yao et al., 2008). The primary treatment for cutaneous melanoma is surgery. Moreover, as with most cancers, the earlier the detection more favorable the prognosis, even though the biopsy is the technique used to confirm the diagnosis. Photoacoustic imaging has been used ex vivo using 680, 800, and 1064 nm wavelength from a heel-pigmented area to diagnose melanoma (Kim et al., 2018; Xu and Wang, 2006). The photoacoustic imaging obtained by the 800 and 1064 nm reveals melanoma details hidden from the surgeon. Based on this result, the photoacoustic imaging was extended to an in vivo study using 700,756, 796, 866, and 900 nm wavelengths. The results are correlated with the histopathological results with a mean absolute error of less than one millimeter.

In the case of Pulsed photothermal radiometry (PPTR), which is sensitive to the distribution of the depth of the absorbents, they are used in some works from 532 nm and are helped by Diffuse reflectance spectroscopy (DRS), allowing the differentiation between different chromophores in the skin, then the color can be evaluated by measuring the total appearance regardless of the surface conditions also in wavelengths in the visible region (Verdel et al., 2019).

Using photoacoustic imaging, Hult and coworkers explored human cutaneous squamous cell cancer (Hult et al., 2020). Photoacoustic imaging's capacity to recognize subtle changes in tissue structure makes it adequate for discerning between healthy and diseased tissue during tumor segmentation. A histological examination of 76 scanned lesions (Hult et al., 2020) confirmed the presence of cancerous cells. The analyzed tissues were placed in a container with buffered saline solution for PA scanning. The samples were excited at a burst rate of 20 Hz and a pulse length of 7 ns. Photoacoustic scanning is a multimodal imaging technology that permits simultaneous high-frequency ultrasound and photoacoustic imaging.

10.7.2 Breast

A photoacoustic mammoscope is reported by Manohar and coworkers (Manohar et al., 2007); an infrared light (1064 nm) is used as an excitation source and a flat array of 2×2 elements with 1 MHz as the central frequency is the US detector. The performed study indicates high absorption that is related to tumor vascularization. According to Manohar, PA imaging has the potential to perceive some breast tumors based on its inherent optical absorption contrast.

Kruger and colleagues report a photoacoustic breast angiography with a cup shape system (Kruger et al., 2010); their approach consists of an array with 128 radially placed US transducers with 5 MHz as the central frequency. The measurements require a coupling fluid, and preliminary tests are performed over a stationary mouse. 3D breast images were reconstructed from a 57-year-old patient, and submillimeter vessels were imaged at a depth of 40 mm. The obtained spatial resolution reaches 250 µm on a 64×64×50 mm field. Kruger's second study with an improved mammoscope was performed employing 512 US transducers having a central frequency of 2 MHz (Kruger et al., 2013). The US central frequency reduction increases the scanning depth but reduces the spatial resolution. Another improvement in the scanning procedure was the incorporation of spiral-like scanning. The resulting scan imaged the breast vasculature over a 1335 mL breast volume performed on four voluntaries.

Kitai proposed a craniocaudal excitation scheme as an alternative to the cup-shaped mammoscope (Kitai et al., 2014). Kitai's prototype uses a tunable laser (700–900 nm wavelength) as excitations sent in a craniocaudal direction, and US detectors are placed on the caudal side. Four different wavelengths are used (756 nm, 797 nm, 825 nm, and 1064 nm), while a 15×23 rectangular US array is used with a central frequency of 1 MHz. A 74% detection rate is reported over a 27-patient population; in Kitai's words, "photoacoustic mammography could provide images of tumor vasculature and oxygenation."

An additional study using a mammoscope proposed by Toi (Toi et al., 1995) was performed on Asian volunteers. The ultrasound detection consists of 512 elements placed in a hemispherical configuration with 2 MHz as a central frequency and two light wavelengths of 755 nm and 795 nm. According to Toi, the photoacoustic technique can describe changes in vascular architecture and haemoglobin saturation because of cancer consequences.

10.7.3 Ovarian

Nandy and collaborators combined photoacoustic tomography and pulse-echo ultrasound to assess ovarian cancer (Nandy et al., 2018). Four different wavelengths were employed (730 nm, 780 nm, 800 nm, and 830 nm), and ultrasound data was recorded between two laser excitations. The multispectral photoacoustic imaging study comprises sixteen voluntaries and consists of photoacoustic images at the four wavelengths captured in a lapse of 12–15 seconds. The averaged hemoglobin (rHbT) and oxygen of the studied areas were calculated and are represented as functional parameters of the analyzed area. The rHbT was effective in identifying ovarian vascular tissue. By integrating the voxels inside the area of interest and dividing them by the total number of voxels, the O_2 mean saturation concentration was obtained. According to the results of Nandy and coworkers, malignant/abnormal ovaries had a more widespread and denser microvessel distribution than benign/normal ovaries (Nandy et al., 2018).

10.7.4 Prostate

Because prostate cancer is one of the most prevalent cancers in men, new techniques for imaging this zone are valuable. Dogra reported studies performed on over 30 male patients with biopsy-confirmed prostate cancer (Dogra et al., 2013). The imaging procedure includes placing *ex vivo* tissue in a saline solution and obtaining the data through a linear sensor array with 32 elements and 5 MHz as the central frequency. The employed light source was a tunable laser (700–1000 nm). The reported sensitivity reaches a value of 81.3%, whereas the specificity is 96.2% for the studied tissues.

The use of transrectal ultrasound combined with photoacoustic imaging is reported by Kothapalli (Kothapalli et al., 2019). The transrectal ultrasound consists of 64 US detectors with 5 MHz as the central frequency and a tunable nanosecond laser (680–950 nm) as the light source. The study was carried out by imaging 20 human prostates with 10 out of them receiving a contrast agent. The reported results mentioned that photoacoustic imaging is reliable for detecting prostate cancer.

10.8 Final words

The use of photoacoustics has permeated more and more in the field of imaging. The higher resolution, compared to classical ultrasound, and the quality of enhancing contrast through the sample's optical properties position photoacoustic imaging is a valuable tool in tumor delineation. The ability of photoacoustic imaging to detect chromophores makes it possible to image angiogenesis and the most diverse organs susceptible to developing cancer. Additionally, the optical properties of haemoglobin combined with the photoacoustic technique make it possible to measure the metabolic rate, which opens new opportunities for using this technique in various biomedical applications.

Acronym/Abbreviation

DRS	Diffuse Reflectance Spectroscopy
EM	Electromagnetic
PA	Photoacoustic
PPTR	Pulsed Photothermal Radiometry
PRF	Pulse Repetition Frequency
rHbT	Averaged Hemoglobin
US	Ultrasound

Reference list

Aguirre, A., Kumavor, P., Ardeshirpour, Y., Sanders, M.M., Brewer et al. (2011). Toward *in vivo* photoacoustic imaging of human ovarian tissue for cancer detection. *Photons Plus Ultrasound: Imaging and Sensing*, 2011.

Aldrich, J.E. (2007). Basic physics of ultrasound imaging. *Critical Care Medicine*, 35(5): S131–S137.

Alqasemi, U., Kumavor, P., Aguirre, A. and Zhu, Q. (2013). Recognizing ovarian cancer from co-registered ultrasound and photoacoustic images. *Photons Plus Ultrasound: Imaging and Sensing*, 2013.

Bell, M.A.L. (2020). Photoacoustic imaging for surgical guidance: Principles, applications, and outlook. *Journal of Applied Physics*, 128(6): 060904. https://doi.org/10.1063/5.0018190.

Blakney, G.B. and Dinwoodie, A.J. (1975). A spectrophotometric scanning technique for the rapid determination of plasma hemoglobin. *Clinical Biochemistry*, 8(1): 96–102. https://doi.org/https://doi.org/10.1016/S0009-9120(75)91005-X.

Borkar, A., Sinha, S., Dhengre, N., Chinni, B., Dogra, V. et al. (2020). Diagnosis of prostate cancer with support vector machine using multiwavelength photoacoustic images. *Proceedings of 3rd International Conference on Computer Vision and Image Processing.*

Chan, V. and Perlas, A. (2011). Basics of ultrasound imaging. *Atlas of Ultrasound-guided Procedures in Interventional Pain Management*, pp. 13–19. Springer.

Diebold, G., Sun, T. and Khan, M. (1991). Photoacoustic monopole radiation in one, two, and three dimensions. *Physical Review Letters*, 67(24): 3384.

Dogra, V.S., Chinni, B.K., Valluru, K.S., Joseph, J.V., Ghazi, A. et al. 2013. Multispectral photoacoustic imaging of prostate cancer: Preliminary ex-vivo results. 3: 41. https://doi.org/10.4103/2156-7514.119139.

García-Figueiras, R., Padhani, A.R., Beer, A.J., Baleato-González, S., Vilanova, J.C. et al. (2015). Imaging of tumor angiogenesis for radiologists—Part 1: Biological and technical basis. *Current Problems in Diagnostic Radiology*, 44(5): 407–424.

Horiguchi, A., Shinchi, M., Nakamura, A., Wada, T., Ito, K. et al. (2017). Pilot study of prostate cancer angiogenesis imaging using a photoacoustic imaging system. *Urology*, 108: 212–219.

Hsu, K.-T., Guan, S. and Chitnis, P.V. (2021). Comparing deep learning frameworks for photoacoustic tomography image reconstruction. *Photoacoustics*, 23: 100271. https://doi.org/https://doi.org/10.1016/j.pacs.2021.100271.

Hughes, S. (2001). Medical ultrasound imaging. *Physics Education*, 36(6): 468.

Hult, J., Dahlstrand, U., Merdasa, A., Wickerström, K., Chakari, R. et al. (2020). Unique spectral signature of human cutaneous squamous cell carcinoma by photoacoustic imaging. *Journal of Biophotonics*, 13(5): e201960212. https://doi.org/https://doi.org/10.1002/jbio.201960212.

Jacques, S.L. (1996, 1996/03/18). Origins of tissue optical properties in the UVA, visible, and NIR regions. *Technical Digest Series*. Advances in Optical Imaging and Photon Migration, Florida.

Jo, J., Tian, C., Xu, G., Sarazin, J., Schiopu, E. et al. (2018). Photoacoustic tomography for human musculoskeletal imaging and inflammatory arthritis detection. *Photoacoustics*, 12: 82–89. https://doi.org/https://doi.org/10.1016/j.pacs.2018.07.004.

John, M.J. and Barhumi, I. (2022). Fast and efficient PAT image reconstruction algorithms: A comparative performance analysis. *Signal Processing*, 201: 108691. https://doi.org/https://doi.org/10.1016/j.sigpro.2022.108691.

Kah, A.C. and Slaney, M. (1999). *Principles of Computerized Tomographic Imaging*. A.C. Kah and Malcolm Slaney.

Kim, J., Kim, Y.H., Park, B., Seo, H.-M., Bang, C.H. et al. (2018). Multispectral *ex vivo* photoacoustic imaging of cutaneous melanoma for better selection of the excision margin. *British Journal of Dermatology*, 179(3): 780–782. https://doi.org/https://doi.org/10.1111/bjd.16677.

Kitai, T., Torii, M., Sugie, T., Kanao, S., Mikami, Y., Shiina, T. et al. (2014). Photoacoustic mammography: Initial clinical results. *Breast Cancer*, 21(2): 146–153. https://doi.org/10.1007/s12282-012-0363-0.

Kothapalli, S.-R., Sonn, G.A., Choe, J.W., Nikoozadeh, A., Bhuyan, A. et al. (2019). Simultaneous transrectal ultrasound and photoacoustic human prostate imaging. *Science Translational Medicine*, 11(507): eaav2169. https://doi.org/doi:10.1126/scitranslmed.aav2169.

Kruger, R.A., Kuzmiak, C.M., Lam, R.B., Reinecke, D.R., Del Rio, S.P. et al. (2013). Dedicated 3D photoacoustic breast imaging. *Medical Physics*, 40(11): 113301. https://doi.org/https://doi.org/10.1118/1.4824317.

Kruger, R.A., Lam, R.B., Reinecke, D.R., Del Rio, S.P., Doyle, R.P. et al. (2010). Photoacoustic angiography of the breast. *Medical Physics*, 37(11): 6096–6100. https://doi.org/https://doi.org/10.1118/1.3497677.

Lara-Hernandez, G., Benavides-Parra, J.C., Arias, N.P., Miranda Hernández, J.G., Gonzalez Moran, C.O. et al. (2017). Influence of poling voltage on optical absorption spectra, thermal properties, and structure of PLZT ceramics. *Ferroelectrics*, 507(1): 159–171. https://doi.org/10.1080/00150193.2017.1283724.

Lin, L., Tong, X., Hu, P., Invernizzi, M., Lai, L. et al. (2021). Clinical photoacoustic computed tomography of breast cancer treated with neoadjuvant chemotherapy. *Photons Plus Ultrasound: Imaging and Sensing*, 2021.

Manohar, S., Vaartjes, S.E., Hespen, J.C.G.v., Klaase, J.M., Engh, F.M.v.d. et al. (2007). Initial results of *in vivo* non-invasive cancer imaging in the human breast using near-infrared photoacoustics. *Optics Express*, 15(19): 12277–12285. https://doi.org/10.1364/OE.15.012277.

Meglinski, I.V. and Matcher, S.J. (2002). Quantitative assessment of skin layers absorption and skin reflectance spectra simulation in the visible and near-infrared spectral regions. *Physiological Measurement*, 23(4): 741–753. https://doi.org/10.1088/0967-3334/23/4/312.

Minghua, X. and Wang, L.V. (2002). Time-domain reconstruction for thermoacoustic tomography in a spherical geometry. *IEEE Transactions on Medical Imaging*, 21(7): 814–822. https://doi.org/10.1109/TMI.2002.801176.

Nandy, S., Mostafa, A., Hagemann, I.S., Powell, M.A., Amidi, E. et al. (2018). Evaluation of ovarian cancer: Initial application of coregistered photoacoustic tomography and US. *Radiology*, 289(3): 740–747. https://doi.org/10.1148/radiol.2018180666.

Neuschmelting, V., Lockau, H., Ntziachristos, V., Grimm, J., Kircher, M.F. et al. (2016). Lymph node micrometastases and in-transit metastases from melanoma: *in vivo* detection with multispectral optoacoustic imaging in a mouse model. *Radiology*, 280(1): 137–150.

Norton, S.J. and Linzer, M. (1981). Ultrasonic reflectivity imaging in three dimensions: exact inverse scattering solutions for plane, cylindrical, and spherical apertures. *IEEE Transactions on Biomedical Engineering*, BME-28(2): 202–220. https://doi.org/10.1109/TBME.1981.324791.

Oraevsky, A., Esenaliev, R., Jacques, S., Thomsen, S., Tittel, F. et al. (1995). *Lateral and Z-axial Resolution in Laser Optoacoustic Imaging with Ultrasonic Transducers* (Vol. 2389). SPIE. https://doi.org/10.1117/12.209986.

Oraevsky, A.A. (1994). A nanosecond acoustic transducer with applications in laser medicine. *IEEE/LEOS Newsletter*, 8(1): 6–17.

Oraevsky, A.A. and Karabutov, A.A. (2000). Ultimate sensitivity of time-resolved optoacoustic detection. *Biomedical Optoacoustics*.

Paltauf, G., Schmidt-Kloiber, H. and Guss, H. (1996). Optical detection of laser-induced stress waves for measurement of the light distribution in living tissue. *Laser-Tissue Interaction and Tissue Optics II*.

Paltauf, G., Schmidt-Kloiber, H. and Guss, H. (1996). Light distribution measurements in absorbing materials by optical detection of laser-induced stress waves. *Applied Physics Letters*, 69(11): 1526–1528.

Park, B., Bang, C.H., Lee, C., Han, J.H., Kim, J. et al. (2020). Wide-field multispectral photoacoustic imaging of human melanomas *in vivo*. *Photons Plus Ultrasound: Imaging and Sensing*, 2020.

Peng, D.-q., Peng, Y.-y., Guo, J. and Li, H. (2016). Laser illumination modality of photoacoustic imaging technique for prostate cancer. *Journal of Physics: Conference Series*.

Piras, D., Xia, W., Heijblom, M., Ten Tije, E., Van Hespen, J. et al. (2011). Breast imaging with the twente photoacoustic mammoscope. *2011 International Workshop on Biophotonics*.

Quiros-Gonzalez, I., Tomaszewski, M.R., Aitken, S.J., Ansel-Bollepalli, L., McDuffus, L.-A. et al. (2018). Optoacoustics delineates murine breast cancer models displaying angiogenesis and vascular mimicry. *British Journal of Cancer*, 118(8): 1098–1106.

Ratto, F., Matteini, P., Rossi, F., Menabuoni, L., Tiwari, N. et al. (2009). Photothermal effects in connective tissues mediated by laser-activated gold nanorods. *Nanomedicine: Nanotechnology, Biology and Medicine*, 5(2): 143–151. https://doi.org/https://doi.org/10.1016/j.nano.2008.10.002.

Salehi, H.S., Kumavor, P.D., Alqasemi, U., Li, H., Wang, T. et al. (2014). High-throughput fiber-array transvaginal ultrasound/photoacoustic probe for ovarian cancer imaging. *Photons Plus Ultrasound: Imaging and Sensing*, 2014.

Sardar, D.K., Yust, B.G., Barrera, F.J., Mimun, L.C., Tsin, A.T.C. et al. (2009). Optical absorption and scattering of bovine cornea, lens and retina in the visible region. *Lasers in Medical Science*, 24(6): 839–847. https://doi.org/10.1007/s10103-009-0677-0.

Shung, K.K. and Thieme, G.A. (1992). *Ultrasonic Scattering in Biological Tissues*. CRC Press.

Shubert, J. and Lediju Bell, M.A. (2018). Photoacoustic imaging of a human vertebra: Implications for guiding spinal fusion surgeries. *Physics in Medicine & Biology*, 63(14): 144001. https://doi.org/10.1088/1361-6560/aacdd3.

Stoffels, I., Jansen, P., Petri, M., Goerdt, L., Brinker et al. (2019). Assessment of nonradioactive multispectral optoacoustic tomographic imaging with conventional lymphoscintigraphic imaging for sentinel lymph node biopsy in melanoma. *JAMA Network Open*, 2(8): e199020–e199020.

Sunar, U., Rohrbach, D.J., Morgan, J. and Zeitouni, N. (2013). Imaging nonmelanoma skin cancers with combined ultrasound-photoacoustic microscopy. *Optical Methods for Tumor Treatment and Detection: Mechanisms and Techniques in Photodynamic Therapy XXII*.

Thomas, T., Dale, P., Weight, R., Atasoy, U., Magee, J. et al. (2008). Photoacoustic detection of breast cancer cells in human blood. *Photons Plus Ultrasound: Imaging and Sensing 2008: The Ninth Conference on Biomedical Thermoacoustics, Optoacoustics, and Acousto-optics*.

Toi, M., Inada, K., Suzuki, H. and Tominaga, T. (1995). Tumor angiogenesis in breast cancer: Its importance as a prognostic indicator and the association with vascular endothelial growth factor expression [Article]. *Breast Cancer Research and Treatment*, 36(2): 193–204. https://doi.org/10.1007/BF00666040.

Verdel, N., Marin, A., Milanič, M. and Majaron, B. (2019). Physiological and structural characterization of human skin *in vivo* using combined photothermal radiometry and diffuse reflectance spectroscopy. *Biomedical Optics Express*, 10(2): 944–960. https://doi.org/10.1364/BOE.10.000944.

Weber, H.J. and Arfken, G.B. (2004). *Essential Mathematical Methods for Physicists, ISE*. Elsevier Science.

Wissmeyer, G., Pleitez, M.A., Rosenthal, A. and Ntziachristos, V. (2018). Looking at sound: Optoacoustics with all-optical ultrasound detection. *Light: Science & Applications*, 7(1): 53. https://doi.org/10.1038/s41377-018-0036-7.

Wu, L., Yao, C., Xiong, Z., Zhang, R., Wang, Z. et al. (2016). The effects of a picosecond pulsed electric field on angiogenesis in the cervical cancer xenograft models. *Gynecologic Oncology*, 141(1): 175–181.

Xu, G., Cheng, Q., Huang, S., Qin, M., Hopkins, T. et al. (2017). Photoacoustic physio-chemical analysis for prostate cancer diagnosis (Conference Presentation). *Photons Plus Ultrasound: Imaging and Sensing*, 2017.

Xu, M. and Wang, L.V. (2006). Photoacoustic imaging in biomedicine. *Review of Scientific Instruments*, 77(4): 041101. https://doi.org/10.1063/1.2195024.

Yao, J., Maslov, K.I. and Wang, L.V. (2012a). *In vivo* photoacoustic tomography of total blood flow and Doppler angle. *Photons Plus Ultrasound: Imaging and Sensing*, 2012.

Yao, J., Maslov, K.I. and Wang, L.V. (2012b). *In vivo* photoacoustic tomography of total blood flow and potential imaging of cancer angiogenesis and hypermetabolism. *Technology in Cancer Research & Treatment*, 11(4): 301–307.

Yao, M., Menda, Y. and Bayouth, J.E. (2008). Chapter 13—Melanoma. pp. 204–215. *In*: Paulino, A.C. and Teh, B.S. (eds.). *PET-CT in Radiotherapy Treatment Planning*. Elsevier. https://doi.org/https://doi.org/10.1016/B978-1-4160-3224-3.50016-5.

CHAPTER 11

Tissue Characterization for Microwave and Ultrasonic Applications

Citlalli J. Trujillo-Romero[1,*] and *Raquel Martínez-Valdez*[2]

In microwave (MW) and ultrasonic (US) thermotherapies such as hyperthermia (42–45°C) or thermal ablation (above 60°C), tissue temperature increases due to its interaction with MW or US energy. Dielectric and ultrasonic properties of biological tissues present temperature and frequency-dependency that must be characterized in order to design new applicators, model EM and US wave-tissue interaction, and improve treatment planning. Electrical conductivity has been modeled as a linear or exponential increase at temperatures lower than 100°C; while, thermal conductivity can be modeled as a linear increase or decrease for temperatures below 100°C. Permittivity describes the interaction between a material and an electric field; hence, it is frequency-dependent. On the other hand, the speed of sound in soft tissues increases as temperature rises; meanwhile, in fatty tissues, it decreases as temperature increments. Moreover, the sound attenuation coefficient rises linearly as frequency increments, while it shows a linear decay for temperatures below 60°C and a non-linear behavior at temperatures higher than 60°C. During thermotherapies,

[1] Division of Medical Engineering Research, National Institute of Rehabilitation LGII. Calz. Mexico Xochimilco No. 289, Col. Arenal de Guadalupe, Mexico City, 14389, Mexico

[2] Biomedical Engineering Program, Polytechnic University of Chiapas. Carretera Tuxtla Gutiérrez – Portillo Zaragoza km 21+500, Col. Las Brisas, Suchiapa, Chiapas, 29150, Mexico.

Email: rmartinez@ib.upchiapas.edu.mx

* Corresponding author: cjtrujillo@inr.gob.mx

tissue properties change along with temperature increase, applicator work frequency, exposure time, and energy intensity.

11.1 Introduction

Biological tissue exposure to electromagnetic (EM), such as microwaves (MW), or ultrasonic (US) energies produces temperature increase in the target region. This increase depends on exposition time and the amount of power delivered which interacts with tissue resulting in absorption, reflection or transmission, whether it is MW or US energy. When it comes to MW applicators, tissue properties, such as electrical conductivity and permittivity must be taken into account during modeling, validation and treatment planning of any specific applicator. On the other hand, when designing US applicators, most common tissue properties to be considered are speed of sound and attenuation. Moreover, tissue properties may be frequency-dependent, temperature-dependent or both. In consequence, tissue characterization must be realized in order to obtain more realistic results during modeling, tissue mimicking or ex-vivo experimental testing, and treatment planning to provide a better approach to the clinical application. In this chapter, EM and US tissue properties are discussed.

11.2 Tissue characterization by using open-ended coaxial probes

As described in Chapter 7, the interaction of the electromagnetic (EM) fields, generated by the thermotherapy applicators, with the body is strongly related to the dielectric properties of the tissues. Dielectric properties define how EM waves are transmitted, absorbed, and reflected by the exposed tissues. Therefore, accurate measurements of dielectric properties such as relative permittivity and electrical conductivity are decisive in several research fields, such as thermotherapies and safety dosimetry. Accurate values of tissue properties are required for the development of computational models to predict the EM energy absorption and consequently the temperature distribution generated by it. Computational models are a potential tool in the development of patient-specific treatment planning systems, especially for thermotherapies. Several authors report the use of tissue properties in the optimization of applicators and treatments for hyperthermia, radiofrequency (RF), and microwave ablation (Kok et al., 2021; Sebek et al., 2021; Tucci et al., 2021). The accuracy of these computational models is related to the accuracy of the tissue properties, and these models can predict the efficiency of the treatment outcome; moreover, the risks can also be predicted.

It is possible to measure the complex permittivity of tissues using a variety of approaches based on wave reflection/transmission characteristics

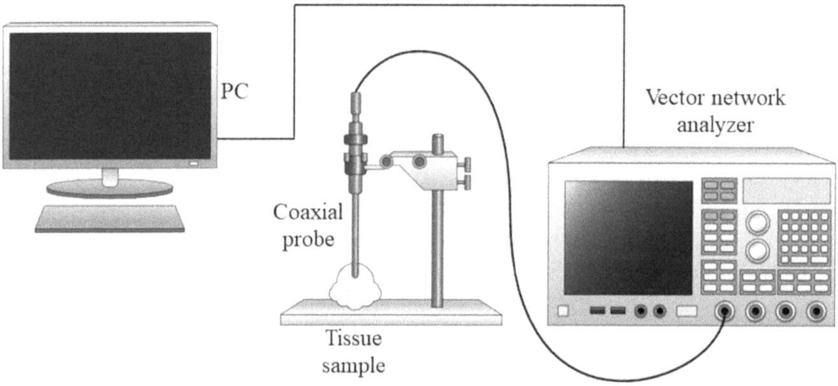

Figure 11.1. Experimental setup to measure tissue dielectric properties by using open-ended coaxial probes.

either in free-space or a transmission line; resonant cavities where the material sample is loaded are also used. (Zheng and Smith, 1991). However, the most promising method is the open-ended coaxial line probe; these probes have been widely used because of their ease of use, wide broad-band response, and ability to conduct noninvasive measurements. It is considered a non-destructive technique for biological tissues; the tissue sample is placed in contact with the open-ended coaxial line, then the frequency-dependence of the complex reflection coefficient is recorded by a vector network analyzer (Misra, 2014). Figure 11.1 shows the common experimental setup to measure tissue properties by means of an open-ended coaxial probe.

To perform the tissue characterization, the vector network analyzer must be calibrated by the OSL method, i.e., open, short, and matched loads must be connected at the end of the cable of the instrument. After this process, the end of the coaxial probe must be in contact with the sample under testing to record the reflection coefficients (S11) related to permittivity and electrical conductivity. The measurements are taken over the frequencies of interest. Permittivity is calculated from the reflection coefficient by using standard inverse techniques (Beruhe, 1996) or numerical models of the coaxial probe. Since the only part of the probe in contact with the tissue is the probe applicator, the accuracy of the measurements is strongly related to the aperture of the probe; literature reports probes with diameters around 1 mm–15 mm (Alanen et al., 1998). Despite the simplicity of this method, errors related either to the equipment or the tissue under test can be presented. Errors due to the equipment are related to measurement uncertainties, calibration, and validation; while the errors due to the tissue are the temperature, contact, and pressure between probe and tissue, tissue heterogeneity and more (La Gioia et al., 2018). The main issues occur since the tissue is considered a homogeneous and isotropic medium.

According to Popovic et al., to obtain precise readings, the tissue being tested must be homogeneous within a sufficient volume so that the measured reflection coefficient will be equal to that of the sample filling the half-space (Hagl et al., 2003). They found that to avoid errors due to boundary reflections, for probes of 3.58 mm and 2.2 mm, the tissue under test must be transversally extended to at least 1.1 cm and 5 mm, respectively. However, the effectiveness of this technique is still under investigation. Several studies about tissue characterization have been reported in the literature; since relative permittivity and electrical conductivity are frequency-dependent, a wide range of frequencies have been studied. Table 11.1 shows a summary of different studies performed by using the open-ended coaxial method to obtain dielectric properties of different kinds of tissues at different frequencies.

11.2.1 Dielectric properties: relative permittivity and electrical conductivity

Oncological thermotherapies have been adopted, in some countries, as a regular treatment of several kinds of tumors. The main goal of these therapies is to heat the biological tissue (tumor) up to temperatures around 44°C (hyperthermia) or higher than 60°C (thermal ablation). Tissue heating is achieved by applicators at different work frequencies. Therefore, to get an accurate prediction of the applicator performance as well as treatment outcome by computational models, precise tissue properties and their behavior due to frequency and temperature must be known. Tissue characterization plays an important role in the design of new devices for thermal therapies, in the modeling stage, and in the patient-specific treatment planning to predict the applicator performance as well as the treatment outcome. Temperature increase can modify several tissue properties such as electrical, thermal and mechanical and consequently, the heat absorbed by the tissues (Bonello et al., 2018). Most of the applicators (RF and MW) used in thermotherapies work at 433 MHz, 915 MHz, 2.45 GHz, and other frequencies. As described in Chapter 7, tissue properties such as relative permittivity and electrical conductivity are strongly related to the interaction between the tissues and the EM energy emitted by the applicators. The equations that model these multi-physical phenomena are mainly defined by the tissue properties and the EM field generated by the applicators. Therefore, the tissues must be characterized electrically and thermally for predicting an accurate treatment outcome by means of computational modeling.

Frequency-dependence of tissue properties has been widely studied (Lau and Gabriel, 1996; Hasgall et al., 2012); moreover, the effect of inaccuracies of tissue properties on the prediction of treatment outcome has been also studied

Table 11.1. Summary of works that report the use of an open-ended coaxial probe to measure dielectric tissue properties in a wide range of frequencies from different tissue samples.

Reference	Frequency and characterized tissues	Tissue samples	Sample size and probe
(Athey et al., 1982)	1.2, 1.8 and 2.4 GHz. Muscle.	*In vivo* measurements in female felines.	At least 5 mm of tissue under the test probe, 8.33 mm probe.
(Schwartz and Mealing, 1985)	200–2000 MHz 2.0–8.0 GHz Muscle, heart, and abdominal skin.	*In vivo* and *in vitro* measurements in a frog.	3 mm diameter probe.
(Smith and Foster, 1985)	10 MHz–1 GHz Subcutaneous adipose tissue, Bone marrow.	*Ex vivo* samples of equine. *Ex vivo* canine tissue. *Ex vivo* tissue from a calf.	Samples of 1 g, 7 mm probe.
(Joines et al., 1994)	50–900 MHz Normal tissue: colon, kidney, liver, lung, mammary, and muscle. Malignant tissues: bladder, colon, kidney, liver, lung, lymph, mammary, spleen, and testes.	Freshly excised human tissues.	Samples larger than 1.5 cm.
(Lau and Gabriel, 1996)	10 Hz–20 GHz Brain, heart muscle, kidney, liver, lung, spleen, muscle, uterus, skin, thyroid, tongue, ovary.	*Ex vivo* tissue from animals (ovine and porcine). Human autopsies Human skin and tongue *in vivo*.	Conical probe, a cube of at least 5 cm linear dimensions.
(Lazebnik et al., 2007)	0.5 GHz–20 GHz Benign, and malignant breast tissues.	*Ex vivo* tissue from breast cancer surgeries.	3 mm probe, sample dimensions 3 mm × 7 mm.
(Peyman et al., 2007)	50 MHz–20 GHz Grey matter, white matter, dura mater, and spinal cord.	*In vivo* and *in vitro* porcine cerebrospinal tissues.	Probes with 1.67 and 2.98 mm internal diameter.
(Hasgall et al., 2012)	10 Hz–100 GHz Dielectric properties of different tissues.	Database from different tissue samples.	–
(Reinecke et al., 2013)	450–500 MHz Thalamus.	Human brain tissue.	–
(Sugitani et al., 2014)	0.5–20 GHz Breast tumor.	*Ex vivo* tissue from breast cancer surgeries.	Breast tissue excised varied from 5 to 30 cm in diameter, probe aperture 2.2 mm.

Table 11.1 contd. ...

...Table 11.1 contd.

Reference	Frequency and characterized tissues	Tissue samples	Sample size and probe
(Azadeh Peyman et al., 2015)	50 MHz–5 GHz Adenocarcinoma of colon. Hemangioma. Hepatocellular carcinoma. Adenocarcinoma of stomach.	Liver tumor samples.	2 cm thick tissue slices, 3.5 mm probe.
(Martellosio et al., 2017)	0.5–50 GHz Normal breast tissue. Breast tumor.	Surgical human breast.	Samples not smaller than 5–6 mm in diameter, Probe with 0.268 mm and 1.6 mm in diameter for inner and outer conductors, respectively.
(Porter et al., 2017)	300 MHz–8.5 GHz Muscle (porcine), fat (porcine), fat (duck).	Freshly excised porcine tissue. *Ex vivo* duck.	2.2 mm probe.
(Di Meo et al., 2018)	0.5–50 GHz Breast, breast tumor.	*Ex vivo* human tissue.	–
(Maenhout et al., 2020)	0.5–20 GHz Liver.	Fresh bovine liver.	A tissue sample of 2 cm × 2 cm × 2 cm.
(Amin et al., 2020)	0.5–8.5 GHz Trabecular bone.	Diseased human bone samples. Bone samples from osteoporotic patients.	Bone samples of 12.7 ± 1.4 mm × 5 ± 0.5 mm × 5 ± 1 mm.
(Fallahi et al., 2021)	915 MHz and 2.45 GHz Liver.	Tissue samples were excised from fresh bovine.	Tissue samples of 2 cm × 2 cm × 2 cm, 2.2 mm probe.
(Gómez-Salazar et al., 2021)	0.5–20 GHz Three different muscle types.	Fresh rabbit meat.	–
(Aydinalp et al., 2021)	0.5–6 GHz Breast tissue, white adipose tissue, wet skin, and tumor.	*Ex vivo* rat breast and skin tissues.	A probe of 2.2 mm aperture diameter.

(Vanni Lopresto et al., 2017). The study of tissue dielectric properties plays a crucial role in the understanding of electromagnetic energy absorption. The interaction of any material with the EM waves is ruled by (1) the displacement, diffusion, and drift of ions and free electrons, (2) dipole moments produced by the polarization of atoms and molecules, and, (3) the orientation of the dipoles in the direction of an applied electric field. In tissues, the movement of the electrons results in conduction currents, as described by Equation 11.1

$$J = \sigma E \tag{11.1}$$

where σ is the tissue conductivity and E is the electric field. In thermotherapies, the dielectric tissue properties are related to the interaction between tissues and EM waves.

The most important tissue properties are relative permittivity and electrical conductivity, which are related by the complex permittivity described by Equation 11.2.

$$\hat{\varepsilon} = \varepsilon_r \varepsilon_0 - j\frac{\sigma}{\omega} \tag{11.2}$$

where ε_r and ε_0 are the relative and the free-space permittivity, respectively, σ is the electrical conductivity and ω is the angular frequency; both parameters, permittivity and conductivity are frequency-dependent. Permittivity is described as the interaction between a material and an electric (E) field, i.e., it measures the opposition offered by the material against the formation of an E field. Permittivity consists of the dielectric constant (real part) and the loss factor (the imaginary part that describes the effect of dielectric loss and electrical conductivity). Therefore, the dielectric constant (relative permittivity) is the amount of energy due to an external field, stored in a material, while the loss factor describes how dissipative a material is to the E field. Moreover, electrical conductivity measures how easily a material allows the flow of an electric current throughout it. Therefore, relative permittivity and electrical conductivity of tissues are required in the design process and modeling of new applicators as well as in patient-specific treatment planning. Table 11.1 describes some of the most relevant works that report the characterization (relative permittivity and electrical conductivity) of different tissues (*ex vivo* and *in vivo*) over a wide range of frequencies.

11.3 Temperature dependence of tissue properties

In thermotherapies, heat is applied to produce either hyperthermia or thermal ablation. The thermal effect produced in the tissues depends on the temperature and exposition time. Moreover, this temperature increase can modify perfusion, electrical, thermal, and mechanical tissue properties. Electromagnetic energy absorption that produces the temperature increase is related to the electrical tissue properties, while thermal tissue properties and blood perfusion affect the heat transfer among the tissues. Consequently, thermal properties also play a key role in the treatment outcome. Tissue properties affect the clinical outcome of thermotherapies; therefore, they have a great impact on treatment planning. To understand the temperature-dependence of tissue properties some experimental studies and mathematical models have been proposed (Bonello et al., 2018; Lau and Gabriel, 1996; Trujillo and Berjano, 2013). This information has been useful to improve results from computational models to predict the treatment outcome (patient-specific treatment planning and applicator design) (Schutt and Haemmerich, 2008; Paulides et al., 2013).

The performance of computational models to predict the outcome of either hyperthermia or thermal ablation is based on the solution of both Maxwell Equations and the Bioheat Transfer Equation. As described in Chapter 7, Maxwell's Equations define the absorption of electromagnetic energy by the body tissues, while the Bioheat Equation describes the heat transfer among them. Electromagnetic models have reached a high level of accuracy; however, thermal models still must be improved. The inaccuracy of thermal models is due to the uncertainties of the thermal tissue properties. Thermal tissue properties vary among patients, tissues, over time, etc. Moreover, some of them are dependent on tissue temperature (Paulides et al., 2013). To improve the accuracy of thermal models, these changes in tissue properties due to temperature increase during thermotherapy must be studied.

11.3.1 Electrical and thermal conductivity

Thermotherapies based on MW, RF, and US energy have been increasingly used in previous years to treat different kinds of tumors, i.e., liver, breast, bone, lung and more (Brace, 2009). Researchers and medical doctors have been contributing to the implementation of clinical trials either for hyperthermia or thermal ablation (Tsuyoshi et al., 2020). Moreover, theoretical models and experimental studies implemented in phantoms, *ex vivo*, and *in vivo* tissues are still under development to fully understand different aspects of the techniques. Theoretical models are widely used in the design of new applicators because they can rapidly predict the applicator performance during treatment. However, the prediction must be as realistic as possible; therefore, the effect of temperature on the dielectric and thermal tissue properties must be considered. To achieve realistic models, different mathematical functions have been proposed to describe temperature-dependence of tissue properties.

Tissue properties such as electrical and thermal conductivity are the most affected by the temperature increase due to water vaporization and the rise in impedance (Trujillo and Berjano, 2013). These phenomena modify the lesion size; therefore, researchers have been studying the thermal-dependence of electrical and thermal conductivity. One of the most relevant studies was made by Trujillo et al. (Trujillo and Berjano, 2013) where different mathematical functions are described and analyzed in order to know how the use of each of them modifies the lesion size produced by the thermotherapy.

Temperature-dependence of tissue properties is mainly important at those temperatures to be able to produce hyperthermia (42°C–46°C) and thermal ablation (60°C–100°C); hence, most of the implemented studies have been done from 37°C to 100°C. Normally, temperatures higher than 100°C are not considered because tissues at those temperatures are already burnt ; therefore, they have lost their structure and properties.

Trujillo et al., report a summary of at least eight cases to model the temperature-dependence of electrical and thermal conductivity for biological tissues. Temperature-dependence of electrical conductivity has been reported with different types and rates of increase and decrease, i.e., linear and exponential increases at a rate of 1.5%/°C and 2%/°C are reported for tissue temperatures lower than 100°C; while for temperatures higher than 100°C drop rates in 2 and 4 orders were reported (Trujillo and Berjano, 2013; Lau et al., 2010; Tungjitkusolmun et al., 2000). On the other hand, thermal conductivity can be modeled by considering either a linear increase rate of 1.5%/°C or a linear drop rate of –1.5%/°C for tissue temperatures lower than 100°C; while for temperatures higher than 100°C, it is considered to be constant . By taking the constant values of electrical and thermal conductivity as a reference for bone, muscle, fat, and skin, Figure 11.2 shows the temperature-dependence of both tissue properties in a temperature range from 37°C–110°C.

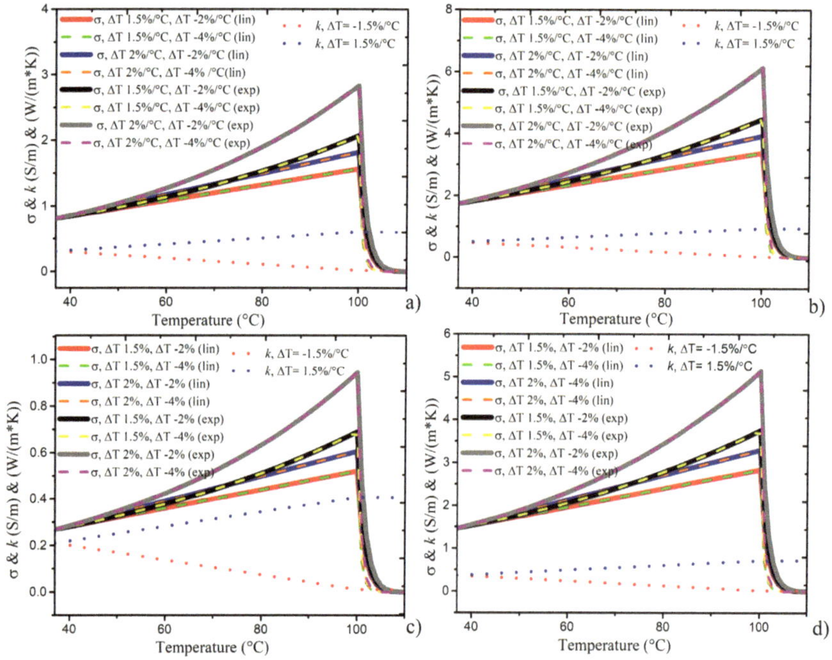

Figure 11.2. Temperature-dependence of electrical (σ) and thermal conductivity (k) for different tissues. (a) bone, (b) muscle, (c) fat, (d) skin.

11.3.2 Blood perfusion

Blood perfusion is a parameter that highly modifies the heat patterns generated by the applicators used for thermotherapies; this is because of the blood circulation during the therapy as well as the heat transfer due to conduction. Therefore, blood flow is crucial in the extension of the thermal damage generated not only in the tumor but also in the surrounding healthy tissues. Blood perfusion of tumors could be quite different, even within tumors with similar histological classification due to their heterogeneous vasculature (Vaupel, 2009). Moreover, during the thermal treatment a temporal variation of blood perfusion occurs due to the response of blood vessels to the thermal damage (Soni et al., 2015). Blood perfusion is a temperature-dependent tissue property. The blood flow in normal tissue increases considerably with temperature increase; in contrast, the blood flow in tumors decreases when the temperature increases (Berlin, 1997). Therefore, the thermal damage produced by heat therapy is mainly because of the reduced ability of the tumors to remove heat due to their poor blood flow. In contrast, less thermal damage in healthy tissue is generated due to its higher blood flow. Heat transfer among tissues in thermotherapies is affected by the thermal tissue properties as well as by blood perfusion (Rossmann and Haemmerich, 2016). Table 11.2 shows a recompilation of the most recent studies performed to measure the temperature-dependence of dielectric and thermal properties, as well as blood perfusion either in healthy tissues or tumors.

Temperature-dependence of blood perfusion has been studied little, mainly because of the difficulty in performing *in vivo* studies during hyperthermia/thermal ablation. Therefore, some computational models have been developed. Drizdal et al. (2010) describe the temperature dependence of perfusion for muscle, fat, and skin according to Equations 11.3–11.5.

$$W_{muscle} * \{0.45 + 3.55\, exp\left(-\frac{(T-45.0)^2}{12.0}\right) \quad T \leq 45°C,\ 4.00 \quad T > 45 \tag{11.3}$$

$$W_{fat} * \{0.036 + 0.036\, exp\left(-\frac{(T-45)^2}{12.0}\right) \quad T \leq 45°C,\ 0.72 \quad T > 45°C \tag{11.4}$$

$$W_{skin}(T) * \{1 + 9.2\left(-\frac{(T-45)^2}{10}\right) \quad T \leq 45°C,\ 10.2 \quad T > 45°C \tag{11.5}$$

As shown in Tables 11.1 and 11.2, tissue properties can be frequency-dependent (relative permittivity and electrical conductivity) and temperature-dependent (thermal and electrical conductivity, perfusion, etc.). Although both dependencies have been studied by different authors, most of these studies are

Table 11.2. Summary of the most recent studies performed to measure the temperature-dependence of dielectric and thermal properties.

Reference	Property	Tissue	Temperature range	Frequency	Variations of the tissue properties
(Jaspard et al., 2003)	The relative permittivity (ε_r) and electrical conductivity (σ).	Cow blood (5 liters) Sheep blood (3 liters) Human blood.	25°C–45°C.	1 MHz–1 GHz.	Conductivity variations: 1%/°C constant in the range frequency. Permittivity variations: from 0.3%/°C (1 MHz) to –0.3%/°C (1 GHz), the change of sign occurs around 50 MHz.
(Zurbuchen et al., 2010)	Electrical conductivity (σ).	Perfused pig liver, 10 × 8 × 5 cm.	30°C–90°C, 5°C steps.	470 kHz.	Perfused liver (coagulation process): 0.36 S/m–0.75 S/m. Perfused liver (cooling process): 0.35 S/m–0.62 S/m.
(Chin and Sherar, 2009)	The relative permittivity (ε_r) and electrical conductivity (σ).	Rat prostate, tissue samples of approximately 4 mm–5 mm.	45°C–75°C.	915 MHz.	Linear coefficients: Conductivity: 1.10 ± 0.11%/°C Relative permittivity: –0.31 ± 0.05%/°C $\varepsilon_r = 62.8 \pm 2.7$ and $\sigma = 1.17 \pm 0.07$ at $T = 23 \pm 1$°C.
(Fu et al., 2014)	The relative permittivity (ε_r) and electrical conductivity (σ).	*Ex vivo* porcine uterus, liver, kidney, urinary bladder, skeletal muscle, and fat.	36°C–60°C, 3°C steps.	43, 64, 128, 170, 298, 400, and 468 MHz.	Uterus: $\varepsilon_{r(43\,MHz)} = 96.88–107$, $\sigma_{(43\,MHz)} = 1.03–1.06$, $\varepsilon_{r(468\,MHz)} = 53.24–57.44$, $\sigma_{(468\,MHz)} = 1.28–1.32$. Liver: $\varepsilon_{r(43\,MHz)} = 121.37–124.77$, $\sigma_{(43\,MHz)} = 0.55–0.57$, $\varepsilon_{r(468\,MHz)} = 49.68–51.68$, $\sigma_{(468\,MHz)} = 0.87–0.88$. Kidney: $\varepsilon_{r(43\,MHz)} = 147.87–152.87$, $\sigma_{(43\,MHz)} = 0.83–0.87$, $\varepsilon_{r(468\,MHz)} = 55.71–59.11$, $\sigma_{(468\,MHz)} = 1.13–1.15$. Urinary bladder: $\varepsilon_{r(43\,MHz)} = 117.13–123.13$, $\sigma_{(43\,MHz)} = 0.87–0.88$, $\varepsilon_{r(468\,MHz)} = 54.93–57.73$, $\sigma_{(468\,MHz)} = 1.07–1.09$. Skeletal muscle: $\varepsilon_{r(43\,MHz)} = 156.08–161.08$, $\sigma_{(43\,MHz)} = 0.97–1.03$, $\varepsilon_{r(468\,MHz)} = 62.81–64.01$, $\sigma_{(468\,MHz)} = 1.22–1.26$. Fat: $\varepsilon_{r(43\,MHz)} = 18.36–21.36$, $\sigma_{(43\,MHz)} = 0.17–0.21$, $\varepsilon_{r(468\,MHz)} = 9.03–11.33$, $\sigma_{(468\,MHz)} = 0.21–0.26$.
(Lazebnik et al., 2006)	The relative permittivity (ε_r) and electrical conductivity (σ).	Animal liver tissue, samples of 5 × 10 × 3 cm.	25.7, 36.4 and 55.5°C.	0.5–20 GHz.	During heating: $\varepsilon_{r(915\,MHz)}$: –0.20 ± 0.28%/°C, $\sigma_{(915\,MHz)}$: 1.33 ± 0.36%/°C, $\varepsilon_{r(2.45\,GHz)}$: –0.17 ± 0.30%/°C, $\sigma_{(2.45\,GHz)}$: 0.20 ± 0.33%/°C. During cooling: $\varepsilon_{r(915\,MHz)}$: –0.13 ± 0.07%/°C, $\varepsilon_{r(915\,MHz)}$: 1.16 ± 0.18%/°C, $\varepsilon_{r(2.45\,GHz)}$: –0.09 ± 0.06%/°C, $\sigma_{(2.45\,GHz)}$: 0.008 ± 0.21%/°C.

(Bhattacharya and Mahajan, 2003)	Thermal conductivity.	Sheep collagen samples greater than 50 mm.	25–80°C.	–	During heating: 0.54–0.79 $Wm^{-1}K^{-1}$ During cooling: 0.62–0.82, 0.54–0.79 $Wm^{-1}K^{-1}$.
(Guntur et al., 2013)	Thermal conductivity.	Fresh porcine liver, a sample of 40 × 40 mm with 20 mm of thickness.	20–90°C.	–	Thermal conductivity decreased by 9.6% from the initial value (20°C) to the turning temperature (35°C), then rose by 45% at 90°C from its minimum.
(Choi et al., 2013)	Thermal conductivity.	*Ex vivo* porcine liver and Cadaveric human liver samples.	25–80°C.	–	*Ex vivo* porcine liver: During heating: It was a steady rise with a 12% increase. During cooling: It was significantly less resulting in an 18% reduction. Cadaveric human liver: The rise in thermal conductivity was lower (5%) and the reduction was 16% after 25°C.
(Lopresto et al., 2019)	Thermal conductivity.	*Ex vivo* bovine liver, cylinders of 7 cm diameter and 6 cm height.	21–113°C.	–	An overall increase of 4.5 times at 99°C with respect to the initial value at 21°C was observed. A dramatic decrease of 70% occurs at 101°C. For 101–113°C a decrease of about 71% was observed.
(Mohammadi et al., 2021)	Thermal conductivity.	*Ex vivo* liver, brain, and pancreas.	20–100°C.	–	Liver: 0.515–1.635 W/m.K (22–97°C) Brain: 0.524–2.005 W/m.K (22–97°C) Pancreas: 0.510–0.524 W/m.K (22–45°C).
(Brown et al., 1992)	Blood perfusion.	Normal and malignant tissue from male mice.	43–46°C.	–	The rise in temperature resulted in a factor of two reductions in heating time for the same microvascular effect. Tumor vasculature is about twice as heat sensitive as the vasculature of normal muscle at 43–46°C.
(Eddy, 1980)	Blood perfusion.	Carcinoma in a Syrian hamster.	41°C, 43°C and 45°C.	–	The pathological changes in the microvasculature of the tumor due to the temperature increase plays an important role in the tumor response and increases the degree of cure due to direct cell killing.

Table 11.2 contd. ...

...Table 11.2 contd.

Reference	Property	Tissue	Temperature range	Frequency	Variations of the tissue properties
(Sturesson et al., 1999)	Blood perfusion.	Twenty-four male Wistar rats (246–277 g).	41°C and 44°C.	–	Heat can produce an increase in local liver perfusion. Moreover, at 41°C liver perfusion increases up to 33%. A gradual decrease of local perfusion occurs at 44°C.
(Vujaskovic et al., 2000)	Blood perfusion.	Thirteen dogs with sarcomas.	44°C.	–	Tumor perfusion was significantly altered after hyperthermia. At temperatures higher than 44°C a significant reduction in oxygenation occurs.
(Akyürekli et al., 1997)	Blood perfusion.	Muscle of 32 healthy pigs (20–54 kg).	44.5°C.	–	Muscle blood flow increases by a factor of 4 (15–30 min, T = 44.5°C). The increase in blood flow cannot be sustained at higher temperatures; therefore, a gradual increase in tissue temperature is presented. The blood flow increase is a local phenomenon.
(Guiot et al., 1998)	Blood perfusion.	Recurrent breast cancer.	Hyperthermia.	–	If tissue is perfused by vessels with a ratio between their physical length and their thermal equilibrium length, much smaller than 1, as in recurrent breast cancer, perfusion notably affects the reached temperature

focused on the characterization of tissues such as liver and muscle. Therefore, a limited number of studies describe the performance of tissues like bone, lungs, and different kinds of tumors. Due to one of the main challenges in thermotherapies is to improve the accuracy of thermal simulations the dielectric and thermal characterization of a wide range of tissues and tumors is still required to cover different temperatures (hyperthermia and thermal ablation).

11.3.3 Speed of sound

Temperature plays an important role in the speed of sound (SoS). Bilanuik and Wonk (Bilaniuk, 1993) proposed a fifth order polynomial to predict the temperature dependence of SoS in pure water that describes sound velocity increments when the temperature rises. *In vivo* human biological tissues present an average temperature of 37°C; then, when subjected to ultrasonic hyperthermia treatments, SoS increases. This behavior is observed in soft tissues; meanwhile, in fatty tissues, the SoS decreases when temperature increases (Johnston et al., 1998). For temperatures higher than 60°C, SoS tends to slowly decay in soft tissues (Techavipoo et al., 2004). The SoS of biological tissue is measured in a temperature regulated water tank (Zeqiri et al., 2010). Literature reports the temperature-dependence of tissues in the hyperthermia temperature range (Bamber and Hill, 1979; Goss, 1978; López-Haro et al., 2010; Martínez-Valdez et al., 2015), but not in the ablation temperature range. Techavipoo et al. (Techavipoo et al., 2004) had made an effort to measure SoS temperature-dependence from 20°C to 95°C (see Figure 11.3a). Although

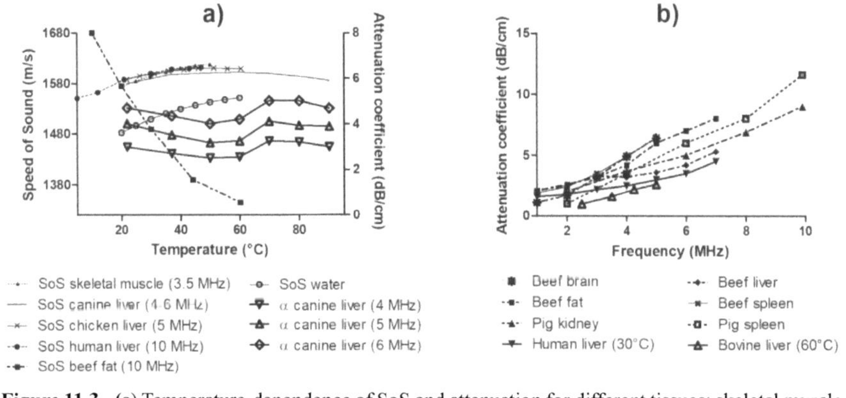

Figure 11.3. (a) Temperature-dependence of SoS and attenuation for different tissues: skeletal muscle (López-Haro et al., 2010), canine liver (Techavipoo et al., 2004), chicken liver (Martínez-Valdez et al., 2015), human liver and beef fat (Goss, 1978), and water (Bilaniuk, 1993); (b) frequency-dependence of attenuation coefficient for different tissues at 37°C: beef brain, beef fat, pig kidney, human liver, beef liver, beef spleen, pig spleen (Goss, 1978) and bovine liver (Gertner et al., 1997).

at temperatures higher than 60°C boiling, cavitation (bubble formation) and microstreaming effects appear and modify the ultrasonic propagation and the tissue changes (Hill et al., 2004; Huang et al., 2015).

11.4 Tissue characterization by acoustic propagation measurements

In clinical applications, ultrasonic waves propagate through different layers of biological tissues with different thicknesses and properties. When developing novel clinical devices based on ultrasonic propagation in the anatomical region the appropriate thickness of layers and the incidenct angle for the acoustic energy delivery must be considered. Acoustic tissue properties such as SoS and attenuation are necessary to understand mechanical wave diffraction, their penetration depth, and thermal effects, among others (Driller and Lizzi, 1987).

11.4.1 Speed of sound

SoS refers to the velocity with which sound travels within a medium; therefore, it depends on the intrinsic properties and structure of the specimen, and its density. Longitudinal ultrasonic waves can propagate in solid, semi-solid and liquid-like media (Zeqiri et al., 2010) such as biological tissues. The measurement of sound velocity is commonly made either by the pulse-echo (PE) technique or the through-transmission (TT) method (Hill et al., 2004; Zeqiri et al., 2010). In the PE technique a single broadband plane transducer is used as emitter and receiver: transceiver. The experimental setup consists of placing the transceiver facing parallel towards a highly reflecting material inside a temperature controlled water tank. Degassed water is considered as a lossless propagating medium; it is commonly used as a reference material because its sound velocity is close to soft tissues (Gutierrez et al., 2012; Hill et al., 2004). The material under test must have parallel faces which are aligned to the transceiver to avoid echo loss, i.e., must be perpendicularly aligned to the acoustic pattern. The acquired echo from the reflecting material in the reference medium is used to locate the echo from the testing material. SoS can be estimated by Equation (11.6),

$$SoS_{tissue} = \frac{2 * D}{TOF} \tag{11.6}$$

where, D is the sample thickness, and TOF is the time-of-flight of the ultrasonic wave propagating in the material under test (Vives, 2008).

The TT technique, also known as the insertion method, consists of placing two plane ultrasonic transducers, with the same operating frequency, aligned and facing each other at a certain distance. The setup is immersed in a temperature-controlled water tank filled with degassed water. Then, the tissue sample is positioned between both transducers; parallelism among the transducers' surfaces and tissue walls must be achieved (Zeqiri et al., 2010). The thickness of the sample is usually known, and the SoS can be obtained from the TOF between the echoes produced from the front and back surfaces of the sample (Braunstein et al., 2022). Also, it can be estimated by a relative measurement between the echo measured in the reference medium (water) and the echo produced by the sample (Zell et al., 2007).

When measuring the temperature-dependence of SoS in water or a tissue sample either by PE or TT methods, the ultrasonic echo can be visually tracked by an oscilloscope (Gertner et al., 1997). As the temperature changes, the ultrasonic echo presents time displacements due to SoS temperature-dependence. Therefore, the specimen must be heated in a water controlled-temperature bath for 20–30 minutes to guarantee that it has attained the desired temperature (Bamber and Hill, 1979; Guiot et al., 1998; Guntur et al., 2013; Maenhout et al., 2020) before measuring the TOF. The sample temperature can also be surveilled with a thermocouple inserted inside the specimen taking care that it does not interfere with the acoustic pattern (Dahis and Azhari, 2020; Martínez-Valdez et al., 2015).

11.4.2 Attenuation

As mentioned before, the ultrasonic waves travel across different tissues such as skin, fat, and muscle, before reaching the target either if it is a diagnostic, physiotherapy, hyperthermia, or ablation application. When the acoustic wave interacts with tissue, the echo energy is reduced along the propagation direction due to losses by absorption, scattering, reflection, refraction, among others (Hill et al., 2004). The combination of these effects results in ultrasonic wave attenuation; however, main energy losses are considered by scattering and absorption (Vives, 2008). Scattering is due to the interaction between the wave and the specimen intrinsic scatterers, which diffract the ultrasonic pattern in different directions relative to the initial one (Driller and Lizzi, 1987). Absorption happens when the acoustic wave energy is converted into heat while propagating in media, and is produced by relaxation processes (Vives, 2008). In tissues liquid-like such as blood, the ultrasonic energy is absorbed less than in tissues with high protein content and lower water content, like bones, ligaments and tendons (Driller and Lizzi, 1987; Watson, 2008). Nevertheless, if blood vessels are exposed to a high acoustic energy level for 1 minute, temperature can rise beyond 100°C causing blood coagulation (Vaezy et al., 1999).

The attenuation coefficient has shown a decreasing trend as temperature increases up to 50°C; for higher temperatures, sound attenuation increases (Bamber and Hill, 1979; Gertner et al., 1997). Its frequency-dependence is linear, i.e., when frequency increases, the attenuation coefficient rises; besides, attenuation increases exponentially with the distance travelled by the ultrasonic wave (Driller and Lizzi, 1987; Sun et al., 2012; Vives, 2008). The attenuation of a plane wave can be expressed as $A = A_0 e^{-\alpha z}$, where A_0 is the amplitude of the echo without attenuation, A is the attenuated echo after the ultrasonic wave has travelled a distance z from the origin, and α is the attenuation coefficient of the wave that propagates in the z direction (Hutchins and Hayward, 1990). The attenuation of the sound wave is commonly expressed in terms of dB/cm, but it can also be found as Np/cm (Driller and Lizzi, 1987; Zeqiri et al., 2010). Therefore, both frequency- and temperature-dependency must be considered during novel application proposals, treatment planning and therapy delivery. Figure 11.3a depicts the temperature-dependence of the attenuation coefficient, while Figure 11.3b shows its frequency-dependence.

Acronym/Abbreviation

dB	Decibels
EM	Electromagnetic
MW	Microwave
N/A	Not applicable
Np	Nepers
OSL	Open-Short-Matched Load
PE	Pulse-Echo
RF	Radiofrequency
SoS	Speed of Sound
TOF	Time-of-Flight
TT	Through-Transmission
US	Ultrasound

Reference list

Akyürekli, D., Gerig, L.H. and Raaphorst, G.P. (1997). Changes in muscle blood flow distribution during hyperthermia. *International Journal of Hyperthermia*, 13(5): 481–496. https://doi.org/10.3109/02656739709023547.

Alanen, E., Lahtinen, T. and Nuutinen, J. (1998). Measurement of dielectric properties of subcutaneous fat with open-ended coaxial sensors. *Physics in Medicine and Biology*, 43(3): 475–485. https://doi.org/10.1088/0031-9155/43/3/001.

Amin, B., Shahzad, A., Farina, L., Parle, E., McNamara, L. et al. (2020). Dielectric characterization of diseased human trabecular bones at microwave frequency. *Medical Engineering and Physics*, 78: 21–28. https://doi.org/10.1016/j.medengphy.2020.01.014.

Athey, T.W., Stuchly, M.A., Stuchly, M.A. and Stuchly, S.S. (1982). Measurement of radio frequency permittivity of biological tissues with an open-ended coaxial line: Part I. *IEEE Transactions on Microwave Theory and Techniques*, 30(1): 82–86. https://doi.org/10.1109/TMTT.1982.1131021.

Aydinalp, C., Joof, S. and Yilmaz, T. (2021). Towards accurate microwave characterization of tissues: Sensing depth analysis of open-ended coaxial probes with *ex vivo* rat breast and skin tissues. *Diagnostics*, 11(2). https://doi.org/10.3390/diagnostics11020338.

Bamber, J.C. and Hill, C.R. (1979). Ultrasonic attenuation and propagation speed in mammalian tissues as a function of temperature. *Ultrasound in Medicine & Biology*, 5(2): 149–157. https://doi.org/10.1016/0301-5629(79)90083-8.

Berlin, I. (1997). *Impact of Nonlinear Heat Transfer Impact of Nonlinear Heat Transfer on Temperature Control in Regional Hyperthermia*, 73(December).

Beruhe, D. (1996). A comparative study of four open-ended coaxial probe models for permittivity measurements of lossy dielectric/biological materials at microwave frequencies. *IEEE Transactions on Microwave Theory and Techniques*, 44(10 Part 2): 1928–1934. https://doi.org/10.1109/22.539951.

Bhattacharya, A. and Mahajan, R.L. (2003). Temperature dependence of thermal conductivity of biological tissues. *Physiological Measurement*, 24: 769–783. https://doi.org/10.16285/j.rsm.2020.1899.

Bilaniuk, N. (1993). Speed of sound in pure water as a function of temperature. *The Journal of the Acoustical Society of America*, 93(3): 1609. https://doi.org/10.1121/1.406819.

Bonello, J., Farrugia, L. and Sammut, C.V. (2018). A review of studies investigating the dielectric properties of biological tissues for application in hyperthermia and microwave thermal ablation. *Xjenza*, 2018(2018): 86. https://doi.org/10.7423/XJENZA.2018.2.02.

Brace, C.L. (2009). Radiofrequency and microwave ablation of the liver, lung, kidney, and bone: What are the differences? *Current Problems in Diagnostic Radiology*, 38(3): 135–143. https://doi.org/10.1067/j.cpradiol.2007.10.001.

Braunstein, L., Brüningk, S.C., Rivens, I., Civale, J., Haar, G. ter et al. (2022). Characterization of acoustic, cavitation, and thermal properties of poly(vinyl alcohol) hydrogels for use as therapeutic ultrasound tissue mimics. *Ultrasound in Medicine & Biology*, 48(6): 1095–1109. https://doi.org/10.1016/J.ULTRASMEDBIO.2022.02.007.

Brown, S.L., Hunt, J.W. and Hill, R.P. (1992). Differential thermal sensitivity of tumour and normal tissue microvascular response during hyperthermia. *International Journal of Hyperthermia*, 8(4): 501–514. https://doi.org/10.3109/02656739209037988.

Chin, L. and Sherar, M. (2009). *Changes in the Dielectric Properties of Rat Prostate ex vivo at 915 Mhz during Heating*, 6736. https://doi.org/10.1080/02656730310001657738.

Choi, J., Morrissey, M. and Bischof, J.C. (2013). Thermal processing of biological tissue at high temperatures: Impact of protein denaturation and water loss on the thermal properties of human and porcine liver in the range 25–80°C. *Journal of Heat Transfer*, 135(6): 1–8. https://doi.org/10.1115/1.4023570.

Dahis, D. and Azhari, H. (2020). Speed of sound and attenuation temperature dependence of bovine brain: *Ex vivo* study. *Journal of Ultrasound in Medicine*, 39(6). https://doi.org/10.1002/jum.15203.

Di Meo, S., Espin-Lopez, P.F., Martellosio, A., Pasian, M., Bozzi, M. et al. (2018). Experimental validation of the dielectric permittivity of breast cancer tissues up to 50 GHz. *2017 IEEE MTT-S International Microwave Workshop Series on Advanced Materials and Processes for RF and THz Applications*, IMWS-AMP 2017, 2018-Janua(September), 1–3. https://doi.org/10.1109/IMWS-AMP.2017.8247408.

Driller, J. and Lizzi, F.L. (1987). Therapeutic applications of ultrasound—A review. *IEEE Engineering in Medicine and Biology Magazine*, 6(4): 33–40.

Drizdal, T., Togni, P., Visek, L. and Vrba, J. (2010). Comparison of constant and temperature dependent blood perfusion in temperature prediction for superficial hyperthermia. *Radioengineering*, 19(2): 281–289.
Eddy, H.A. (1980). Alterations in tumor microvasculature during hyperthermia. *Radiology*, 137(2): 515–521. https://doi.org/10.1148/radiology.137.2.7433685.
Fallahi, H., Sebek, J. and Prakash, P. (2021). Broadband dielectric properties of *ex vivo* bovine liver tissue characterized at ablative temperatures. *IEEE Transactions on Biomedical Engineering*, 68(1): 90–98. https://doi.org/10.1109/TBME.2020.2996825.
Fu, F., Xin, S.X. and Chen, W. (2014). Temperature-and frequency-dependent dielectric properties of biological tissues within the temperature and frequency ranges typically used for magnetic resonance imaging-guided focused ultrasound surgery. *International Journal of Hyperthermia*, 30(1): 56–65. https://doi.org/10.3109/02656736.2013.868534.
Gertner, M.R., Wilson, B.C. and Sherar, M.D. (1997). Ultrasound properties of liver tissue during heating. *Ultrasound in Medicine and Biology*, 23(9): 1395–1403.
Gómez-Salazar, J.A., Alvarado-Iglesias, R., Kaur, T., Corona-Chávez, A., Olvera-Cervantes, J.L. et al. (2021). Dielectric properties of fresh rabbit meat in the microwave range. *Journal of Food Science*, 86(3): 952–959. https://doi.org/10.1111/1750-3841.15631.
Goss, S.A. (1978). Comprehensive compilation of empirical ultrasonic properties of mammalian tissues. *The Journal of the Acoustical Society of America*, 64(2): 423. https://doi.org/10.1121/1.382016.
Guiot, C., Madon, E., Allegro, D., Piantà, P.G., Baiotto, B. et al. (1998). Perfusion and thermal field during hyperthermia. Experimental measurements and modelling in recurrent breast cancer. *Physics in Medicine and Biology*, 43(10): 2831–2843. https://doi.org/10.1088/0031-9155/43/10/012.
Guntur, S.R., Lee, K.Il, Paeng, D.G., Coleman, A.J. and Choi, M.J. et al. (2013). Temperature-dependent thermal properties of *ex vivo* liver undergoing thermal ablation. *Ultrasound in Medicine and Biology*, 39(10): 1771–1784. https://doi.org/10.1016/j.ultrasmedbio.2013.04.014.
Gutierrez, M.I., Martinez, R., Vera, A. and Leija, L. (2012). Technology in ultrasonic hyperthermia. pp. 41–83. *In*: Gao, X.-H. and Chen, H.-D. (eds.). *Hyperthermia: Recognition, Prevention and Treatment* (Issue 2). NOVA Publishers.
Hagl, D.M., Popovic, D., Hagness, S.C., Booske, J.H., Okoniewski, M. et al. (2003). Sensing volume of open-ended coaxial probes for dielectric characterization of breast tissue at microwave frequencies. *IEEE Transactions on Microwave Theory and Techniques*, 51(4 I): 1194–1206. https://doi.org/10.1109/TMTT.2003.809626.
Hasgall, P., Neufeld, E., Gosselin, M., Klingenbock, A., Kuster, N. et al. (2012). *IT'IS Database for Thermal and Electromagnetic Parameters of Biological Tissues*.
Hill, C.R., Bamber, J.C. and Haar, G.R. ter. (2004). *Physical Principles of Medical Ultrasonics* (2nd ed.). John Wiley & Sons, Ltd.
Huang, C.W., Sun, M.K., Chen, B.T., Shieh, J., Chen, C.S. et al. (2015). Simulation of thermal ablation by high-intensity focused ultrasound with temperature-dependent properties. *Ultrasonics Sonochemistry*, 27: 456–465. https://doi.org/10.1016/J.ULTSONCH.2015.06.003.
Hutchins, D.A. and Hayward, G. (1990). The radiated field of ultrasonic transducers. pp. 1–80. *In*: Thurston, R.N. and Pierce, A.D. (eds.). *Physical Acoustics*, 19. Academic Press.
Jaspard, F., Nadi, M. and Rouane, A. (2003). Dielectric properties of blood: An investigation of haematocrit dependence. *Physiological Measurement*, 24(1): 137–147. https://doi.org/10.1088/0967-3334/24/1/310.
Johnston, R.L., Dunn, F. and Goss, S.A. (1998). Compilation of empirical ultrasonic properties of mammalian tissues. II. *The Journal of the Acoustical Society of America*, 68(1): 93. https://doi.org/10.1121/1.384509.

Joines, W.T., Yang, Z., Li, C. and Jirtle, R.L. (1994). The measured electrical properties of normal and malignant human tissues from 50 to 900 MHz. *Medical Physics*.

Kok, H.P., van der Zee, J., Guirado, F.N., Bakker, A., Datta, N.R. et al. (2021). Treatment planning facilitates clinical decision making for hyperthermia treatments. *International Journal of Hyperthermia*, 38(1): 532–551. https://doi.org/10.1080/02656736.2021.1903583.

La Gioia, A., Porter, E., Merunka, I., Shahzad, A., Salahuddin, S. et al. (2018). Open-ended coaxial probe technique for dielectric measurement of biological tissues: Challenges and common practices. *Diagnostics*, 8(2): 40. https://doi.org/10.3390/diagnostics8020040.

Lau, M., Hu, B., Werneth, R., Sherman, M., Oral, H. et al. (2010). A theoretical and experimental analysis of radiofrequency ablation with a multielectrode, phased, duty-cycled system. *PACE - Pacing and Clinical Electrophysiology*, 33(9): 1089–1100. https://doi.org/10.1111/j.1540-8159.2010.02801.x.

Lau, R.W. and Gabriel, C. (1996). The dielectric properties of biological tissues: II. Measurements in the frequency range 10 Hz to 20 GHz. *Phys. Med. Biol.*, 41(11): 2251–2269. http://stacks.iop.org/0031-9155/41/i=11/a=002.

Lazebnik, M., Converse, M.C., Booske, J.H. and Hagness, S.C. (2006). Ultrawideband temperature-dependent dielectric properties of animal liver tissue in the microwave frequency range. *Physics in Medicine and Biology*, 51(7): 1941–1955. https://doi.org/10.1088/0031-9155/51/7/022.

Lazebnik, M., Popovic, D., McCartney, L., Watkins, C.B., Lindstrom et al (2007). A large-scale study of the ultrawideband microwave dielectric properties of normal, benign and malignant breast tissues obtained from cancer surgeries. *Physics in Medicine and Biology*, 52(20): 6093–6115. https://doi.org/10.1088/0031-9155/52/20/002.

López-Haro, S.A., Leija, L., Favari, L. and Vera, A. (2010). Measurement of ultrasonic properties into biological tissues in the hyperthermia temperature range. *Physics Procedia*, 3(1): 551–558. http://www.sciencedirect.com/science/article/B8JJ4-4YHJVRN-2G/2/de3289e28b58aae6087fa01d25ce65d1.

Lopresto, V., Argentieri, A., Pinto, R. and Cavagnaro, M. (2019). Temperature dependence of thermal properties of *ex vivo* liver tissue up to ablative temperatures. *Phys. Med. Bio*l., 27(xxxx): 0–31. https://doi.org/10.1016/j.jare.2020.01.010%0Ahttps://doi.org/10.1016/j.nano.2021.102426%0Ahttps://doi.org/10.1080/03008207.2019.1617280%0Ahttp://dx.doi.org/10.1038/s41598-019-38972-2%0Ahttps://doi.org/10.1016/j.matpr.2019.12.188%0Ahttps://doi.org/10.1016/.

Lopresto, Vanni, Pinto, R., Farina, L. and Cavagnaro, M. (2017). Microwave thermal ablation: Effects of tissue properties variations on predictive models for treatment planning. *Medical Engineering and Physics*, 46: 63–70. https://doi.org/10.1016/j.medengphy.2017.06.008.

Maenhout, G., Markovic, T., Ocket, I. and Nauwelaers, B. (2020). Effect of open-ended coaxial probe-to-tissue contact pressure on dielectric measurements. *Sensors (Switzerland)*, 20(7): 1–13. https://doi.org/10.3390/s20072060.

Martellosio, A., Pasian, M., Bozzi, M., Perregrini, L., Mazzanti, A. et al. (2017). Dielectric properties characterization from 0.5 to 50 GHz of breast cancer tissues. *IEEE Transactions on Microwave Theory and Techniques*, 65(3): 998–1011. https://doi.org/10.1109/TMTT.2016.2631162.

Martínez-Valdez, R., Contreras, M.V.H., Vera, A. and Leija, L. (2015). Sound speed measurement of chicken liver from 22°C to 60°C. *Physics Procedia*, 70: 1260–1263. https://doi.org/10.1016/j.phpro.2015.08.280.

Misra, D.K. (2014). Permittivity measurement. *In*: Webster, H. and Eren, J.G. (eds.). *Measurement, Instrumentation, and Sensors Handbook: Electromagnetic, Optical,Radiation, Chemical, and Biomedical Measurement* (2nd ed.). CRC Press. https://doi.org/https://doi.org/10.1201/b15664.

Mohammadi, A., Bianchi, L., Asadi, S. and Saccomandi, P. (2021). Measurement of *ex vivo* liver, brain and pancreas thermal properties as function of temperature. *Sensors*, 21(12). https://doi.org/10.3390/s21124236.

Paulides, M.M., Stauffer, P.R., Neufeld, E., Maccarini, P.F., Kyriakou, A. et al. (2013). Simulation techniques in hyperthermia treatment planning. *International Journal of Hyperthermia*, 29(4): 346–357. https://doi.org/10.3109/02656736.2013.790092.

Peyman, A., Holden, S.J., Watts, S., Perrott, R. and Gabriel, C. et al. (2007). Dielectric properties of porcine cerebrospinal tissues at microwave frequencies: *In vivo*, *in vitro* and systematic variation with age. *Physics in Medicine and Biology*, 52(8): 2229–2245. https://doi.org/10.1088/0031-9155/52/8/013.

Peyman, Azadeh, Kos, B., Djokić, M., Trotovšek, B., Limbaeck-Stokin, C., Serša, G. et al. (2015). Variation in dielectric properties due to pathological changes in human liver. *Bioelectromagnetics*, 36(8): 603–612. https://doi.org/10.1002/bem.21939.

Porter, E., Gioia, A. La, Santorelli, A. and O'Halloran, M. (2017). Modeling of the dielectric properties of biological tissues within the histology region. *IEEE Transactions on Dielectrics and Electrical Insulation*, 24(5): 3290–3301. https://doi.org/10.1109/TDEI.2017.006690.

Reinecke, T., Hagemeier, L., Schulte, V., Klintschar, M., Zimmermann, S. et al. (2013). Quantification of edema in human brain tissue by determination of electromagnetic parameters. *Proceedings of IEEE Sensors*. https://doi.org/10.1109/ICSENS.2013.6688477.

Rossmann, C. and Haemmerich, D. (2016). Review of temperature dependence of thermal properties, dielectric properties, and perfusion of biological tissues at hyperthermic and ablation temperatures. *Critical Reviews in Biomedical Engineering*, 42(6): 467–492.

Schutt, D.J. and Haemmerich, D. (2008). Effects of variation in perfusion rates and of perfusion models in computational models of radio frequency tumor ablation. *Medical Physics*, 35(8): 3462–3470. https://doi.org/10.1118/1.2948388.

Schwartz, J.L. and Mealing, G.A.R. (1985). Dielectric properties of frog tissues *in vivo* and *in vitro*. *Physics in Medicine and Biology*, 30(2): 117–124. https://doi.org/10.1088/0031-9155/30/2/001.

Sebek, J., Taeprasartsit, P., Wibowo, H., Beard, W.L., Bortel, R. et al. (2021). Microwave ablation of lung tumors: A probabilistic approach for simulation-based treatment planning. *Medical Physics*, 48(7): 3991–4003. https://doi.org/10.1002/mp.14923.

Smith, S.R. and Foster, K.R. (1985). Dielectric properties of low-water-content tissues. *Phys. Med. Biol.*

Soni, S., Tyagi, H., Taylor, R.A. and Kumar, A. (2015). The influence of tumour blood perfusion variability on thermal damage during nanoparticle-assisted thermal therapy. *International Journal of Hyperthermia*, 31(6): 615–625. https://doi.org/10.3109/02656736.2015.1040470.

Sturesson, C., Ivarsson, K., Andersson-Engels, S. and Tranberg, K.G. (1999). Changes in local hepatic blood perfusion during interstitial laser-induced thermotherapy of normal rat liver measured by interstitial laser Doppler flowmetry. *Lasers in Medical Science*, 14(2): 143–149. https://doi.org/10.1007/s101030050036.

Sugitani, T., Kubota, S.I., Kuroki, S.I., Sogo, K., Arihiro, K. et al. (2014). Complex permittivities of breast tumor tissues obtained from cancer surgeries. *Applied Physics Letters*, 104(25): 1–6. https://doi.org/10.1063/1.4885087.

Sun, C., Pye, S.D., Browne, J.E., Janeczko, A., Ellis, B. et al. (2012). The speed of sound and attenuation of an IEC agar-based tissue-mimicking material for high frequency ultrasound applications. *Ultrasound in Medicine and Biology*, 38(7): 1262–1270. https://doi.org/10.1016/j.ultrasmedbio.2012.02.030.

Techavipoo, U., Varghese, T., Chen, Q., Stiles, T.A., Zagzebski, J.A. et al. (2004). Temperature dependence of ultrasonic propagation speed and attenuation in excised canine liver tissue measured using transmitted and reflected pulses. *The Journal of the Acoustical Society of America*, 115(6): 2859–2865. http://www.ncbi.nlm.nih.gov/pubmed/15237809.

Trujillo, M. and Berjano, E. (2013). Review of the mathematical functions used to model the temperature dependence of electrical and thermal conductivities of biological tissue in radiofrequency ablation. *International Journal of Hyperthermia*, 29(6): 590–597. https://doi.org/10.3109/02656736.2013.807438.

Tsuyoshi, H., Inoue, D., Kurokawa, T. and Yoshida, Y. (2020). Hyperthermic intraperitoneal chemotherapy (HIPEC) for gynecological cancer. *Journal of Obstetrics and Gynaecology Research*, 46(9): 1661–1671. https://doi.org/10.1111/JOG.14391.

Tucci, C., Trujillo, M., Berjano, E., Iasiello, M., Andreozzi, A. et al. (2021). Pennes' bioheat equation vs. porous media approach in computer modeling of radiofrequency tumor ablation. *Scientific Reports*, 11(1). https://doi.org/10.1038/s41598-021-84546-6.

Tungjitkusolmun, S., Woo, E.J., Cao, H., Tsai, J.Z., Vorperian, V.R. et al. (2000). Thermal-electrical finite element modelling for radio frequency cardiac ablation: Effects of changes in myocardial properties. *Medical and Biological Engineering and Computing*, 38(5): 562–568. https://doi.org/10.1007/BF02345754.

Vaezy, S., Martin, R., Kaczkowski, P., Keilman, G., Goldman, B., Yaziji, H., Carter, S., Caps, M. and Crum, L. (1999). Use of high-intensity focused ultrasound to control bleeding. *Journal of Vascular Surgery*, 29(3): 533–542.

Vaupel, P. (2009). *Pathophysiology of Solid Tumors*, 51–92. https://doi.org/10.1007/978-3-540-74386-6_4.

Vives, A.A. (2008). Piezoelectric transducers and applications. *Piezoelectric Transducers and Applications*, 1–532. https://doi.org/10.1007/978-3-540-77508-9/COVER.

Vujaskovic, Z., Poulson, J.M., Gaskin, A.A., Thrall, D.E. Page et al. (2000). Temperature-dependent changes in physiologic parameters of spontaneous canine soft tissue sarcomas after combined radiotherapy and hyperthermia treatment. *International Journal of Radiation Oncology Biology Physics*, 46(1): 179–185. https://doi.org/10.1016/S0360-3016(99)00362-4.

Watson, T. (2008). Ultrasound in contemporary physiotherapy practice. *Ultrasonics*, 48(4): 321–329.

Zell, K., Sperl, J.I., Vogel, M.W., Niessner, R., Haisch, C. et al. (2007). Acoustical properties of selected tissue phantom materials for ultrasound imaging. *Physics in Medicine and Biology*, 52(20): 475–484. https://doi.org/10.1088/0031-9155/52/20/N02.

Zeqiri, B., Scholl, W. and Robinson, S.P. (2010). Measurement and testing of the acoustic properties of materials: A review. *Metrologia*, 47(2). https://doi.org/10.1088/0026-1394/47/2/S13.

Zheng, H. and Smith, C.E. (1991). Permittivity measureents using a short open-ended coaxial line probe. *IEEE Microwave and Guided Wave Letters*, 1: 337–339.

Zurbuchen, U., Holmer, C., Lehmann, K.S., Stein, T., Roggan, A. et al. (2010). Determination of the temperature-dependent electric conductivity of liver tissue *ex vivo* and *in vivo*: Importance for therapy planning for the radiofrequency ablation of liver tumours. *International Journal of Hyperthermia*, 26(1): 26–33. https://doi.org/10.3109/02656730903436442.

CHAPTER 12

Nanotheranostics in Cancer

Eunice Vargas-Viveros, Dora-Luz Flores and *Dayanira Paniagua**

This chapter begins with classifying nanomaterials to visualize the variety of materials and their characteristics at the nanometric scale, according to their dimensions and properties. The following section discusses the characteristics of nanoparticles that are applied for diagnosing and treating cancer. The section on mechanisms for diagnosis and therapy explains the primary mechanisms by which diagnosis and therapy are given simultaneously through processes assisted by nanostructured systems. Functionalization is a section that deals with how nanomaterials are adapted to be applied in these medical application processes thanks to the properties they have. The relevant properties in their study are mentioned in the characterization of the nanoparticles used in these nanostructured systems. Finally, in applications, the formulation is related to the mechanism of action of these nanostructured multifunctional systems for cancer diagnosis and therapy, known as nanotheranostics. For the applications, we used the Prisma methodology through the Covidence platform to summarize and evaluate the most relevant results with different nanotheranostics in cancer, focusing on characterization, mechanisms, and different nanoparticles.

Facultad de Ingeniería, Arquitectura y Diseño, Universidad Autónoma de Baja California, Ensenada, B. C. 22860 México

Emails: eunice@uabc.edu.mx; dflores@uabc.edu.mx

* Corresponding author: paniagua.dayanira@uabc.edu.mx

12.1 Introduction

Nowadays, cancer is still a threat to people's lives. These malignant tumors have not been identified early using traditional medical diagnostic techniques. So once cancerous tumors are diagnosed, they are already in advanced stages. Current chemical (drug) and physical (radiation) approaches to cancer treatment are imprecise and sometimes have irreversible side effects. In addition, it could also trigger other problems, such as recurrence or metastasis after treatment. One of the proposals to overcome these limitations is the development of more precise and efficient theranostic systems against cancer that will be powerful for improving the survival rate of patients and fighting cancer (Dai et al., 2022).

According to Dai et al., 2022, nanotheranostics refers to nanoparticles targeting cancer cells for diagnostic purposes and simultaneously carrying out therapeutic activities—synthesizing these nanosystems to offer targeted therapy while minimizing side effects and increasing treatment efficacy. In recent years, nanotheranostics has been a strategy for cancer treatment using intelligent multifunctional structured nanomaterials that include therapeutic and imaging functionalities.

12.2 Fundamentals of nanomaterials

12.2.1 Nanomaterials classification

Nanotechnology within recent decades is a discipline that has attracted great interest as emerging research is substantiated with the exploration of nanosized materials (i.e., nanoparticles, nanofibers, nanowires nanomembranes, nanotubes, and so on) in several fields. Nanomaterials (NMs) have many advantages over conventional materials, including target, selectivity, plasticity and high stability. Multiple biotic and abiotic materials can be used in the nanomaterials synthesis process through various methods (Shahcheraghi et al., 2022).

Nanobiotechnology is the union of two disciplines: nanotechnology and biotechnology as well as its application in biological fields. In order to enhance conventional biotechnological methods and defeating their limitations, nano-based approaches are developed. These limitations are for instance, the side effects caused by conventional therapies. Therapeutic focusing is expected to be functional in prevention and treatment of different diseases, such as cancers, neurodegenerative diseases, diabetes, inflammatory diseases and genetic disorders. Talking about cancers, traditional therapies (irradiation, chemotherapy and surgery) might provoke harsh aftereffects with no efficient disease treatment often. Researchers have recently developed drugs using nanomaterials through drug delivery systems that reduce side effects (Shahcheraghi et al., 2022).

Nanomaterials can be classified by considering different criteria such as dimensionality, chemical composition, state, and morphology. A material is considered on a nanometric scale if its size is between 1 and 100 nm. There are four classes of nanomaterials based on their dimensionality and shape. When they have all their dimensions in nanoscale they are named zero-dimensional (0D), i.e., below 100 nm. Examples include hollow sphere, nanorod, cube, spherical, metal, polygon, core–shell nanomaterials and quantum dots (QDs). Materials classified as one-dimensional (1D) are those with two dimensions in nanoscale but one dimension not in nanoscale; they include nanotubes, nanofibers, polymeric metallics, nanorod filaments or fibers, nanowires and ceramics. Materials with only one dimension in nanoscale and two otherwise are called two-dimensional nanomaterials (2D); they can include nanoplates, crystalline or amorphous, single-layered or multi-layered, thin films and nanocoatings. Materials with various dimensions above 100 nm are three-dimensional materials (3D); they mix numerous nanocrystals with different directions. For instance, carbon nanobuds, polycrystals, nanotubes, honeycombs, foams, layer skeletons, fibers, pillars and fullerenes are included in this group (Saleh, 2020).

On the other hand, nanomaterials are important due to the properties they possess, such as chemical and electromagnetic. They can exist in dispersed forms, suspensions, and colloids, as well as agglomerates, i.e., nanoparticles (NPs). The latter tend to form clusters unless they can be functionalized from their surface. Nanomaterials might be classified also based on their chemical composition in various types such as individuals composed as nanocomposites and more types of NPs within. Carbonaceous nanomaterials are composed of carbon, e.g., carbon nanotubes (CNTs), graphenes and fullerenes. Metallic nanomaterials are those from metals such as zinc, titanium, iron, silver, copper, silica gold and silver. Branched dendrimers form other kinds of nanomaterials with structures similar to branches with nano dimensions. When nanoparticles are incorporated to a standard materials grid they are called nanocomposites in order to enhance mechanical toughness or electrical conductivity. Quantum dots (QDs) are few nanometer semiconductors that have electronic and optical properties quite different from those of bigger particles because of quantum mechanics. These properties consist of absorbing white or UV light and reemitting it as a particular wavelength. Here either valence band holes, conduction band electrons or excitons are imprisoned to three spatial dimensions (Gavas et al., 2021).

Finally, diagnosis of cancer nanomaterials can be grouped as contrasting agents (i.e., gold nanoparticles and iron oxide among others) as well as fluorescent agents (quantum dots). There are some carriers in nanoscale that have built-in optical properties (e.g., gold, magnetic nanoparticles and carbon nanotubes). These nanocarriers can be transformed to a high energy for cell destruction and assist as nanotheranostics (Shahcheraghi et al., 2022).

12.2.2 Nanoparticles in cancer

Nanoparticles have many advantages in cancer diagnosis due to their characteristics, such as dissimilarity, high surface/volume ratio, nanometric size, and improved targeting system.

Nanopartice systems can selectively accumulate at the tumor site as a consequence of the enhanced permeation and retention effect (EPR) (Jiao et al., 2022). This preferential distribution is due to the presence of fenestrations at the tumor blood vessels sized from 200 to 2000 nm, and the poor lymphatic drainage (Gavas et al., 2021). Taking advantage of these features, the encapsulation of chemotherapeutics is routinely performed to reduce side effects, improve pharmacokinetics, and provide tumor selectivity. Also, to improve the release rate of drugs, particle polymer characteristics can be manipulated. (Gavas et al., 2021).

The different types of NPs used in cancer can have an organic or inorganic nature exhibiting different properties and advantages. Organic NPs are made from organic materials that mimic cellular membrane structures or polymeric biodegradable substances which increase their biocompatibility and biodegradability. Next, we describe the most common organic NPs:

- Polymeric nanoparticles (PNPs): They are “colloidal macromolecules” formed by different biodegradable polymers with a specific architecture such as: chitosan, polylactic acid, albumin, poly(amino acids) or alginate. Because of their nature, they reduce toxicity, enhance biocompatibility and drug release (Gavas et al., 2021).
- Dendrimers: They are hyperbranched polymeric macromolecules with a spherical structure (Gavas et al., 2021).
- Specific antibody–drug conjugates (ADCs). Monoclonal antibodies (mAb) combined with NPs. They are extensively used in cancer treatment because of their well-known targeting abilities (Gavas et al., 2021).
- Extracellular vesicles (EVs): They are double-layered phospholipid vesicles that mimic natural vesicles like exosomes, microvesicles and apoptotic bodies. They are used to deliver anti-tumor drugs to the target sites. NPs combined with exosomes are used because of their intrinsic

biocompatibility, chemical stability, and intracellular communications. Besides, they internalized very quickly within the cancer cells and escape the immune system (Gavas et al., 2021).

- Liposomes: They are spherical vesicles of uni-lamellar or multi-lamellar phospholipids that are used to encapsulate drug molecules. They are biologically inert, have low toxicity and weak immunogenicity. The most common liposome structure has a hydrophobic phospholipid bilayer and a hydrophilic core. This unique architecture makes them suitable for entrapping hydrophilic and hydrophobic drugs and protects them effectively from environmental degradation during systemic circulation. (Gavas et al., 2021).
- Solid Lipid Nanoparticles (SLN): Colloidal nanocarriers from 1–100 nm size, made of a phospholipid monolayer (from triglycerides, fatty acids, waxes, steroids, and PEGylated lipids), emulsifier and water, all together form a "micelle-like structure". The lipid component makes it possible to entrap drugs in a non-aqueous core (Gavas et al., 2021).
- Nanoemulsions: These NPs range from 10–1000 nm. They can be made in water-in-oil systems,oil-in-water systems, and bi-continuous nanoemulsions. They need high temperatures and pressure and instruments like homogenizers and microfluidizers, making their clinical application challenging (Gavas et al., 2021).
- Cyclodextrin Nanosponges: Tiny, mesh-like structures that stabilize the NPs and increase their drug loading capacity (Gavas et al., 2021).

Inorganic NPs have also been explored for cancer treatment since they exhibit specific properties that are not found in organic components:

- Carbon Nanoparticles: Carbon NPs include carbon nanotubes, graphene, carbon nanohorns, fullerenes, and graphene. Each of these has different structures, eliciting different tumor cell biological activities. Graphene oxide (GO) can target irregular angiogenesis and hypoxia in the tumor microenvironment (TME). Alternatively, fullerenes are the most widely studied nanocarriers. They exhibit triple yield, generate ROS and have the ability to absorb light. These features make them suitable for photodynamic therapy (PDT) and exhibit specific structural, physical, chemical, and electrical properties (Gavas et al., 2021).
- Quantum dots: Nano-semiconductors used in biological imaging because of their broad absorption spectrum, narrow emission bands, and high photostability (Gavas et al., 2021).
- Metallic NPs: Due to their remarkable photothermal, magnetic, and optical properties, they are commonly explored in "biological imaging" and as

targeted drug delivery systems. They are used as targeted drug carriers because their surface properties are easily controlled. Gold, silver, iron, and copper are the most widely used metallic nanoparticles (Gavas et al., 2021).

- Magnetic NPs: They are widely used in biomedical applications as delivery systems, visualisation with magnetic spectra for magnetic resonance imaging (MRI) or even for magnetic hyperthermia, generating thermal ablation of cancer cells. These NPs are from iron, gadolinium, manganese, and other metals. However, most magnetic NPs are iron oxide based (Griaznova et al., 2022). To enhance stability and biocompatibility, these NPs are covered with polymers or fatty acids (Gavas et al., 2021).
- Calcium phosphate NPs: Since they are biologically compatible, biodegradable, and do not cause adverse reactions, they are used as delivery systems. These NPs have been combined with viral or non-viral vectors for cellular gene transfer (Gavas et al., 2021).
- Mesoporous silica NPs: Has been widely used in immunotherapy. They also exhibit good pharmacokinetic properties making them one of the best drug carriers (Gavas et al., 2021).
- Conjugated Polymers (CP): These polymers are made of a carbon backbone with alternate single and double bounds which result in delocalized π-electrons. Due to their unique properties, they have been explored as a promising alternative for minimally invasive therapies such as phototherapy (Shete et al., 2022; Wei et al., 2021).

12.2.3 Mechanisms for diagnostics and therapy

Given that NPs can be retained in the tumor due to the EPR effect described above, being able to detect this accumulation leads to exploring non-invasive techniques for diagnosis. In this sense, many NPs have been used as contrast agents, such as for magnetic resonance imaging (MIR), NPs based on Gadolinium, Manganese, iron oxide, and Superparamagnetic iron–platinum particles (SIPPs) are the most common; for Fluorescence imaging (FLI) gold NPs, lanthanide NPs, QDs, nanotubes, for photoacoustic imaging (PAI) nanotubes and gold NPs; nanobubbles, silica NPs, and carbon nanotubes have been used as CAs for Ultrasound imaging (USI); nanoparticles based on heavy atoms such as gold, tantalum, lanthanides, and even Radiolabel QDs (Ga 68 or Iodine 125) have been used for nuclear imaging techniques such as Computed Tomography (CT), Positron Emission Tomography (PET) or Single Photon Emission Computed Tomography (SPECT (Smith and Gambhir, 2017).

The imaging techniques are based on the irradiation of the NPs with electromagnetic or mechanical energy to generate the emission of the detected signal. This irradiation also generates a thermal or chemical response in many of the nanomaterials, offering the possibility that the same nanoparticles can be used as both therapeutic agents and contrast media allowing monitoring of biodistribution and therapy simultaneously (Debbage and Jaschke, 2008).

The photothermal response gives place to Photothermal Therapy (PTT), which has shown minimal side effects and high performance in tumor ablation. High-energy PTT, which induces necrosis, can increase debris generation and trigger damage-associated molecular patterns (DAMPs), causing inflammation and, in many cases inducing secondary tumor growth. In contrast, low-energy PTT induces cellular apoptosis, leading to beneficial immunogenic responses that remove tumors and induce T-cell infiltration that mimics the effect of a vaccine (Han and Choi, 2021). Essentially, photothermal agents should have a strong extinction coefficient and high photothermal conversion efficiency (Wei et al., 2021).

Photothermal NPs include organic nanomaterials (such as porphyrin-lipid conjugates or organic semiconducting polymeric NPs); carbon-based nanomaterials (such as graphene oxide or carbon nanotubes); and inorganic materials such as metal NPs, based on metal oxides or gold and also QDs (Han and Choi, 2021).

Organic nanomaterials generally show good biocompatibility and are readily biodegradable, but they also show some limitations as low photothermal conversion efficiency, complicated synthesis, or poor photothermal stability (Han and Choi, 2021). Meanwhile, inorganic nanomaterials are known for their toxicity and being non-biodegradable; however, they have advantages such as excellent NIR light absorption, narrow emission spectra, high photothermal efficiency and photostability, and ease of synthesis (Han and Choi, 2021).

Another therapy mechanism that is photosensitive and has become trending is photodynamic therapy (PDT). The action mechanism is the photoactivation of different materials (photosensitizers), which can be activated by the absorption of visible light and induce photochemical reactions in the presence of reactive oxygen species (ROS). ROS induce severe hypoxia in tumor cells due to oxygen depletion and trigger ROS-mediated oxidative stress and unscheduled cell apoptosis. The composition of the components of nanomaterials can trigger the Fenton or Fenton-like reaction, which is catalyzed by transition metal ions such as Fe or Cu ions (Li et al., 2022). Dynamic therapy can also be activated with ultrasound stimulation, known as sonodynamic therapy (SDT) (Zhang et al., 2022).

Combined PTT treatment with other therapies like photodynamic therapy (PDT), immunotherapy, chemotherapy, or radiotherapy, has shown significant treatment outcomes (Han and Choi, 2021).

12.3 Multifunctional nanomaterials

12.3.1 Functionalization

To enhance the effectiveness of cancer therapy, it is necessary to develop delivery systems with the ability to target the specific site and discriminate cancer cells from healthy cells, shielding them from cytotoxicity (Gavas et al., 2021).

NPs are well-known due to their nonselective distribution and high toxicity. However, to be effective, NPs should have the ability to remain stable in the vascular system, penetrate the tumor fluid, reach the target and interact exclusively with tumor cells, and escape the different clearance systems (Gavas et al., 2021; Shete et al., 2022).

Functionalization refers to the modification of NP surfaces, includes conjugation of biomolecules or chemicals, and offers the possibility of a design on demand to improve their properties and characteristics, giving place to NPs with multimodal characteristics.

The use of organic molecules as functionalities or matrix reduces toxicity, among other advantages: particles with different fluorescent properties, functionalized with tryptophan, increase the lifetime of excited states since tryptophan displays an excitation and emission spectrum in the UV region. NPs can be functionalized for excellent stability with chitosan molecules, allowing further functionalization with other biomolecules. The surface of silica NPs can be altered by linking various functional molecules. Especially mesoporous particles. To enhance biodistribution, NPs have been functionalized with organic polymers like starch, dextran, citrate, or polyethylene glycol (PEG), which also provide stabilization and reduce toxicity; since functionalization modifies surface properties of NPs, their pharmacokinetics also modified (Thiruppathi et al., 2017).

The components of the nanotheranostics could be customized for tracking by different imaging modalities into one nanoprobe, such as MRI, PET, USI and PAI (Jiao et al., 2022). Signal generators incorporated into NPs include gadolinium, iron oxide, iodine, fluorine, bismuth, quantum dots, radionuclides, metal nanoclusters and Conjugated Polymers (Debbage and Jaschke, 2008; Wei et al., 2021).

Another advantage of functionalization is that it allows the loading of therapeutic drugs into the NPs, like chemotherapeutics, giving place to drug delivery systems, working synergistically with PTT, PDT, immunotherapy, or radiotherapy, which has shown significant treatment outcomes (Han and Choi, 2021). Stimulus-responsive NDDS, sensitive to external factors such as light, ultrasound, or magnetic field, can controllably release drugs while being tracked by imaging techniques, helping prevent unwanted drug leakage during the administration of the delivery process (Gavas et al., 2021). It has also been observed that when responding to multiple stimuli, a superior tumor-specific drug release yield occurs compared to those responding to a single stimulus (Gou et al., 2022).

Another hallmark that has been explored as an advantage for Stimuli-responsive nanomaterials is the unique physiological property of solid human tumors, which consists in a set of metabolic processes that give place to a unique microenvironment, different from healthy cells, known as Tumor Microenvironment (TME). These characteristics include, low pH, high inflammatory, severe hypoxia, elevated glutathione (GSH) levels, excessive H_2O_2, elevated hyper permeability vessel and overexpression of different receptors or enzymes that are associated with specific cancer types.

NDDS that are sensitive to these conditions offer the possibility of specific delivery at tumor lesions, avoiding systemic leakage, reducing side effects of treatment (Gavas et al., 2021). Some examples include the use of pH-sensitive NPs that release drugs at low pH (Gavas et al., 2021). The use of Cu ion NPs instead of Fe ions with Fe ions is preferred since they catalyzed Fenton-like reactions more efficiently within wide pH ranges (Li et al., 2022).

Another improvement in cancer therapy with NPs is targeting through functionalization. As mentioned, tumors can be targeted by taking advantage of the TME and EPR effects. This strategy is known as "passive" targeting. For this purpose, NPs should size between 10 and 200 nm. That specific size can be reached by functionalizing with hydrophilic polymers. The most commonly used hydrophilic polymer is PEG (Shete et al., 2022). However, this approach still has some limitations due to the insufficient diffusion into the tumor, no EPR effect in the early stage of tumors, and retention in high-fenestrated organs like the liver or kidneys (Jiao et al., 2022).

To improve targeting, specific ligands with high affinity against overexpressed receptors are used in NPs functionalization giving place to an "active targeting" approach. Most widely used targeting moieties are listed and described in Table 12.1 (Shete et al., 2022). The NPs functionalized with these specific ligand molecules not only increase specificity and retention in the TME, but this ligand-target interaction also triggers internalization of NPs via receptors-mediated endocytosis (Gavas et al., 2021).

Table 12.1. Targeting molecules.

Molecule	**Properties**	**Cancer types**	**Advantages**
Folic acid (FA)	FA, known as vitamin B9, is a water-soluble molecule that has shown promising target capacities due to its high affinity to the folate receptor (FR).	FRs are overexpressed on the membranes of the tumor cells, like meningiomas, non-Hodgkin's lymphomas, choriocarcinomas, uterine sarcomas, osteosarcomas, and ovarian carcinomas.	FA is relatively abundant, non-toxic, economical, easily conjugate to the carriers, non-immunogenic, keep high binding affinity, is small-sized, and stable. Then, the NPs release the drug inside the tumor cells, at the cytoplasm, where the drugs finally interact with the intracellular components.
Tumor factor (Tf)	Tf is a membrane glycoprotein that functions as an iron transporter, which binds selectively to the CD71 receptors.	Tumor factor Receptors (TfR) are overexpressed on their membrane, while healthy cells express it at low levels.	Tf is internalized by receptor-mediated endocytosis and delivers the NP accumulating a high amount inside the tumor cells.
Aptamers	These are molecules such as ssDNA or RNA between 6 kDa and 26 kDa. They are developed by the Systematic Evolution of Ligands by Exponential Enrichment (SELEX) and are designed to recognize specific targets based on their three-dimensional structure.	They can be developed against different cancer markers. One of the most promising aptamers is AS1411 used against metastatic renal cell carcinoma.	The development of aptamers is an easy scale-up process. They have low immunogenicity and are smaller than antibodies, allowing them to get easily internalized into the cells and interact with tiny proteins on the cell surface.
Peptides	Peptides are specific amino acid sequences that can bind to the receptors inside the tumors. The most common peptide motif is Arginyl-glycyl-aspartic acid, known as RGD peptide, which is responsible for cell adhesion to the extracellular matrix. It binds to the endothelial cell receptor integrin $\alpha v \beta 3$.	Peptides have different targets. Integrin $\alpha v \beta 3$ is a crucial component of angiogenesis and is overexpressed on invasive tumors like glioblastomas and malignant melanomas.	Peptides are non-infectious, easier to produce and more stable than antibodies. They exert good biocompatibility, are economical, and easy to scale up. Peptides are associated with limitations like the risk of enzymatic degradation. Also, non-specific targeting and cellular uptake have been reported, maybe due to their weak receptors affinity.

Table 12.1 contd. ...

...Table 12.1 contd.

Molecule	Properties	Cancer types	Advantages
Angiopep-2 (ANG)	Peptides derived from the kunitz domain, they bind to the low-density lipoprotein receptor-related proteins (LRPs).	LRPs are overexpressed on the blood-brain barrier (BBB) and GBM or glioma tumor cells.	The parenchymal accumulation and high transcytosis capability of the angiopep-2 allows the functionalized NP to cross the BBB, accumulate and target the glioma within the brain.
Monoclonal antibodies (mAbs)	Therapeutically active mAbs can recognize and target hematological tumor-associated antigens, which are glycoproteins related to cluster of differentiation (CD) groups (CD20, CD30, CD33, and CD52).	These mAbs are used to treat different hematological cancer types and solid tumors.	Based on their humanness they are named with the following suffixes: for fully humanized mAbs -umab, for humanized mAbs -zumab, for chimeric mAbs -ximab and for murine mAbs -omab.

12.3.2 Characterization of functionalized nanoparticles

The importance of using nanostructured materials has been based on the study of their physical and chemical properties, since they are very different from those of conventional materials, even with the same chemical composition. This has been allowed by the use of tools and techniques that provide information on materials at the nanoscale. One of the most important constituents of nanomaterials are metallic nanoparticles, which provide these magical-like properties and significantly improve their performance.

As it is well known, in Nanotechnology, nanomaterials with specific properties are designed and synthesized, in order to apply them in the solution of a particular problem or optimization of existing processes. Table 12.2 lists the most common characterization techniques to evaluate and analyze the physical and chemical properties, which are important in nanotheranostic systems for cancer.

12.4 Applications

Various applications of multifunctional nanostructured systems act as biomarkers, drug nanocarriers, magnetics, and hotters that have used different techniques for synthesizing and characterizing these nanomaterials, some of which are listed in Table 12.3.

The literature was systematically reviewed to describe the usefulness of the newest contributions of nanotheranostics in cancer. The review focused

Table 12.2. Characterization techniques of different nanoparticles used for PPT.

Technique	**Properties of interest**	**Reference**
Atomic Force Microscopy (AFM)	pH-triggered formation of actives	(Huang et al., 2022)
Circular Dichroism (CD)	pH-responsiveness	(Huang et al., 2022)
Confocal Microscopy	Location of nanoparticles and activation of their functionalization	(Liu et al., 2022) (García-Hevia et al., 2022)
Differential Scanning Calorimetry (DSC)	Determination of the melting point	(García-Hevia et al., 2022)
Dynamic Light Scattering (DLS)	Hydrodynamic diameter, stability of the nanoparticle and ζ-potential values of some formulations	(Zhao et al., 2022) (García-Hevia et al., 2022) (Zhang et al., 2022) (Liu et al., 2022) (Kong et al., 2022) (Li et al., 2022) (Jiao et al., 2022) (Li et al., 2022) (Li et al., 2022) (Gou et al., 2022)
Energy Dispersive X ray Spectroscopy (EDX, EDS, EDAX)	Element distribution on the surface or on specific NP sites, elemental analysis, chemical composition	(Esmaeili et al., 2022) (Jiao et al., 2022) (Zhai et al., 2022) (Zhang et al., 2022)
Electronic Supplementary Material (ESM)	Average aspect ratio	(Zhang et al., 2022)
Infrared diode laser	Up-conversion spectrum, photothermal analysis of materials and thermal images	(Liu et al., 2022)
Field Emission-Scanning Electron Microscopy (FE-SEM)	Ionic distribution of the nanoconjugates	(Esmaeili et al., 2022)
Fluorescence spectrophotometry	pH-response performance of the proposed strategy, fluorescence spectrum	(Huang et al., 2022) (Li et al., 2022) (Dai et al., 2022) (Li et al., 2022)
Fourier-Transform Infrared Spectroscopy (FTIR)	Chemical composition and interfacial interaction of the nanosystems	(Esmaeili et al., 2022) (Liu et al., 2022) (Wang et al., 2022) (Li et al., 2022) (Gou et al., 2022)
Scanning Transmission Electron Microscopy (HAADF-STEM)	Morphology	(Zhai et al., 2022)
High Performance Liquid Chromatography (HPLC)	Composition and physicochemical properties	(García-Hevia et al., 2022) (Li et al., 2022)
High Resolution Transmission Electron Microscopy (HRTEM)	Morphology, textural properties, interplanar distances of metals planes, lattice spacing	(Esmaeili et al., 2022) (Zhang et al., 2022) (Jiao et al., 2022)

Table 12.2 contd. ...

...Table 12.2 contd.

Technique	Properties of interest	Reference
Inductively Coupled Plasma-Atomic Emission Spectroscopy (ICP-AES)	Metal ions identification	(Zhai et al., 2022)
Inductively Coupled Mass Spectrometry (ICP MS)	Mass ratio of metals, concentrations of metals in NPs, elemental content percentage	(Griaznova et al., 2022) (Kong et al., 2022)
Near-Infrared Spectroscopy (NIR)	Reduction-sensitive property	(Zhang et al., 2022) (Liu et al., 2022) (Kong et al., 2022) (Dai et al., 2022) (Jiao et al., 2022), (Li et al., 2022)
Infrared Thermal (IRT)	Imaging capacities	(Li et al., 2022) (Jiao et al., 2022) (Li et al., 2022)
Nitrogen adsorption	Specific surface area	(Esmaeili et al., 2022) (Wang et al., 2022) (Zhai et al., 2022)
Polydispersity Index (PDI)	NPs size distribution	(Gou et al., 2022)
Scanning Electron Microscopy (SEM)	Surface morphology and the elemental composition of NPs, Nanoparticle images and mean size of the bimetal core satellites	(Wang et al., 2022) (Kong et al., 2022) (Gou et al., 2022) (Griaznova et al., 2022)
Magnetic Resonance (MR)	Imaging	(Li et al., 2022)
Transmission Electron Microscopy (TEM)	NP size, morphological, structural and magnetic characterization, NP diameters	(Zhao et al., 2022) (Zhang et al., 2022) (García-Hevia et al., 2022) (Liu et al., 2022) (Wang et al., 2022) (Kong et al., 2022) (Li et al., 2022) (Jiao et al., 2022) (Li et al., 2022) (Zhai et al., 2022) (Li et al., 2022)
Thermogravimetry (TG, TGA)	Amount of PEG loaded onto Ti3C2@TiO2-x hybrid was about 3% as quantitatively determined	(Wang et al., 2022) (Zhang et al., 2022)
UV-Vis	Gold NPs plasmon	(Griaznova et al., 2022) (Liu et al., 2022) (Gou et al., 2022)
Uv-Vis and fluorescence spectrophotometry	Photothermal ability	(Li et al., 2022)
UV-Vis, Visible and NIR spectroscopy	PDT and PTT capabilities, optical absorption	(Zhao et al., 2022) (Zhang et al., 2022) (Wang et al., 2022) (Zhai et al., 2022) (Gou et al., 2022)

Table 12.2 contd. ...

...Table 12.2 contd.

Technique	Properties of interest	Reference
UV-Vis spectrometer and HPLC	The loading and encapsulation efficiency	(Li et al., 2022)
VSM LakeShore 7407 Series magnetometry	Field dependence of magnetization	(Griaznova et al., 2022)
X-ray Photoelectron Spectroscopy (XPS)	Elemental composition and oxygen vacancies, chemical composition	(Zhang et al., 2022) (Wang et al., 2022) (Li et al., 2022) (Jiao et al., 2022) (Zhai et al., 2022)
X Ray Diffraction (XRD)	Crystalline structure	(Esmaeili et al., 2022) (Zhang et al., 2022) (Liu et al., 2022) (Li et al., 2022) (Zhai et al., 2022)
Zeta potential	Particle size distribution and colloidal stability	(Esmaeili et al., 2022) (Zhao et al., 2022) (Liu et al., 2022) (Wang et al., 2022) (Gou et al., 2022) (Zhang et al., 2022)

on the different nanomaterials used as thermal therapy for different types of cancer, as well as the mechanism used in each of the formulations. For the review of articles, the Prisma methodology was implemented. The use of the Covidence platform was integrated to assist in elaboration of reviews.

Literature search strategies were developed using text words related to nanotheranostics and cancer. ScienceDirect (Elsevier, n.d.) and Springer databases were searched, and searches were limited to English. Data were collected through a search taken from various publications, including the words "nanotheranostic" AND/OR "nanotheranostics", AND/OR "cancer".

Articles from peer-reviewed journals were included if they were published between the period 2021–2022, written in English, and had no reviews. The literature search results were uploaded into a collaborative file, and eligibility criteria and citation abstracts were attached to a database. Titles and abstracts produced by the search were independently screened against eligibility criteria. Complete reports were obtained for all titles that appeared to meet the eligibility criteria or where there was some uncertainty. We then examined the full texts and verified that they met the eligibility criteria.

According to the search criteria, a total of 66 articles were found in the databases. Of these, 4 were duplicated, 25 were eliminated because they did not meet the eligibility criteria, and 23 were eliminated because they did not present a methodology or objectives for this research. In total, 14 articles were included for analysis (see Figure 12.1).

Table 12.3. Applications using PRISMA methodology.

Formulation	**Mechanism**	**References**
Nanoparticles (NPs) based on regenerated silk fibroin, loaded with Doxorubicin (DOX), with surface functionalization by photosensitizer (N770). The NP was labeled as N770-DOX@ NPs.	They were designed to treat tumoral, lung, and breast cancer. The authors designed a nanoscale drug delivery system with quadruple responsiveness (pH, ROS, glutathione, and hyperthermia), capable of oxygen self-production, mitochondrial phototherapy, chemotherapy, dual-level targeted drug delivery, and multimodal imaging. Demonstrating that the NPs functionalized with N770 achieved the endocytosis process.	(Gou et al., 2022)
Concave octahedral PtCu nanoframes (COPtCu-Ns) loaded with Vitamin C (VC), modified with PEG and FA for tumor targeting.	VC has shown to be a promising pro-oxidant for the generation of H2O2, causing selective toxicity against tumor cells. Meanwhile, the COPtCu-Ns induce a Fenton-like reaction in the acidic TME generating hydroxyl radical, together with the photothermal effect of the NPs, the chemical efficiency would be promoted, and the output of OH radicals enhanced, which would improve the therapeutic effect of CDT.	(Zhai et al., 2022)
The authors first synthesized MRP (a NIR-II fluorescent probe) and JPC (a reduction-activatable prodrug of JQ1). Then, T7-PEG5k-DSPE (T7-modified phospholipid) was subsequently self-assembled with MRP and JPC due to hydrophobic interaction generating a micellar NP named TNP@JQ1/MRP.	The NP design could cross BBB due to T7 and deliver JPC and MRP in GBM. Then, JPC will give place to JQ1 that would suppress interferon gamma-inducible PDL1 expression in the tumor cells; meanwhile, MRP could generate NIR-II fluorescence to navigate 808 nm lasers and induce a photothermal effect to trigger *in situ* antigen release. The PTT, together with the JQ1 effect, would elicit antitumor immunogenicity.	(Li et al., 2022)
The authors designed a theranostic agent that is activated with NIR-II laser. It is constituted by Pluronic F127, with PEG moieties (amphiphilic polymers), coated with PTTBT (a NIR-II-absorbing conjugated polymer) and bF as NO donor. The nanotheranostic was named CP-bF@PEG which worked as a NO nanogenerator and a NIR-II photothermal inducer.	CP-bF@PEG showed high-contrast NIR-II fluorescence imaging, the ability to identify tumors accurately, and excellent photothermal effects under deep tissue penetration of NIR-II laser irradiation. The NP also generated NO via glutathione activation in a controllable manner at the TME, achieving synergistic therapeutic effects.	(Li et al., 2022)

Table 12.3 contd. ...

...Table 12.3 contd.

Formulation	Mechanism	References
CD doped with Gadolinium and conjugated with AS1411 aptamers. The NP mas named AS1411-Gd-CD.	AS1411-Gd-CD provided comprehensive diagnostic information by showing FLI/MRI high sensitivity and superior spatial resolution. Also elicited high photothermal conversion when irradiated by 808 nm laser achieving high PTT therapeutic efficacy and tumor targeting, since AS1411 aptamers display high affinity against nucleotin, which is highly expressed in several cancer cell types.	(Jiao et al., 2022)
The authors engineered a novel NP with poly(styrene-co-chloromethyl styrene)- graft-poly(ethylene glycol), used as matrix, coupled with boron di-fluoride formazanate (BDF), a new type of electron acceptor, and a novel organic small molecular NIR-II dye named BDF1005.	BDF1005 NPs got accumulated at the tumor site due to EPR effect. The NPs presented a great photothermal conversion attributed to the high molar absorption coefficient of BDF1005 at 808 nm.	(Dai et al., 2022)
A theranostic nanoplatform was designed using hollow mesoporous organosilica as a nanoplatform where a near-infrared carbocyanine dye (DIR) and Gd_2O_3 were integrated. The NP was named HMON@CuS/Gd_2O_3.	HMON@CuS/Gd_2O_3 NPs was able to generate oxidative stress elevation at tumor lesions, good photothermal stability under NIR irradiation and mild PTT effect, while achieving real-time visualization by MRI/NIR fluorescence multimodality imaging-guided PT for tumor identification and elimination.	(Li et al., 2022)
Polydopamine NPs doped with Gadolinium (PDA@Gd) were functionalized with RGD2 (a RGD-derived peptide RGDfRGDfC) for targeted delivery to bone tumors and D8 (a bone targeting peptide of eight aspartic acids).	The RGD peptide is known to be essential in the bone remodeling process. It recognizes integrin receptors and inhibits the activation of osteoclast. Meanwhile the delivered PDA@Gd NPs allow dual-modal imaging (PAI/MRI) with good biocompatibility and excellent photothermal properties. The nanotheranostic showed good biocompatibility and excellent photothermal properties. The results showed a promising new nanoplatform for treatment of osteosarcoma and related osteolysis.	(Kong et al., 2022)

Table 12.3 contd. ...

...Table 12.3 contd.

Formulation	Mechanism	References
The authors report the design of a mesoporous silica nanoparticle (PMSN) decorated with polydopamine (PDA). The pores of PMSN were loaded with Fe^{3+} and coated with Lauric acid (LA) to prevent the undesired leakage of iron. The NP was called PMFLB. Then, Bovine serum albumin (BSA) was used to provide colloidal stability.	Under the NIR laser, the PDA (a light-sensitive carbon polymer) produced enough heat to kill the colorectal cancer cells through hyperthermia. Meanwhile, the heat induced the phase change of LA, and the iron was released. Fe3+ reacts with the endogenous H2S of the colorectal TME transforming it into Fe2+, triggering the Fenton reaction with endogenous H2O2, and generating hydroxyl radical. Finally, the Fe2+ regenerated into Fe3+, which enhanced the chemodynamic therapy. The innovation relies on the thermal response of LA, which reacts as a doorkeeper preventing the iron from leaking before reaching the TME of cancer cells.	(Wang et al., 2022)
UCNPs@mSiO_2-Au-Cys are reported as nanomotors utilized in combined therapy as tumor multimode imaging and photothermal (PTT)/ photodynamic (PDT). At first, by upconversion nanoparticles (UCNPs), photothermal agent Au NPs and photosensitizer Cy-S-Ph-NH_2 (Cys) are loaded with the mesoporous silica shell.	The synergistic treatment of PTT and PDT by UCNPs@mSiO2-Au-Cys + NIR eradicated the tumor after one dose, and no recurrence/ regeneration was observed after ten days. Due to their asymmetric distribution on UCNPs, Au loaded on the NPs provided PTT heat and acted as a driving source of nanomotors. UCNPs@mSiO2-Au-Cys can produce autonomous motion and become upconversion nanomotors under an 808 nm laser irradiation. Autonomous movement improves the tumor permeability of therapeutic agents. Moreover, it can overcome the problem of uneven photothermal effect in the tumor. Furthermore, AuNPs enhance the ROS mechanism improving PDT, which is done while Cys is triggered by NIR light and cooperates to provide NIR photothermal imaging capability. UCNPs of nanomotors under 980 nm excitation can produce two emissions at 659 nm and 800 nm, and this can be used to provide real-time online monitoring of singlet oxygen level generated in an organic body; that means real-time monitoring of PDT treatment effect.	(Liu et al., 2022)

Table 12.3 contd. ...

...Table 12.3 contd.

Formulation	Mechanism	References
In this work, Fe-Au nanoparticles were synthetized and coated with polyacrylic acid (PAA) for colloidal stability (Fe-Au@PAA)	These NPs showed strong photoconversion when irradiated at 808 nm, which belongs to the first biological transparency window (650–950 nm), where tissues show low absorption and scattering of light; this provides the highest light penetration depth. The delivery was reached by the EPR effect and assisted with a neodymium magnet placed near the tumor to increase NP accumulation.	(Griaznova et al., 2022)
The authors reported the synthesis of magnetic lipid nanocomposite vehicles (mLNVs) that are formed by a hydrophobic matrix of Carnauba wax, which encapsulate magnetic nanoparticles (MNPs) of Fe3O4 as imaging reporters for MRI, and effectors for magnetic hyperthermia. The nanocomposites also encapsulated doxorubicin (a potent chemotherapeutic), and a lipophilic commercial fluorescent dye (DiO).	The nanpocomposites showed low toxicity due to their organic nature. The proposed NPs were able to deliver doxorubicin after inducing magnetic hyperthermia due to the Carauba wax melting. The high melting point of the wax makes it a good candidate for drug delivery, minimizing pre-leakage of the drug. With this formulation the NPs showed a synergic therapy, combining the antimitotic effect with the thermal ablation of solid tumor cells. The imaging capability made these NPs trackable during therapy. The authors also observed that the NPs accumulated at the tumor despite the lack of specific targeting. The authors also address that compared to liposomal nanocarriers, their formulation shows high encapsulation efficiency, high stability, low cost, and enables a close control over the drug delivery.	(Garcia-Hevia et al., 2022)
Heterogeneous Pd-Au nanorods by depositing palladium nanoclusters on gold nanorods and coating them with BSA (BSA-Pd-Au NRs).	The nanorods showed extended light absorption from the NIR-I region into the NIR-II region and good photothermal conversion. That induced not only PTT but heterogeneous dual catalytic activities: bioorthogonal chemistry, which led to the *in situ* conversion of 5-fluoro-1-propargyluracil (Pro-5-Fu) into 5-fluorouracil (5-Fu) when coadministered, and decomposition of hydrogen peroxide (H2O2) into hydroxyl radical, providing PDT y chemotherapy synergistically.	(Zhang et al., 2022)

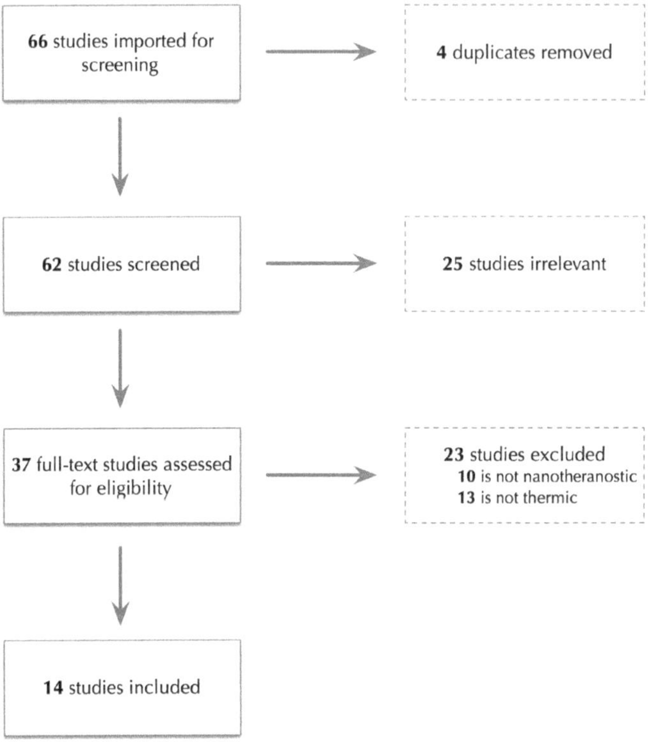

Figure 12.1. PRISMA flow diagram for systematic review.

Acronym/Abbreviation

ADC	Antibody–Drug Conjugate
bF	Benzofuroxan
BBB	Blood Brain Barrier
CD	Carbon Dots
CDT	Chemodynamic Therapy
CNT	Carbon Nanotubes
CP	Conjugated Polymers
CT	Computerized Tomography
DAMP	Damage-Associated Molecular Pattern
DOX	Doxorubicin
EPR	Enhanced Permeation and Retention
EV	Extracellular Vesicles
FA	Folic Acid
FLI	Fluorescence Imaging

GMB	Glioblastoma
GO	Graphene Oxide
GSH	Glutathione
mAb	Monoclonal Antibodies
MRI	Magnetic Resonance Imaging
N770	Heptamethine Cyanine-Based Photosensitizers
NIR	Near Infrared
NM	Nanomaterial
NP	Nanoparticle
PA	Photoacoustic
PDT	Photodynamic Therapy
PEG	Polyethylene Glycol
PET	Positron Emission Tomography
PNP	Polymeric Nanoparticles
PTT	Photothermal Therapy
QD	Quantum Dots
ROS	Reactive Oxygen Species
SLN	Solid Lipid Nanoparticles
TME	Tumor Microenvironment
VC	Vitamin C

Reference list

Dai, H., Cheng, Z., Zhang, T., Wang, W., Shao, J. et al. (2022). Boron difluoride formazanate dye for high-efficiency NIR-II fluorescence imaging-guided cancer photothermal therapy. *In Chinese Chemical Letters*, 33(5): 2501–2506. https://doi.org/10.1016/j.cclet.2021.11.079.

Debbage, P. and Jaschke, W. (2008). Molecular imaging with nanoparticles: Giant roles for dwarf actors. *Histochemistry and Cell Biology*, 130(5): 845–875. https://doi.org/10.1007/s00418-008-0511-y.

Esmaeili, Y., Khavani, M., Bigham, A., Sanati, A., Bidram, E. et al. (2022). Mesoporous silica@chitosan@gold nanoparticles as "on/off" optical biosensor and pH-sensitive theranostic platform against cancer. *International Journal of Biological Macromolecules*, 202(December 2021): 241–255. https://doi.org/10.1016/j.ijbiomac.2022.01.063.

García-Hevia, L., Casafont, Í., Oliveira, J., Terán, N., Fanarraga, M.L. et al. (2022). Magnetic lipid nanovehicles synergize the controlled thermal release of chemotherapeutics with magnetic ablation while enabling non-invasive monitoring by MRI for melanoma theranostics. *Bioactive Materials*, 8(October 2020): 153–164. https://doi.org/10.1016/j.bioactmat.2021.06.009.

Gavas, S., Quazi, S. and Karpiński, T.M. (2021). Nanoparticles for cancer therapy: Current progress and challenges. *Nanoscale Research Letters*, 16(1). https://doi.org/10.1186/s11671-021-03628-6.

Gou, S., Chen, N., Wu, X., Zu, M., Yi, S. et al. (2022). Multi-responsive nanotheranostics with enhanced tumor penetration and oxygen self-producing capacities for multimodal synergistic cancer therapy. *Acta Pharmaceutica Sinica B*, 12(1 PG-406–423): 406–423. https://doi.org/https://doi.org/10.1016/j.apsb.2021.07.001.

Griaznova, O.Y., Belyaev, I.B., Sogomonyan, A.S., Zelepukin, I.V., Tikhonowski, G.V. et al. (2022). Laser synthesized core-satellite Fe-Au nanoparticles for multimodal *in vivo* imaging and *in vitro* photothermal therapy. *Pharmaceutics*, 14(5). https://doi.org/10.3390/pharmaceutics14050994.

Han, H.S. and Choi, K.Y. (2021). Advances in nanomaterial-mediated photothermal cancer therapies: Toward clinical applications. *Biomedicines*, 9(3): 1–15. https://doi.org/10.3390/biomedicines9030305.

Huang, J., Wu, Y.C., He, H., Ma, W.J., Liu, J.B. et al. (2022). Acidic microenvironment triggered *in situ* assembly of activatable three-arm aptamer nanoclaw for contrast-enhanced imaging and tumor growth inhibition *in vivo*. *Theranostics*, 12(7 PG-3474–3487): 3474–3487. https://doi.org/10.7150/thno.72028.

Jiao, M., Wang, Y., Wang, W., Zhou, X., Xu, J. et al. (2022). Gadolinium doped red-emissive carbon dots as targeted theranostic agents for fluorescence and MR imaging guided cancer phototherapy. *Chemical Engineering Journal*, 440(PG-135965): 135965. https://doi.org/https://doi.org/10.1016/j.cej.2022.135965.

Kong, Y., Zhou, L., Liao, S., Wang, C., Chen, J. et al. (2022). Dual peptide-engineered and gadolinium-doped polydopamine particles as targeted nanotheranostics for the treatment of osteosarcoma and related osteolysis. *Chemical Engineering Journal*, 444(February): 136516. https://doi.org/10.1016/j.cej.2022.136516.

Li, F., Lai, Y., Ye, J., Saeed, M., Dang, Y. et al. (2022). Dual-targeting prodrug nanotheranostics for NIR-II fluorescence imaging-guided photo-immunotherapy of glioblastoma. *Acta Pharmaceutica Sinica B*, 12(9): 3486–3497. https://doi.org/10.1016/j.apsb.2022.05.016.

Li, J., Xie, L., Sang, W., Li, W., Wang, G. et al. (2022). NIR-II-absorbing conjugated polymer-based theranostic agent for NIR-II fluorescence imaging-guided photothermal therapy acting synergistically with tumor microenvironment-responsive nitric oxide therapy. *ChemPhysMater.*, 1(1 PG-51–55): 51–55. https://doi.org/https://doi.org/10.1016/j.chphma.2021.09.004.

Li, P.F., Lin, B.Q., Chen, Z.A., Liu, P., Liu, J.Q. et al. (2022). Biodegradable hollow mesoporous organosilica nanotheranostics (HMONs) as a versatile platform for multimodal imaging and phototherapeutic-triggered endolysosomal disruption in ovarian cancer. *Drug Delivery*, 29(1 PG-161–173): 161–173. https://doi.org/10.1080/10717544.2021.2021322.

Liu, H.X., Zhang, J.Z., Jia, Y.N., Liu, X., Chen, X.Q. et al. (2022). Theranostic nanomotors for tumor multimode imaging and photothermal/ photodynamic synergistic therapy. *Chemical Engineering Journal*, 442(PG-). https://doi.org/10.1016/j.cej.2022.135994.

Saleh, T.A. (2020). Nanomaterials: Classification, properties, and environmental toxicities. *Environmental Technology and Innovation*, 20: 101067. https://doi.org/10.1016/j.eti.2020.101067.

Shahcheraghi, N., Golchin, H., Sadri, Z., Tabari, Y., Borhanifar, F. et al. (2022). Nano-biotechnology, an applicable approach for sustainable future. *3 Biotech.*, 12(3): 1–24. https://doi.org/10.1007/s13205-021-03108-9.

Shete, M.B., Patil, T.S., Deshpande, A.S., Saraogi, G., Vasdev, N. et al. (2022). Current trends in theranostic nanomedicines. *Journal of Drug Delivery Science and Technology*, 71(December 2021): 103280. https://doi.org/10.1016/j.jddst.2022.103280.

Smith, B.R. and Gambhir, S.S. (2017). Nanomaterials for *in vivo* imaging. *Chemical Reviews*, 117(3): 901–986. https://doi.org/10.1021/acs.chemrev.6b00073.

Thiruppathi, R., Mishra, S., Ganapathy, M., Padmanabhan, P., Gulyás, B. et al. (2017). Nanoparticle functionalization and its potentials for molecular imaging. *Advanced Science*, 4(3). https://doi.org/10.1002/advs.201600279.

Wang, S., Yang, Y., Wu, H., Li, J., Xie, P. et al. (2022). Thermosensitive and tum or microenvironment activated nanotheranostics for the chemodynamic/photothermal therapy of colorectal tumor. *Journal of Colloid and Interface Science*, 612: 223–234. https://doi.org/10.1016/j.jcis.2021.12.126.

Wei, J., Liu, Y., Yu, J., Chen, L., Luo, M. et al. (2021). Conjugated polymers: Optical toolbox for bioimaging and cancer therapy. *Small*, 17(43): 1–31. https://doi.org/10.1002/smll.202103127.

Zhai, J., Gao, J., Zhang, J., Liu, D., Gao, S. et al. (2022). Concave octahedral PtCu nanoframes mediated synergetic photothermal and chemodynamic tumor therapy. *Chemical Engineering Journal*, 442(P2): 136172. https://doi.org/10.1016/j.cej.2022.136172.

Zhang, L., Wang, W., Ou, M., Huang, X., Ma, Y. et al. (2022). NIR-II photothermal therapy for effective tumor eradication enhanced by heterogeneous nanorods with dual catalytic activities. *Nano Research*, 15(5): 4310–4319. https://doi.org/10.1007/s12274-022-4096-x.

Zhao, H., Guo, Y., Yuan, A., Xia, S., Gao, Z. et al. (2022). Nature-inspired nanothylakoids for multimodal cancer therapeutics. *Science China Materials*, 65(7 PG-1971–1979): 1971–1979. https://doi.org/10.1007/s40843-021-2003-5.

CHAPTER 13

Magneto Hyperthermia

Christian Chapa González

This chapter deals with magneto hyperthermia from a perspective that allows understanding the fundamentals associated with nanoparticles. First, it covers some clinical bases on the effect of hyperthermia on cancer cells. Next, it describes the mechanisms associated with the temperature increase in nanoparticle systems that causes the increase in temperature. Finally, some factors affecting the design of the formulations are described with an emphasis on the properties of the nanoparticles. Among these factors, the chemical composition, the synthesis method, and the surface modification that has driven scientific research in recent years for the design of magneto-hyperthermia treatments stand out. Finally, a reflection on the scope of nanomedicine systems developed for magneto hyperthermia therapy in the context of clinical studies and leaving open the question of whether we are close to seeing this treatment as a daily clinical practice against cancer.

13.1 Introduction

Hyperthermia is an alternative thermal treatment used against cancer, which consists of raising the temperature in the tumor region to cause damage to the tumor or destroy it. A new method to achieve this treatment is by applying an alternating magnetic field to a suspension of magnetic particles in a fluid. The problem with this treatment is controlling the generation of heat so that it only

Autonomous University of Ciudad Juarez. Av. del Charro 450 nte. Ciudad Juárez, Chihuahua, México. C.P. 32310.
Email: christian.chapa@uacj.mx

affects the cancer cells without damaging healthy cells due to the magnetic anisotropy of the material used in the suspension. Although the benefits of heat as a medicinal agent have been recognized since ancient times, the clinical application of heat continues to improve with technological and scientific advances. The ways to take advantage of clinical hyperthermia have advanced and are varied from hot water heating pads to high-density nano radio frequency antennas for flexible hyperthermia arrays. More recently, we intend to leverage our knowledge in the control of nanoscale materials to develop nanomedicine systems capable of converting outside energy into heat for targeted therapy and drug delivery in a safe and effective manner.

It is an encouraging scenario to get patients to recover more quickly and safely using magnetic nanoparticles designed to concentrate or convert external energy into heat with ever-improving precision. This would overcome the serious limitations of current medicine, such as the side effects of chemotherapies. At the same time, this undoubtedly represents the consolidation of a truly multidisciplinary field. It is difficult to find other disciplines of human endeavor that integrate material sciences, biology, chemistry, physics, electrical engineering, among others in a clinical practice that has given rise to disruptive technology. This disruptive technology and growing industry are expected to significantly impact society's economy and healthcare. However, as the scope of Nanomedicine applications, such as magneto hyperthermia, evolves it is important to simultaneously recognize the fundamentals and obstacles relevant to its clinical practice. Biomedical engineering-driven research must be harnessed in order to achieve the maximum potential impact on the health of individuals and populations.

It is undeniable that the evolution of magneto hyperthermia research is extraordinary since it explored, perhaps for the first time, iron oxide particles and electromagnetic radiation. This fact was reported in the Annals of Surgery in 1957 by R.K. Gilchrist et al.

> *When injections of a particulate matter were made into the lymphatics of tumor in man, it was possible to fill the normal parts of the node with particles one micron or less in diameter. These studies show that it is possible to obtain appreciable differential heating of nodes containing 5 mg of selected* Fe_2O_3 *per one gram of tissue. This heating is dependent on the radio frequency, magnetic field strength and the coercive force of the magnetic particles.*

It is noteworthy that this work was carried out between physicians and engineers. Nowadays, it is more common for research teams to be made up of an interdisciplinary and multidisciplinary group. Nowadays, it is common for biomedical engineers, physicians and chemists to be involved in the study or development of new therapeutic strategies, such as hyperthermia mediated by magnetic nanoparticles. In the article mentioned above, Dr. R.K. Gilchrist indicated the following, referring to the future implications that the engineer Earl Ballantine mentioned to him:

> *This is the way this will work, and there is no question in the world but what this can be done. It's only a matter of whether you can get someone to make these metals fine enough for you, and secondly, someone who will make the field.*

In these few lines, Gilchrist and Ballantine probably foretold the birth of a disruptive discipline that would forever change medical practice by studying new cancer treatment options. The existing literature currently has a vast amount of information related to the control of the size of the materials used in magneto hyperthermia. We also find extensive literature that talks about other aspects that were not mentioned at that time, such as the shape of nanoparticles, surface modification and biocompatibility. On the other hand, there is extensive literature that deals with anisotropy, magnetic relaxation, magnetic moments, and coercivity of the nanoparticles used for this biomedical application. Not only that, however, the reader will also learn about the scope of magnetic nanoparticles in other areas such as drug and gene delivery, imaging and biomolecule separation. Thanks to interdisciplinary research we have achieved some understanding of the clinical basis of induced hyperthermia, mechanisms of magnetic nanomaterials-based hyperthermia, and factors influencing the design of formulations for magneto hyperthermia-based therapy. In this chapter, we offer an approach to these topics in a narrative form, so that both experienced and neophyte readers will enjoy the reading in the same way.

13.2 Clinical basis of induced hyperthermia

Magneto-hyperthermia treatment can be more efficient with the help of superparamagnetic particles. The particle suspension can convert the energy it receives from an oscillating magnetic field into heat, which can be employed *in vivo* to raise the temperature of tumor tissue and kill cancer cells. Magneto hyperthermia offers significant benefits compared to local and whole-body hyperthermia.

Magneto hyperthermia is based on the increase of temperature in tissue by the application of an external variable magnetic field to a magnetic fluid. As cancer cells are more sensitive to temperature than normal cells neoplastic cells are less tolerable to temperatures in the range of 42 to 45°C in which cancer cells start to die due to photothermal instability of proteins, low availability of oxygen and nutrients in the tumor region (Figure 13.1).

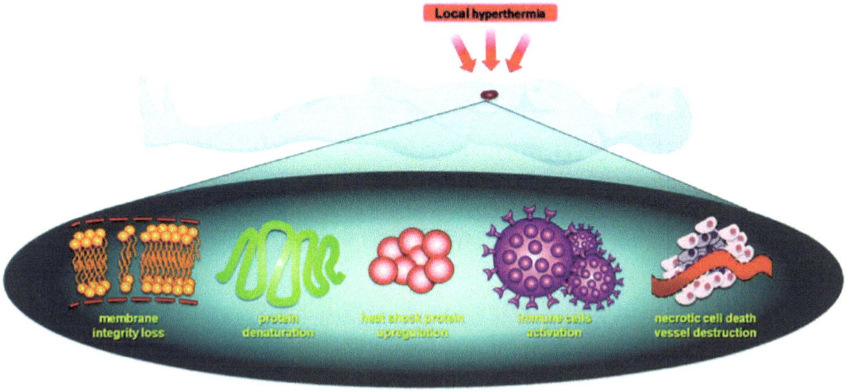

Figure 13.1. There are several mechanisms by which local hyperthermia can affect cells and promote antitumor responses. Reprinted from OpenAccess Publication (Clavel et al., 2015).

Hyperthermia or thermoablation may occur depending on the rate of heating. At temperatures above 42°C, cancer cells begin to die by apoptosis due to some factors. On one hand, a tumor is inefficiently irrigated because when the temperature rises, the tumor cannot receive the necessary nutrients and the cancer cells are in an acidic environment with low oxygen levels. On the other hand, healthy tissues present an ideal vascularization that allows them to resist a few degrees more of heat. Another factor is that the heat generated has a cytotoxic effect on malignant cells since heat inhibits some proteins that serve as signaling for cell proliferation of cancer cells, in addition to inducing apoptosis.

Currently, nanostructured composite materials have been extensively explored to obtain or modulate properties such as magnetization. The applications of the materials depend on the properties they possess, and, in turn, these properties are governed by the process of obtaining them. Thus, control over the properties of nanostructured materials, such as particle morphology, size distribution, surface chemistry, and consequently biocompatibility, is achieved by manipulating variables in the process of obtaining a nanostructured composite material.

Magnetic nanoparticle drug delivery systems can be manipulated through an external magnetic field and if combined with hyperthermia result in a more effective treatment. Hyperthermia treatment and drug delivery mechanisms by magnetic nanoparticle systems are related to the dissipation of energy in the form of heat. The possibility of converting magnetic dissipation energy into thermal energy gave rise to the application of magnetic materials in biomedicine. We should consider that the applied magnetic field must preserve the integrity of the tissues. On this basis, there are two main aspects. On the one hand, hyperthermia uses a moderate temperature rise (42°C for

60 min) that does not exceed a critical temperature (between 51 and 55°C). The viability of healthy cells is directly related to the thermostability of vital proteins.

13.3 Mechanisms of magnetic nanomaterials-based hyperthermia

The biomedical applications of single-domain magnetic nanoparticles, particularly hyperthermia, are possible because these materials can transmit energy in the form of heat. For several decades, it has been known that malignant cells *in vivo* can be destroyed at temperatures of 41–43°C (Overgaard, 1977). The heat transfer of magnetic nanoparticles is explained by the dissipation energy in alternating magnetic fields. The size and magnetic characteristics of the particles have a significant impact on the source of magnetic heating. Hysteresis losses cause heat to be produced for particles with multidomains that are either Ferro- or ferrimagnetic. A superparamagnetic or single-domain particle produces heat when an alternating magnetic field is applied through two relaxation processes known as Néel relaxation and Brown relaxation.

The dissipation energy of magnetic nanoparticles in a magnetic field is given by the Equation (13.1):

$$P = \pi\mu_0\chi_0 H_0^2 f \frac{2\pi f\tau}{1+(2\pi f\tau)^2} \tag{13.1}$$

where μ_0, χ_0, H_0, and f represent the free space permeability, equilibrium susceptibility, amplitude, and frequency of the alternating magnetic field, respectively. While τ symbolizes the relaxation time given by Equation (13.2):

$$\tau^{-1} = \tau_N^{-1} + \tau_B^{-1} \quad \text{where} \tag{13.2}$$

τ_N is the Néel relaxaion, while τ_B is the Brownian relaxation time. These terms are described mathematically with the Equations (13.3) and (13.4), respectively:

$$\tau_N = \tau_0 e^{\left(\frac{\Delta E}{kT}\right)} \tag{13.3}$$

$$\tau_B = \frac{3\eta V_H}{kT} \quad \text{where,} \tag{13.4}$$

τ_0 denotes the relaxation time, usually, 10^{-9} s, η is the viscosity, V_H is the hydrodynamic volume, k is Bolzmann's constant (1.38×10^{23} J·K^{-1}), T is the temperature, and V_M is the volume of the nanoparticles.

When nanoparticles are dispersed in fluids the relaxation behavior is strongly influenced by heat release and the Brownian mechanism occurs. As can be seen, the temperature can be cleared, the resulting Equation is (13.5):

$$T = \frac{3\eta V_H}{k\tau_R} \tag{13.5}$$

The volume of the MPNs is expressed as V_M (13.6), and the hydrodynamic volume (including the overlay of the MPNs) is expressed by the term V_H (13.7).

$$V_M = \frac{\pi D^3}{6} \tag{13.6}$$

$$V_H = \frac{\pi(D + 2\delta)^3}{6} \tag{13.7}$$

D is the diameter of the nanoparticles, and δ is the thickness of the coating.

For smaller nanoparticles τ_N is smaller than τ_B and the relaxation will take effect by the rotation of the magnetic moment within the nanoparticle. The energy barrier, which is determined by the product of the magnetic core volume and the magnetic anisotropy constant, exponentially affects the Néel relaxation time. In contrast, Brownian relaxation takes place when nanoparticles undergo rotational diffusion when suspended in a liquid medium (Figure 13.2). That is, Brownian relaxation occurs predominantly in fluid systems.

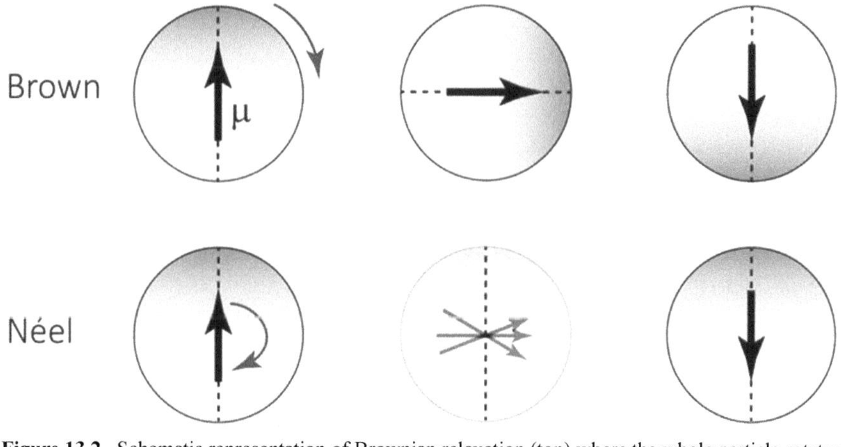

Figure 13.2. Schematic representation of Brownian relaxation (top) where the whole particle rotates in the fluid, while in Néel relaxation (bottom) it is the magnetic moment that rotates. Taken from (Ilg and Kröger, 2020) under CC BY license.

13.4 Factors influencing the design of formulations for magneto hyperthermia-based therapy

In recent years, research on the relationship between chemical composition, size, shape, and surface modification of nanoparticles to find the optimum properties for designing magneto hyperthermia-based therapy has increased.

The development of magnetic nanoparticles is constantly advancing in relation to their potential uses in the biomedical field. This is because of one of their main characteristics, the temperature increase effect induced by the magnetic field (Dan et al., 2015a; Suto et al., 2009). Magnetic nanoparticle heat generation in clinical and experimental settings increases with AMF frequency (f) and field strength (H) (Maier-Hauff et al., 2007; Motoyama et al., 2008; Thiesen and Jordan, 2008). The experiments, however, are constrained to a biologically safe and physiologically acceptable range (Ortega and Pankhurst, 2012; Spirou et al., 2018). The magnetic nanoparticles lose heat through relaxation (Obaidat et al., 2015; Sohail et al., 2017). As a result, it is necessary to control the properties of the nanoparticles under an AMF that can be used in clinical settings in order to increase their heating efficiency.

It is possible to rationally develop well-defined magnetic materials by comprehending the intricate relationship between composition, microstructure, and magnetic characteristics. Undoubtedly, the most explored materials in formulations for magneto hyperthermia-based therapy are iron oxides and their derivatives (Blanco-Andujar et al., 2016; Dulińska-Litewk et al., 2019; Low et al., 2022; Sangaiya and Jayaprakash, 2018; Wu et al., 2015). To date, they remain a platform for the development of hyperthermia systems for various types of cancer such as breast cancer (Cędrowska et al., 2020; Onyekanne et al., 2022), prostate cancer (Albarqi et al., 2020; Johannsen et al., 2010; Peralta and González, 2021), lung cancer (Theodosiou et al., 2022), pancreatic cancer (Engelmann et al., 2018; Palzer et al., 2021; Zhu et al., 2022), brain cancer or gliomas (Dan et al., 2015b; Dhar et al., 2022; Maier-Hauff et al., 2011; Schwake et al., 2022), ovarian cancer (Bai et al., 2021; Mérida et al., 2020; Taratula et al., 2013), cervical cancer (Wen et al., 2021), liver cancer (Chan et al., 2021; Chen et al., 2021; Gordon et al., 2014; Yan et al., 2014).

13.4.1 Chemical composition

Iron oxides are naturally occurring compounds that can be easily synthesized in the laboratory. In total there are 16 types of iron oxides, which can be oxides, hydroxides or oxyhydroxides. Among the iron oxides, magnetite nanoparticles are probably the most extensively studied material. Magnetite is the name assigned to the iron oxide whose chemical formula is Fe_3O_4.

Due to the presence of both iron ions (trivalent and divalent) in its minimal formula, $Fe^{2+}(Fe^{3+}O_2)_2$ it stands out from other iron oxides. Magnetite adopts an inverse spinel structure. The backbone structure corresponds to a mixture of metal oxides whose general chemical composition is AB_2O_4. Normally, A is a divalent atom whose atomic radius is between 80 and 110 pm, such as Cu, Fe, Mg, Mn, and Zn. On the other hand, B represents a trivalent cation whose atomic radius is between 75 and 90 pm, such as Al, Co, Fe and Ti. The structure consists of a matrix of 32 oxide ions forming 64 tetrahedral and 32 octahedral holes in a unit cell. In turn, the unit cell contains eight formula units $(AB_2O_4)_8$.

There are two types of subcells in the backbone structure. The subcell of A, in these two tetrahedral sites, is filled with 1/8 of the unit cell. The arrangement of these two cubic unit cells gives a total of 16 occupied octahedral sites. All trivalent cations in a typical backbone structure are found in the center of the octahedral sites, while all divalent cations take up 1/8 of the tetrahedral sites. The crystallographic unit of Fe_3O_4 is composed of eight formula units. That is, it consists of 24 iron atoms and 32 oxygen atoms. Magnetite has a cubic crystallographic system with a lattice parameter of 8.3941 Å. When the Fe^{+2} ions are replaced by divalent ions of some transition metal with its 3d orbital almost complete within the spinel structure it will be possible to change the magnetic properties of the material. Some elements that are not ferromagnetic when exchanged for the Fe^{+2} ions in the spinel structure may have contributed in the magnetic properties of these ferrites. This can be achieved with several synthesis methods among which the coprecipitation method stands out for its simplicity. The coprecipitation synthesis method allows the production of magnetic nanoparticles in a short period of time and in a reproducible manner in the laboratory, in addition to the fact that the synthesis parameters are very easy to control, such as pH, agitation, temperature, and precursor concentration.

Magnetite has a high saturation magnetization. The material exceeds the thermal energy released required in hyperthermia. Hence the need to control its anisotropy to manipulate its magnetic properties. Cobalt and nickel can modify the anisotropy in the magnetite since these are ferromagnetic materials with a magnetic moment lower than that of the iron present in the magnetite (Flores-Urquizo et al., 2017). By varying stoichiometry, we can modify the relaxation mechanisms of the suspended particles and thus manipulate the heat generated in the ferrofluid.

The Ms value can be tailored if the magnetic anisotropy is modified. This can be achieved by partially or totally replacing the Fe^{2+} with the Ni^{2+} ion in octahedral sites of the inverse spinel structure of the $Fe^{3+}(Fe^{3+}Fe^{2+})O_4$, which further affects the magnetic properties of the ferrite nanocrystals (Carta et al., 2009). According to recent publications, nickel ferrite nanoparticles were obtained, modified, and subjected to cytotoxicity tests for biological uses (Amiri et al., 2018; Ibraheem et al., 2019; Khan et al., 2019; Sharifi et al., 2012). In addition, the *in vitro* cytotoxicity of the $NiFe_2O_4$ nanoparticles causes a significant dose-dependent reduction in the cell number as compared to Fe_3O_4 nanoparticles.

The chemical composition is certainly one of the main factors influencing the design of formulations for magneto-hyperthermia-based therapy. The compositions of the materials fabricated for hyperthermia may contain elements such as cobalt (Phong et al., 2021), zinc (Jiang et al., 2011; Kahmei et al., 2021), manganese (Iacovita et al., 2019), yttrium (Gordon et al., 2014), lanthanum-strontium (Chen et al., 2016; Haghniaz et al., 2016; Shlapa et al., 2018; Usković et al., 2006), barium (Farrokhtakin et al., 2013). However, some compositions are too elaborate despite possessing very interesting properties such as the combination of luminescence and hyperthermia. They have the disadvantage of their complex formulation such as $Mn_{0.5}Fe_{2.5}O_4@YVO_4:Eu^{3+}$ (Ningombam et al., 2018) and $NaYF_4:Yb, Er@PE_3@Fe_3O_4$ (Wang et al., 2019).

13.4.2 Method of synthesis

In order that the factors affecting the formulation of hyperthermia systems can be appreciated, we will concentrate on methods of synthesis for nickel ferrites. The reader will be able to realize the vast possibility of experimental parameter control when designing formulations for hyperthermia. In the literature, it is described how the synthesis factors affect the Ms of nickel ferrite nanoparticles. In this way, the Ms of $NiFe_2O_4$ nanoparticles (28 nm), obtained by mechanical milling and sintering, was 47.0 emu/g, the results in (Ghayour et al., 2017) proved that the Ms is very important for the heating rate of zinc doped with nickel ferrite nanoparticles. By using the sol-gel process, $NiFe_2O_4$ coercivity, saturation magnetization, and particle size all increased as the calcination temperature rose. (500 to 800°C) (Liu and Gao, 2012). Another research group also showed that with sol-gel the Ms of the $NiFe_2O_4$ NPs (35 to 40 emu/g) and the crystallite size (30.75 to 42.32 nm) increases with increasing temperature (900 to 1100°C) (Asiri et al., 2018). While, by the hydrothermal (to 125 from 200°C) process (Karaagac et al., 2017), it was reported that the reaction time also influences the size of $NiFe_2O_4$ (6.6 to 12 nm) and (Ms) increases (45.6 to 51.7 emu/g) with time. It has been demonstrated that the nickel ferrites' coating affects the Ms more so than their particle size. According to the findings in (Kurosawa et al., 2011), the Ms was reduced by 5.4% using oleic acid as a coating of $NiFe_2O_4$ nanoparticles, while changing the diameter affected the Ms by 0.3%.

With any particle composition, the heating takes place just before the transition from a superparamagnetic to a single domain state (Starsich et al., 2018). The heating effectiveness of magnetic nanoparticles depends on their saturation magnetization (Ms), among other things like anisotropy. Because of their biocompatibility and adjustable surface modification, superparamagnetic Fe_3O_4 NPs are one of the best materials capable of inducing thermal effects in the biomedical field. As a result, numerous research groups have developed flexible methods to create them (Ling and Hyeon, 2013; Mehta, 2017; Sohail et al., 2017). Researchers have noted that altering the particle size, content, and coatings can change the Ms of Fe_3O_4 nanoparticles. In the study of (Motoyama et al., 2008) the researchers discovered a substantial correlation between heat generation and the Ms (38.1 to 82.5 emu/g) and size (10 to 120 nm) of nanoparticles. In a similar vein, the results reported in (Li et al., 2017) indicated that independent of particle shape, the Ms rises with particle size, from 54.7 emu/g (9.6 nm) to 84.7 emu/g (287 nm). The suspensions of Fe_3O_4 nanoparticles (20 $mg \cdot mL^{-1}$) can reach temperatures of up to 90°C in an AMF (f = 80 Hz and H = 30 $kA \cdot m-^{-1}$), according to the study presented by (Zhao et al., 2006), also showed that the Ms of Fe_3O_4 nanoparticles decreases as Fe^{3+}/Fe^{2+} molar ratio increases from 1.8:1 (65.53 emu/g) to 2.2:1 (32.16 emu/g).

13.4.3 Surface modification

As can be seen, particle formulations for hyperthermia have been developed thanks to their size-dependent superparamagnetic property. However, the surface area/volume ratio in small particles is too high and they tend to reduce their surface energy. Consequently, magnetic nanoparticles are stabilized by the formation of agglomerates or by the adsorption of surrounding molecules. For this reason, steric stabilization is pursued by experimental methods using various inorganic or organic molecules as a coating. Such steric stabilization mechanisms are based on the repulsion between molecules adsorbed on the surface of nanoparticles. At the same time, the coatings modify the physicochemical properties of the magnetic nanoparticles and can improve biocompatibility, dispersion in aqueous media, cellular uptake, and enhanced biodistribution.

Several studies show that magnetic particle properties vary with coatings for example the hydrodynamic size (Chapa González et al., 2021), including how they act with respect to biomolecules (Villegas-Serralta et al., 2018), the interaction with cells (Chapa Gonzalez et al., 2014; Roacho-Pérez et al., 2020; Urquizo et al., 2022), so another factor to be considered is the dependence of the properties of the magnetic particles on the coating. The temperatures of suspensions (30 mg/ml) containing Fe_3O_4 nanoparticles (67.06 emu/g) or PEG-coated Fe_3O_4 nanoparticles (64.11 emu/g) were 89.2°C and 72.2°C, respectively. Similarly, the Ms of carboxymethyl dextran and folate-coated magnetite nanoparticles were 35 emu/g and 30 emu/g, respectively, in (Jiang et al., 2014), which were significantly lower than that of bare Fe_3O_4 nanoparticles. Concentrations of up to 55 mg/mL were required to achieve temperatures above 40°C and frequencies of 80 Hz. The Ms of Fe_3O_4 (51.68 emu/g) in the study by (Shete et al., 2014) was found to be greater than that of the chitosan-coated Fe_3O_4 nanoparticles (49.96 emu/g), and they also found that the temperature rises as the suspension concentration rises (from 2 to 10 mg/mL). In contrast, when the strength of the magnetic field was increased (in a range of 167.6–335.2 Oe), the temperature was higher in coated nanoparticles. The results show how crucial it is to regulate magnetic characteristics in both uncoated and coated nanoparticles.

Together, the magnetic nanoparticles and the coating form homogeneous suspensions whose physicochemical properties (size distribution, polydispersity, and surface charge or zeta potential) must meet the criteria for stability. The nanoparticles should have a unimodal distribution with a low polydispersity index. The method usually employed for the determination of the size distribution is photon correlation spectroscopy (PCS). Light scattering occurs by particles with Brownian motion; this is the basis of the method; hence it is known as dynamic light scattering (DLS). Laser light is used to illuminate the particles, and the intensity of the variations in the scattered light is examined to determine the particles' sizes. In principle, suspended particles scatter incident light due to random motion, rotation, and the morphology factor. The rate of motion is inversely proportional to the particle size; that is, smaller particles have greater motion or scattering. The ratio of particle size to incident wavelength determines the strength of the scattered electromagnetic field. Thus, DLS is used to determine molecular weight, morphology, diffusion coefficients, particle aggregation or decay, electrophoretic mobility, as well as particle size by means of the hydrodynamic diameter from the Stokes-Einstein Equation (13.8):

$$d_H = \frac{kT}{3\pi\eta D} \tag{13.8}$$

where k represents the Boltzmann's constant, T and η represent the temperature and viscosity, respectively, while D denotes the diffusion coefficient.

The uniformity of the size distribution can be characterized by the polydispersity index. The polydispersity index is a measure of the width of a size distribution derived from the analysis of cumulative DDL data for a single Gaussian population with standard deviation (σ), and average size

(χ_{DDL}), thus polydispersity index is the relative variance of the distribution in the Equation (13.9):

$$PI = \frac{\sigma^2}{\chi_{DDL}^{\;2}} \tag{13.9}$$

PI, polydispersity index is a dimensionless value calculated from correlation data or cumulant analysis.

The polydispersity index value can be used as a measure of the non-uniformity of the measurements. A value of polydispersity index < 0.02 corresponds to monodisperse particles. While a value between 0.02 and 0.08 represents a rather narrow size distribution. Also, as a rule of thumb, a value of 0.04 can be considered for monodisperse suspensions. Similarly, the rate of aggregation or hydrolytic breakdown depends on an initial value of the polydispersity index. The maximum value of the polydispersity index is arbitrary and is limited to 1.0; this value indicates that the sample contains a very broad size distribution and/or contains agglomerates and that sedimentation may eventually occur.

On the other hand, the measurement of the zeta potential can be used to determine the physicochemical stability of samples in suspension. Zeta potential is an indicator of the extent of aggregation. For a charged particle in suspension, the electric potential decreases as it moves away from the particle surface. Polar aqueous diluent molecules form an imbalanced region of charge around the particles called a "double layer". The diluent molecules are bound together and successfully move with the particle over the solution. The electric potential at that point is the zeta potential. Typically, measuring devices use Henry's equation to first determine the electrophoretic mobility and then use that information to calculate the zeta potential. Electrophoretic mobility is the term used to describe a particle's speed in an electric field during electrophoresis.

Electrophoresis occurs as the response of charged particles suspended in an electrolyte to the application of an electric field. The particles migrate towards the oppositely charged electrode while the viscosity of the medium opposes the movement. These two forces reach an equilibrium, and the particles move at a constant velocity. This is related to parameters like the magnitude of the applied electric field or the voltage differential, the medium's electrical conductivity, the fluidity, and the surface charge. Thus, the zeta potential can be known from the Equation (13.10):

$$U_E = \frac{2\varepsilon z f(ka)}{3\eta} \tag{13.10}$$

where ε, U_E, z, and f(ka) represent the dielectric constant, electrophoretic mobility, zeta potential, and the Henry function, which generally uses two approximation values of 1.5 or 1.0.

The repulsive forces between the particles are promoted by a high zeta potential. The particles in suspension acquire a greater separation distance from each other. As a result, there is less aggregation brought on by Van der Waals interactions. Particles in the samples will have the tendency to form agglomerates over time when the zeta potential is far from ±30 mV. This zeta potential value is dependent on the electrical charge of the particle surface. Both the pH and the ionic strength of the medium can alter the adsorption of diluent molecules on the particle surface and modify the zeta potential value.

13.5 Performance of nanomedicine systems developed for magnetic hyperthermia therapy, clinical phase studies

Magnetic nanoparticle-mediated hyperthermia is a disruptive medical technology in clinical practice, although it is still at an early stage of development and the evidence for its effectiveness in cell culture and animal models is overwhelming. Despite potentially offering a safer and more effective means of treatment, the lack of evidence derived from prospective cohort studies and other observational studies or randomized clinical trials does not provide sufficient knowledge to represent a comprehensive understanding derived from the action of magnetic nanoparticles on the tumor. For example, investigators in MAGNABLATE I, a phase 0 study, have already conducted much of the preclinical work to develop this type of treatment; however, this study was limited to evaluating the retention of nanoparticles in the prostate, the site of injection. The researchers did not propose to heat the tissue, but rather to observe the pathology samples to see if the nanoparticles remain in the prostate or travel to other organs. (Hashim Uddin, 2017). Very little is known about the *in vivo* events that occur after the administration of nanomaterials. The study of the phenomena presented by nanoparticles in the reticuloendothelial system is still a very active and developing area of research. The mechanism by which specific cellular responses occur is not yet fully understood, although proteins and blood serum components are well known. The lack of clinical studies may be attributable to the systems' inherent variability, which has complicated our understanding of these mechanisms and led to contradictory results.

On the other hand, ferumoxytol, a compound of iron oxide particles with sorbitol approved by the FDA as a treatment for anemia in adults with chronic kidney disease, has shown promising results for brain cancer. This compound showed in humans a dose-dependent plasma clearance with a half-life of approximately 16 hours. Originally the use of ferumoxytol was started as a contrast agent to observe the behavior of tumors. Ferumoxytol penetrates the macrophages of the endothelial reticular system of the liver, spleen, and bone marrow, and in doing so stimulates their phagocytic capacity. When magnetic resonance images obtained before and after the doses administered were compared, a decrease in tumor size was observed. This aroused the scientists' interest in evaluating it as a possible treatment for brain cancer. In a pilot study with 14 patients (Gahramanov et al., 2011) a comparison was made between the contrast agent gadoteridol and ferumoxytol. This study showed that ferumoxytol could be a strong contender as a contrast agent and showed promise for immune system response. Another study with 26 patients (Dósa et al., 2011) related the presence of the nanoparticles with an inflammatory component since the phagocytic cells capture the ferumoxytol. This gives hope for its use as a treatment because it stimulates the immune system and could be an indicator of the progression of grade III tumors.

In another study with 4 patients (Qiu et al., 2012) with low-grade brain tumors, ferumoxytol was used to measure the blood volume of the human brain and the authors opened the possibility that it could be used to decrease blood flow in brain tumors. We believe that these findings and the accumulating evidence of its effectiveness in preclinical models open the possibility of employing magnetic nanoparticles for cancer hyperthermia in clinical practice soon (Figure 13.3).

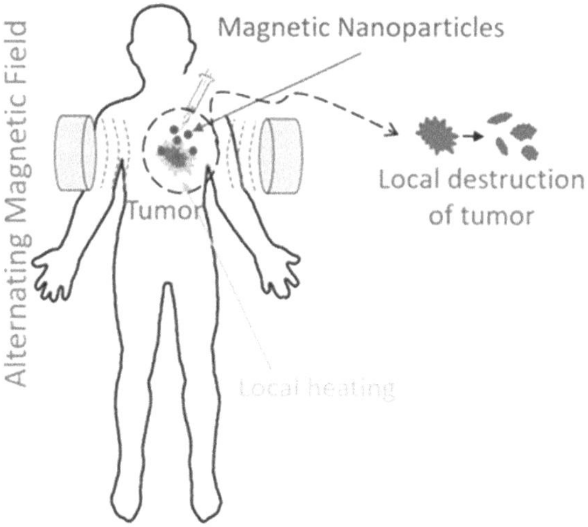

Figure 13.3. Diagram of magneto-hyperthermia where a local increase in temperature caused by nanoparticles exposed to the alternating magnetic field causes cell death. Taken from (Benos et al., 2022) under CC BY license.

Acronym/Abbreviation

Å	Angstrom
Al	Aluminum
AMF	Alternating Magnetic Field
Co	Cobalt
Cu	Coper
d	Diameter of the Nanoparticle
D	Diffusion Coefficient
DLS	Dynamic Light Scattering
emu	Electromagnetic Units
Er	Erbium
Eu	Europium
f	Frequency
F	Fluor
f(ka)	Henry function
FDA	Food and Drug Administration
Fe	Iron
Fe_2O_3	Iron(III) oxide or ferric oxide
Fe_3O_4	Magnetite
g	Grams
H0	Amplitude
k	Bolzmann's constant
kA	Kiloampere
m	Meter
MAGNABLATE I	MAGnetic NAnoparticle thermoaBLATion - Retention and Maintenance in prostatE
Mg	Magnesium
mg	Miligram
min	Minutes
mL	Mililiter
Mn	Manganese
Ms	Saturation magnetization
mV	Milivolts
Na	Sodium
Ni	Nickel
$NiFe_2O_4$	Nickel ferrite
nm	Nanometers
NPs	Nanoparticles
O	Oxygen
PCS	Photon Correlation Spectroscopy

pH	a scale used to specify the acidity or basicity of an aqueous solution
PI	Polydispersity Index
T	Temperature
Ti	Titanium
UE	Electrophoretic Mobility
V	Vanadium
VH	Hydrodynamic Volume
VM	Volume of the Nanoparticle
Y	Yttrium
Yb	Ytterbium
z	Zeta Potential
Zn	Zinc
δ	Thickness of the Coating
ε	Dielectric Constant
η	Viscosity
μ0	Free Space Permeability
σ	Standard Deviation
τ	Relaxation Time
τB	Brownian Relaxation Time
τN	Néel Relaxation Time
χ0	Equilibrium Susceptibility

Reference list

Albarqi, H.A., Demessie, A.A., Sabei, F.Y., Moses, A.S., Hansen, M.N. et al. (2020). Systemically delivered magnetic hyperthermia for prostate cancer treatment. *Pharmaceutics*, 12(11): 1–14. https://doi.org/10.3390/PHARMACEUTICS12111020.

Amiri, M., Pardakhti, A., Ahmadi-Zeidabadi, M., Akbari, A., Salavati-Niasari, M. et al. (2018). Magnetic nickel ferrite nanoparticles: Green synthesis by Urtica and therapeutic effect of frequency magnetic field on creating cytotoxic response in neural cell lines. *Colloids and Surfaces B: Biointerfaces*, 172: 244–253. https://doi.org/10.1016/J.COLSURFB.2018.08.049.

Asiri, S., Sertkol, M., Güngüneş, H., Amir, Md., Manikandan, A. et al. (2018). The temperature effect on magnetic properties of $NiFe_2O_4$ nanoparticles. *Journal of Inorganic and Organometallic Polymers and Materials*, 28(4): 1587–1597. https://doi.org/10.1007/s10904-018-0813-z.

Bai, M.Y., Yu, M.H., Wang, T.T., Chen, S.H., Wang, Y.C. et al. (2021). Plate-like alginate microparticles with disulfiram-spio-coencapsulation: An *in vivo* study for combined therapy on ovarian cancer. *Pharmaceutics*, 13(9). https://doi.org/10.3390/PHARMACEUTICS13091348.

Benos, L., Ninos, G., Polychronopoulos, N.D., Exomanidou, M.-A., Sarris, I. et al. (2022). Natural convection of Blood–Magnetic iron oxide bio-nanofluid in the context of hyperthermia treatment. *Computation*, 10(11): 190. https://doi.org/10.3390/COMPUTATION10110190.

Blanco-Andujar, C., Walter, A., Cotin, G., Bordeianu, C., Mertz, D. et al. (2016). Design of iron oxide-based nanoparticles for MRI and magnetic hyperthermia. *Nanomedicine*, 11(14): 1889–1910. https://doi.org/10.2217/nnm-2016-5001.

Carta, D., Casula, M.F., Falqui, A., Loche, D., Mountjoy, G. et al. (2009). A structural and magnetic investigation of the inversion degree in ferrite nanocrystals MFe_2O_4 (M = Mn, Co, Ni). *The Journal of Physical Chemistry C*, 113(20): 8606–8615. https://doi.org/10.1021/jp901077c.

Cędrowska, E., Pruszyński, M., Gawęda, W., Zuk, M., Krysiński, P. et al. (2020). Trastuzumab conjugated superparamagnetic iron oxide nanoparticles labeled with 225Ac as a perspective tool for combined α-radioimmunotherapy and magnetic hyperthermia of HER2-positive breast cancer. *Molecules (Basel, Switzerland)*, 25(5). https://doi.org/10.3390/MOLECULES25051025.

Chan, M.H., Lu, C.N., Chung, Y.L., Chang, Y.C., Li, C.H. et al. (2021). Magnetically guided theranostics: Montmorillonite-based iron/platinum nanoparticles for enhancing *in situ* MRI contrast and hepatocellular carcinoma treatment. *Journal of Nanobiotechnology*, 19(1). https://doi.org/10.1186/S12951-021-01052-7.

Chapa González, C., Navarro Arriaga, J.U. and García Casillas, P.E. (2021). Physicochemical properties of chitosan–magnetite nanocomposites obtained with different pH. *Polymers and Polymer Composites*, 29(9): S1009–S1016. https://doi.org/10.1177/09673911211038461/ASSET/IMAGES/LARGE/10.1177_09673911211038461-FIG2.JPEG.

Chapa Gonzalez, C., Roacho Pérez, J.A., Martínez Pérez, C.A., Olivas Armendáriz, I., Jimenez Vega, F., Castrejon Parga, K.Y. and Garcia Casillas, P.E. (2014). Surface modified superparamagnetic nanoparticles: Interaction with fibroblasts in primary cell culture. *Journal of Alloys and Compounds*, 615(S1): S655–S659. https://doi.org/10.1016/J.JALLCOM.2014.01.155.

Chen, B.W., Chiu, G.W., He, Y.C., Huang, C.Y., Huang, H.T. et al. (2021). Extracellular and intracellular intermittent magnetic-fluid hyperthermia treatment of SK-Hep1 hepatocellular carcinoma cells based on magnetic nanoparticles coated with polystyrene sulfonic acid. *PloS One*, 16(2). https://doi.org/10.1371/JOURNAL.PONE.0245286.

Chen, Y., Wang, Y., Liu, X., Lu, M., Cao, J. et al. (2016). LSMO nanoparticles coated by hyaluronic acid for magnetic hyperthermia. *Nanoscale Research Letters*, 11(1). https://doi.org/10.1186/S11671-016-1756-3.

Clavel, C.M., Nowak-Sliwinska, P., Pəunescu, E. and Dyson, P.J. (2015). Thermoresponsive fluorinated small-molecule drugs: A new concept for efficient localized chemotherapy. *MedChemComm.*, 6(12): 2054–2062. https://doi.org/10.1039/C5MD00409H.

Dan, M., Bae, Y., Pittman, T.A. and Yokel, R.A. (2015a). Alternating magnetic field-induced hyperthermia increases iron oxide nanoparticle cell association/uptake and flux in blood-brain barrier models. *Pharmaceutical Research*, 32(5): 1615–1625. https://doi.org/10.1007/s11095-014-1561-6.

Dan, M., Bae, Y., Pittman, T.A. and Yokel, R.A. (2015b). Alternating magnetic field-induced hyperthermia increases iron oxide nanoparticle cell association/uptake and flux in blood-brain barrier models. *Pharmaceutical Research*, 32(5): 1615–1625. https://doi.org/10.1007/s11095-014-1561-6.

Dhar, D., Ghosh, S., Das, S. and Chatterjee, J. (2022). A review of recent advances in magnetic nanoparticle-based theranostics of glioblastoma. *Nanomedicine (London, England)*, 17(2): 107–132. https://doi.org/10.2217/NNM-2021-0348.

Dósa, E., Guillaume, D.J., Haluska, M., Lacy, C.A., Hamilton, B.E. et al. (2011). Magnetic resonance imaging of intracranial tumors: Intra-patient comparison of gadoteridol and ferumoxytol. *Neuro-Oncology*, 13(2): 251–260. https://doi.org/10.1093/NEUONC/NOQ172.

Dulińska-Litewka, J., Łazarczyk, A., Hałubiec, P., Szafrański, O., Karnas, K. et al. (2019). Superparamagnetic iron oxide nanoparticles-current and prospective medical applications. *Materials (Basel, Switzerland)*, 12(4). https://doi.org/10.3390/MA12040617.

Engelmann, U.M., Roeth, A.A., Eberbeck, D., Buhl, E.M., Neumann, U.P. et al. (2018). Combining bulk temperature and nanoheating enables advanced magnetic fluid hyperthermia efficacy on pancreatic tumor cells. *Scientific Reports*, 8(1). https://doi.org/10.1038/S41598-018-31553-9.

Farrokhtakin, E., Ciofani, G., Puleo, G.L., de Vito, G., Filippeschi, C. et al. (2013). Barium titanate core–gold shell nanoparticles for hyperthermia treatments. *International Journal of Nanomedicine*, 8: 2319–2331. https://doi.org/10.2147/IJN.S45654.

Flores-Urquizo, I.A., García-Casillas, P. and Chapa-González, C. (2017). Desarrollo de nanopartículas magnéticas Fe+32X+21O4(X= Fe, Co y Ni) recubiertas con amino silano. *Revista Mexicana de Ingenieria Biomedica*, 38(1): 402–411. https://doi.org/10.17488/RMIB.38.1.36.

Gahramanov, S., Raslan, A.M., Muldoon, L.L., Hamilton, B.E., Rooney et al (2011). Potential for differentiation of pseudoprogression from true tumor progression with dynamic susceptibility-weighted contrast-enhanced magnetic resonance imaging using ferumoxytol vs. gadoteridol: A pilot study. *International Journal of Radiation Oncology Biology Physics*, 79(2): 514–523. https://doi.org/10.1016/j.ijrobp.2009.10.072.

Ghayour, H., Abdellahi, M., Ozada, N., Jabbrzare, S. and Khandan, A. et al. (2017). Hyperthermia application of zinc doped nickel ferrite nanoparticles. *Journal of Physics and Chemistry of Solids*, 111: 464–472. https://doi.org/10.1016/J.JPCS.2017.08.018.

Gilchrist, R.K., Medal, R., Shorey, W.D., Hanselman, R.C., Parrott, J.C. and Taylor, C.B. (1957). Selective inductive heating of lymph nodes. Annals of Surgery, 146(4): 596. https://doi.org/10.1097/00000658-195710000-00007.

Gordon, A.C., Lewandowski, R.J., Salem, R., Day, D.E., Omary, R.A. et al. (2014). Localized hyperthermia with iron oxide-doped yttrium microparticles: Steps toward image-guided thermoradiotherapy in liver cancer. *Journal of Vascular and Interventional Radiology: JVIR*, 25(3): 397–404. https://doi.org/10.1016/J.JVIR.2013.10.022.

Haghniaz, R., Umrani, R.D. and Paknikar, K.M. (2016). Hyperthermia mediated by dextran-coated La0.7Sr0.3MnO3 nanoparticles: *In vivo* studies. *International Journal of Nanomedicine*, 11: 1779–1791. https://doi.org/10.2147/IJN.S104617.

Hashim Uddin, A. (2017). Magnetic Nanoparticle Thermoablation—Retention and Maintenance in the Prostate: A Phase 0 Study in Men (MAGNABLATE I). *Clinical Trials. Gov.* https://www.clinicaltrials.gov/ct2/show/record/NCT02033447?term=Hyperthermia%2C+magnetic&draw=3&rank=20.

Iacovita, C., Florea, A., Scorus, L., Pall, E., Dudric, R. et al. (2019). Hyperthermia, cytotoxicity, and cellular uptake properties of manganese and zinc ferrite magnetic nanoparticles synthesized by a polyol-mediated process. *Nanomaterials (Basel, Switzerland)*, 9(10). https://doi.org/10.3390/NANO9101489.

Ibraheem, F., Aziz, M.H., Fatima, M., Shaheen, F., Ali, S.M. et al. (2019). *In vitro* cytotoxicity, MMP and ROS activity of green synthesized nickel oxide nanoparticles using extract of Terminalia chebula against MCF-7 cells. *Materials Letters*, 234: 129–133. https://doi.org/10.1016/J.MATLET.2018.09.075.

Ilg, P. and Kröger, M. (2020). Dynamics of interacting magnetic nanoparticles: Effective behavior from competition between Brownian and Néel relaxation. *Physical Chemistry Chemical Physics*, 22(39): 22244–22259. https://doi.org/10.1039/D0CP04377J.

Jiang, Q.L., Zheng, S.W., Hong, R.Y., Deng, S.M., Guo, L. et al. (2014). Folic acid-conjugated Fe_3O_4 magnetic nanoparticles for hyperthermia and MRI *in vitro* and *in vivo*. *Applied Surface Science*, 307: 224–233. https://doi.org/10.1016/J.APSUSC.2014.04.018.

Jiang, Y., Ou, J., Zhang, Z. and Qin, Q.H. (2011). Preparation of magnetic and bioactive calcium zinc iron silicon oxide composite for hyperthermia treatment of bone cancer and repair of bone defects. *Journal of Materials Science. Materials in Medicine*, 22(3): 721–729. https://doi.org/10.1007/S10856-010-4225-Z.

Johannsen, M., Thiesen, B., Wust, P. and Jordan, A. (2010). Magnetic nanoparticle hyperthermia for prostate cancer. *International Journal of Hyperthermia*, 26(8): 790–795. https://doi.org/10.3109/02656731003745740.

Kahmei, R.D.R., Seal, P. and Borah, J.P. (2021). Tunable heat generation in nickel-substituted zinc ferrite nanoparticles for magnetic hyperthermia. *Nanoscale Advances*, 3(18): 5339–5347. https://doi.org/10.1039/D1NA00153A.

Karaagac, O., Atmaca, S. and Kockar, H. (2017). A facile method to synthesize nickel ferrite nanoparticles: Parameter effect. *Journal of Superconductivity and Novel Magnetism*, 30(8): 2359–2369. https://doi.org/10.1007/s10948-016-3796-4.

Khan, S., Ansari, A.A., Malik, A., Chaudhary, A.A., Syed, J.B. et al. (2019). Preparation, characterizations and *in vitro* cytotoxic activity of nickel oxide nanoparticles on HT-29 and SW620 colon cancer cell lines. *Journal of Trace Elements in Medicine and Biology*, 52: 12–17. https://doi.org/10.1016/J.JTEMB.2018.11.003.

Kurosawa, R., Suzuki, T., Nakayama, T., Suematsu, H. and Niihara, K. et al. (2011). Change in magnetic characteristics of $NiFe_2O_4$ nanoparticles upon organic matter adsorption and desorption. *Japanese Journal of Applied Physics*, 50(1S2): 01BE11. https://doi.org/10.1143/JJAP.50.01BE11.

Li, Q., Kartikowati, C.W., Horie, S., Ogi, T., Iwaki, T. et al. (2017). Correlation between particle size/domain structure and magnetic properties of highly crystalline Fe_3O_4 nanoparticles. *Scientific Reports*, 7(1): 9894. https://doi.org/10.1038/s41598-017-09897-5.

Ling, D. and Hyeon, T. (2013). Chemical design of biocompatible iron oxide nanoparticles for medical applications. *Small*, 9(9-10): 1450–1466. https://doi.org/10.1002/smll.201202111.

Liu, X. and Gao, W.-L. (2012). Preparation and magnetic properties of $NiFe_2O_4$ nanoparticles by modified Pechini method. *Materials and Manufacturing Processes*, 27(9): 905–909. https://doi.org/10.1080/10426914.2011.610082.

Low, L.E., Lim, H.P., Ong, Y.S., Siva, S.P., Sia, C.S. et al. (2022). Stimuli-controllable iron oxide nanoparticle assemblies: Design, manipulation and bio-applications. *Journal of Controlled Release*, 345: 231–274. https://doi.org/10.1016/j.jconrel.2022.03.024.

Maier-Hauff, K., Rothe, R., Scholz, R., Gneveckow, U., Wust, P. et al. (2007). Intracranial thermotherapy using magnetic nanoparticles combined with external beam radiotherapy: Results of a feasibility study on patients with glioblastoma multiforme. *Journal of Neuro-Oncology*, 81(1): 53–60. https://doi.org/10.1007/s11060-006-9195-0.

Maier-Hauff, K., Ulrich, F., Nestler, D., Niehoff, H., Wust, P. et al. (2011). Efficacy and safety of intratumoral thermotherapy using magnetic iron-oxide nanoparticles combined with external beam radiotherapy on patients with recurrent glioblastoma multiforme. *Journal of Neuro-Oncology*, 103(2): 317–324. https://doi.org/10.1007/S11060-010-0389-0.

Mehta, R.V. (2017). Synthesis of magnetic nanoparticles and their dispersions with special reference to applications in biomedicine and biotechnology. *Materials Science and Engineering: C*, 79: 901–916. https://doi.org/10.1016/J.MSEC.2017.05.135.

Mérida, F., Rinaldi, C., Juan, E.J. and Torres-Lugo, M. (2020). *In vitro* ultrasonic potentiation of 2-phenylethynesulfonamide/magnetic fluid hyperthermia combination treatments for ovarian cancer. *International Journal of Nanomedicine*, 15: 419–432. https://doi.org/10.2147/IJN.S217870.

Motoyama, J., Hakata, T., Kato, R., Yamashita, N., Morino, T. et al. (2008). Size dependent heat generation of magnetite nanoparticles under AC magnetic field for cancer therapy. *Biomagnetic Research and Technology*, 6: 4. https://doi.org/10.1186/1477-044X-6-4.

Ningombam, G.S., Ningthoujam, R.S., Kalkura, S.N. and Singh, N.R. (2018). Induction heating efficiency of water-dispersible Mn0.5Fe2.5O4@YVO4:Eu3+ magnetic-luminescent nanocomposites in an acceptable ac magnetic field: hemocompatibility and cytotoxicity studies. *The Journal of Physical Chemistry. B*, 122(27): 6862–6871. https://doi.org/10.1021/ACS.JPCB.8B02364.

Obaidat, I.M., Issa, B. and Haik, Y. (2015). Magnetic properties of magnetic nanoparticles for efficient hyperthermia. *Nanomaterials (Basel, Switzerland)*, 5(1): 63–89. https://doi.org/10.3390/nano5010063.

Onyekanne, C.E., Salifu, A.A., Obayemi, J.D., Ani, C.J., Ashouri Choshali, H. et al. (2022). Laser-induced heating of polydimethylsiloxane-magnetite nanocomposites for hyperthermic inhibition of triple-negative breast cancer cell proliferation. *Journal of Biomedical Materials Research. Part B, Applied Biomaterials*, 110(12). https://doi.org/10.1002/JBM.B.35124.

Ortega, D. and Pankhurst, Q.A. (2012). Magnetic hyperthermia. pp. 60–88. *In*: O'Brien, P. (ed.). *Nanoscience: Volume 1: Nanostructures through Chemistry*. RCS Publishing. https://doi.org/10.1039/9781849734844-00060.

Overgaard, J. (1977). Effect of hyperthermia on malignant cells *in vivo*. A review and a hypothesis - Forskning - Aarhus Universitet. *Cancer*, 39(6): 2637–2646. https://pubmed.ncbi.nlm.nih.gov/872062/.

Palzer, J., Mues, B., Goerg, R., Aberle, M., Rensen, S.S. et al. (2021). Magnetic fluid hyperthermia as treatment option for pancreatic cancer cells and pancreatic cancer organoids. *International Journal of Nanomedicine*, 16: 2965–2981. https://doi.org/10.2147/IJN.S288379.

Peralta, J.A.P. and González, C.C. (2021). Prostate cancer tumor growth inhibition in nanoparticle mediated magnetic hyperthermia on preclinical trials: A scoping review. *Memorias Del Congreso Nacional de Ingeniería Biomédica*, 8(1): 164–167. https://doi.org/10.24254/CNIB.21.26.

Phong, L.T.H., Manh, D.H., Nam, P.H., Lam, V.D., Khuyen, B.X. et al. (2021). Structural, magnetic and hyperthermia properties and their correlation in cobalt-doped magnetite nanoparticles. *RSC Advances*, 12(2): 698–707. https://doi.org/10.1039/D1RA07407E.

Qiu, D., Zaharchuk, G., Christen, T., Ni, W.W., Moseley, M.E. et al. (2012). Contrast-enhanced functional blood volume imaging (CE-fBVI): Enhanced sensitivity for brain activation in humans using the ultrasmall superparamagnetic iron oxide agent ferumoxytol. *NeuroImage*, 62(3): 1726–1731. https://doi.org/10.1016/J.NEUROIMAGE.2012.05.010.

Roacho-Pérez, J.A., Ruiz-Hernandez, F.G., Chapa-Gonzalez, C., Martínez-Rodríguez, H.G., Flores-Urquizo, I.A. et al. (2020). Magnetite nanoparticles coated with PEG 3350-Tween 80: *In vitro* characterization using primary cell cultures. *Polymers*, 12(2): 300. https://doi.org/10.3390/POLYM12020300.

Sangaiya, P. and Jayaprakash, R. (2018). A review on iron oxide nanoparticles and their biomedical applications. *Journal of Superconductivity and Novel Magnetism*, 31(11): 3397–3413. https://doi.org/10.1007/S10948-018-4841-2.

Schwake, M., Müther, M., Bruns, A.-K., Zinnhardt, B., Warneke, N. et al. (2022). Combined fluorescence-guided resection and intracavitary thermotherapy with superparamagnetic iron-oxide nanoparticles for recurrent high-grade glioma: Case series with emphasis on complication management. *Cancers*, 14(3): 541. https://doi.org/10.3390/CANCERS14030541.

Sharifi, I., Shokrollahi, H. and Amiri, S. (2012). Ferrite-based magnetic nanofluids used in hyperthermia applications. *Journal of Magnetism and Magnetic Materials*, 324(6): 903–915. https://doi.org/10.1016/J.JMMM.2011.10.017.

Shete, P.B., Patil, R.M., Thorat, N.D., Prasad, A., Ningthoujam, R.S. et al. (2014). Magnetic chitosan nanocomposite for hyperthermia therapy application: Preparation, characterization and *in vitro* experiments. *Applied Surface Science*, 288: 149–157. https://doi.org/10.1016/J.APSUSC.2013.09.169.

Shlapa, Y., Solopan, S., Belous, A. and Tovstolytkin, A. (2018). Effect of synthesis method of La1 - xSr x MnO3 manganite nanoparticles on their properties. *Nanoscale Research Letters*, 13(1). https://doi.org/10.1186/S11671-017-2431-Z.

Sohail, A., Ahmad, Z., Bég, O.A., Arshad, S., Sherin, L. et al. (2017). A review on hyperthermia via nanoparticle-mediated therapy. *Bulletin Du Cancer*, 104(5): 452–461. https://doi.org/10.1016/J.BULCAN.2017.02.003.

Spirou, S.V., Basini, M., Lascialfari, A., Sangregorio, C., Innocenti, C. et al. (2018). Magnetic hyperthermia and radiation therapy: radiobiological principles and current practice†. *Nanomaterials (Basel, Switzerland)*, 8(6). https://doi.org/10.3390/nano8060401.

Starsich, F.H.L., Eberhardt, C., Boss, A., Hirt, A.M., Pratsinis, S.E. et al. (2018). Coercivity determines magnetic particle heating. *Advanced Healthcare Materials*, 7(19). https://doi.org/10.1002/ADHM.201800287.

Suto, M., Kosukegawa, H., Maruta, K., Ohta, M., Tohji, K. et al. (2009). Heat diffusion characteristics of magnetite nanoparticles dispersed hydro-gel in alternating magnetic field. *Journal of Magnetism and Magnetic Materials*, 321(20): 3483–3487. https://doi.org/10.1016/J.JMMM.2009.06.067.

Taratula, O., Dani, R.K., Schumann, C., Xu, H., Wang, A. et al. (2013). Multifunctional nanomedicine platform for concurrent delivery of chemotherapeutic drugs and mild hyperthermia to ovarian cancer cells. *International Journal of Pharmaceutics*, 458(1): 169–180. https://doi.org/10.1016/J.IJPHARM.2013.09.032.

Theodosiou, M., Sakellis, E., Boukos, N., Kusigerski, V., Kalska-Szostko, B. et al. (2022). Iron oxide nanoflowers encapsulated in thermosensitive fluorescent liposomes for hyperthermia treatment of lung adenocarcinoma. *Scientific Reports*, 12(1). https://doi.org/10.1038/S41598-022-12687-3.

Thiesen, B. and Jordan, A. (2008). Clinical applications of magnetic nanoparticles for hyperthermia. *International Journal of Hyperthermia*, 24(6: 467–474. https://doi.org/10.1080/02656730802104757.

Urquizo, I.A.F., García, T.C.H., Loredo, S.L., Galindo, J.T.E., Casillas, P.E.G. et al. (2022). Effect of aminosilane nanoparticle coating on structural and magnetic properties and cell viability in human cancer cell lines. *Particle & Particle Systems Characterization*, 2200106. https://doi.org/10.1002/PPSC.202200106.

Uskoković, V., Košak, A. and Drofenik, M. (2006). Preparation of silica-coated lanthanum-strontium manganite particles with designable curie point, for application in hyperthermia treatments. *International Journal of Applied Ceramic Technology*, 3(2): 134–143. https://doi.org/10.1111/j.1744-7402.2006.02065.x.

Villegas-Serralta, E., Zavala, O., Flores-Urquizo, I.A., García-Casillas, P.E., Chapa González, C. et al.(2018). Detection of HER2 through antibody immobilization is influenced by the properties of the magnetite nanoparticle coating. *Journal of Nanomaterials*, 2018. https://doi.org/10.1155/2018/7571613.

Wang, X., Kang, C., Pan, Y. and Jiang, R. (2019). Photothermal effects of NaYF4:Yb,Er@PE3@Fe3O4 superparamagnetic nanoprobes in the treatment of melanoma. *International Journal of Nanomedicine*, 14: 4319–4331. https://doi.org/10.2147/IJN.S203077.

Wen, S., Xing, W., Gao, L. and Zhao, S. (2021). Effect of superparamagnetic DMSO@ γ-Fe_2O_3 combined with carmustine on cervical cancer. *Journal of Nanoscience and Nanotechnology*, 21(12): 6196–6204. https://doi.org/10.1166/JNN.2021.18596.

Wu, W., Wu, Z., Yu, T., Jiang, C., Kim, W.S. et al. (2015). Recent progress on magnetic iron oxide nanoparticles: Synthesis, surface functional strategies and biomedical applications. *Science and Technology of Advanced Materials*, 16(2). https://doi.org/10.1088/1468-6996/16/2/023501.

Yan, S.Y., Chen, M.M., Fan, J.G., Wang, Y.Q., Du, Y.Q. et al. (2014). Therapeutic mechanism of treating smmc-7721 liver cancer cells with magnetic fluid hyperthermia using Fe_2O_3 nanoparticles. *Brazilian Journal of Medical and Biological Research*, 47(11): 947–959. https://doi.org/10.1590/1414-431X20143808.

Zhao, D., Zeng, X., Xia, Q. and Tang, J. (2006). Inductive heat property of Fe_3O_4 nanoparticles in AC magnetic field for local hyperthermia. *Rare Metals*, *25*(6 SUPPL. 1): 621–625. https://doi.org/10.1016/S1001-0521(07)60159-4.

Zhu, L., Mao, H. and Yang, L. (2022). Advanced iron oxide nanotheranostics for multimodal and precision treatment of pancreatic ductal adenocarcinoma. *Wiley Interdisciplinary Reviews. Nanomedicine and Nanobiotechnology*, 14(4). https://doi.org/10.1002/WNAN.1793.

Index

U

V